Foundations of Relational Realism

Contemporary Whitehead Studies

Series Editors:

Roland Faber (Claremont Lincoln University)
and Brian G. Henning (Gonzaga University)

Contemporary Whitehead Studies, co-sponsored by the Whitehead Research Project, is an interdisciplinary book series that publishes manuscripts from scholars with contemporary and innovative approaches to Whitehead studies by giving special focus to projects that:

- explore the connections between Whitehead and contemporary Continental philosophy, especially sources, like Heidegger, or contemporary streams like poststructuralism,
- reconnect Whitehead to pragmatism, analytical philosophy and philosophy of language,
- explore creative East/West dialogues facilitated by Whitehead's work,
- explore the interconnections of the mathematician with the philosopher and the contemporary importance of these parts of Whitehead's work for the dialogue between sciences and humanities,
- reconnect Whitehead to the wider field of philosophy, the humanities, the sciences and academic research with Whitehead's pluralistic impulses in the context of a pluralistic world,
- address Whitehead's philosophy in the midst of contemporary problems facing humanity, such as climate change, war & peace, race, and the future development of civilization.

Titles in the Series:

Butler on Whitehead: On the Occasion, edited by Roland Faber, Michael Halewood, and Deena Lin

Foundations of Relational Realism

A Topographical Approach to Quantum Mechanics and the Philosophy of Nature

Michael Epperson and Elias Zafiris

LEXINGTON BOOKS
Lanham • Boulder • New York • Toronto • Plymouth, UK

Published by Lexington Books
A wholly owned subsidiary of The Rowman & Littlefield Publishing Group, Inc.
4501 Forbes Boulevard, Suite 200, Lanham, Maryland 20706
www.rowman.com

10 Thornbury Road, Plymouth PL6 7PP, United Kingdom

British Library Cataloguing in Publication Information Available

Library of Congress Cataloging-in-Publication Data
Epperson, Michael.
Foundations of relational realism : a topoglogical approach to quantum mechanics and the philosophy of nature / Michael Epperson and Elias Zafiris.
pages cm
Includes bibliographical references and index.
ISBN 978-0-7391-8032-7 (cloth : alk. paper) — ISBN 978-0-7391-8033-4 (electronic)
1. Quantum logic. 2. Quantum theory—Philosophy. 3. Philosophy of nature. I. Zafiris, Elias, 1970- II. Title.
QC174.17.M35E67 2013
530.1201—dc23 2013010287

™ The paper used in this publication meets the minimum requirements of American National Standard for Information Sciences—Permanence of Paper for Printed Library Materials, ANSI/NISO Z39.48-1992.

Printed in the United States of America

For Anna and Kitty,
our wives,
to whom we are forever grateful.

CONTENTS

Preface xiii

Part I: Philosophical Foundations of Quantum Relational Realism
Michael Epperson

Elementary quantum relational events, logical causality, mereotopological extension, and the internal relation of the local to the global in quantum mechanical predication 1

Chapter 1—Introduction:
Relational Realism: A Program in Speculative Philosophy 3

1.1 Bipolar Dualism: Reducing Contrast to Mutually Exclusive Relata 5

1.2 Bipolar Dualism and the Scientific Method: The Example of Physical and Philosophical Cosmology 10

1.3 The Fallacy of Misplaced Concreteness 13

1.4 Dipolar Duality: Contrasts of Mutually Implicative Relata 22

Chapter 2:
Substance and Logic in Quantum Mechanics 29

2.1 'Liberating' Science from Boolean Logic: A Sisyphean Adventure 32

2.2 Abandoning the Logic of One World for the Logic of Many Worlds: An Unlikely Liberation 39

2.3 Quantum Mechanics and Boolean Logic 43

2.4 Internal Relations in Quantum Mechanics 52

2.5 The Compatibility Condition for Logical Causality in Quantum Mechanics 58

2.6 Inducing the Global from the Local 63

2.7 Decoherence: Causal Relation or Logical Integration? 68

2.8 EPR and Quantum Nonlocality 71

2.9 Topological Relations vs. Metrical Relations: Quantum Mechanics and Spacetime 80

2.10 Substance in Quantum Mechanics: Relational Realism and the Philosophy of Alfred North Whitehead 86

Chapter 3: Predication in Quantum Mechanics 103

3.1 EPR and Decoherence 106

3.2 The Problem of Self-Reference in Quantum Systems 110

3.3 Quantum Mechanics and the Theory of Logical Types 116

3.4 Relations as Relata: From Set-Theoretic Classical Material Objects to Category-Theoretic Quantum Relational Events 119

3.5 Summary 132

Chapter 4:
Logical Causality in Quantum Mechanics:
A Relational Realist Ontology 139

4.1 Internal Relation and Logical Implication in Quantum Mechanics 140

4.2 The Compatibility Condition Revisited 148

4.3 The Evolution of Potentiality to Probability 157

4.4 Quantum Events as Dipolar Units of Relation: Their Subjective and Objective Features Are Mutually Implicative 163

4.5 Quantum Mechanics Presupposes Logically Related Actual Occasions 168

4.6 Summary 170

Chapter 5:
Integrating Logical Relation and Extensive Relation:
Mereotopology and Quantum Mechanics 179

5.1 Mereotopological Notions of Internal Relation and Logical Implication 185

5.2 Mereotopological Extensive Relations 190

5.3 Summary 194

Interlude 199

Part II: Mathematical Foundations of Quantum Relational Realism
Elias Zafiris

A sheaf-theoretic description of internally related quantum event structures in terms of Boolean covering systems 211

Chapter 6:
Notion of Localization Processes 213

6.1 Localization in Physical Theories 213

6.2 Localization Schemes 216

6.3 The Conceptual Framework of Category Theory 220

6.3.1 The Necessity of a Categorical Framework 220

6.3.2 Categorical Principles and Terminology 222

6.3.3 Universality and Equivalence 227

6.4 The Methodology of Uniform Fibrations: Variable Set Presheaves 232

6.5 The Localization Role of Topology: Covering Sieves and Sites 233

Chapter 7:
Sheaves of Germs: The Topological Case 243

7.1 Localization over a Topological Space 243

7.2 Uniform Localization of Observables 244

7.3 Gluing Sections and Local-to-Global Compatibility 246

7.4 Contextuality: Germs and Stalks of Observable Sheaves 248

7.5 Completion and Functionalization: Display Bundles 252

7.6 Events: From Set-Theoretic to Sheaf-Theoretic Semantics 253

7.7 Covering Sieves as Representable Functors 257

Chapter 8:
Sheaves of Boolean Germs: The Quantum Topological Case 261

8.1 Quantum Logical Event Algebras and Observables 261

8.2 The Topological Significance of Boolean Contexts 266

8.3 Logical Internal Relation: Coherent Local-to-Global Ontological Formation 267

8.4 Functor of Boolean Reference Frames 270

8.5 Functorial Boolean-Quantum Internal Relation 273

8.6 Quantum Localization: Sheaf of Boolean Reference Frames 277

8.7 Boolean Localization Systems and Boolean Germs 279

8.8 Gluing Isomorphism of Local Boolean Reference Frames 281

Chapter 9:
Functorial Entanglement and Logical Classification 287

9.1 Composite Systems and Entanglement 287

9.2 Functorial Entanglement 295

9.3 Interpretation of Quantum Events 303

9.4 EPR Correlations 307

9.5 Logical Classification of Quantum Events 313

9.6 Truth Valuations in Boolean Localization Systems 318

9.7 Boolean Germ Classification of Measurement Events 323

Chapter 10:
Quantum Localization in a Broader Conceptual Perspective 327

10.1 Boolean Covers and Physical Contexts 327

10.2 Partial Congruence and Adjunction 328

10.3 Classical to Quantum: From Groups to Categories 330

10.4 Topological Localization as Genetic Analysis 333

10.5 Boolean Localization and Decoherence 334

10.6 Sheaf-Theoretic Quantum Relational Realism 345

Chapter 11:
Recapitulation: A Semantic Bridge Between Process Metaphysics and Quantum Theory via Sheaves 353

11.1 General Conceptual Dimension 353

11.2 Subtleties of Quantum Theory 356

11.3 Localization in the Quantum Domain 360

11.4 Relational Localization via Sheaves 364

11.5 Germs and Extensive Connection 372

11.6 The Road Ahead: Differential Extensive Connection and Dynamics with Applications to Quantum Gravity 376

Bibliography 389
Index 399
About the Authors 421

PREFACE

If there is a central conceptual framework that has reliably borne the weight of modern physics as it ascends into the twenty-first century, it is the framework of quantum mechanics. Because of its enduring stability in experimental application, physics has today reached heights that not only inspire wonder, but arguably exceed the limits of intuitive vision, if not intuitive comprehension. Indeed, it is the dizzying aspect of the ascent that many physicists and philosophers have taken as quantum theory's emblematic mark of achievement; the theoretical fruits of its complex formalism are, by this inclination, selectively interpreted as nothing less than the scientific discovery of a heretofore concealed and counterintuitive physical reality—a strange new world spanned by the extremes of quantum singularity on one end, and multiverse on the other.

For a great many other physicists and philosophers, however, this currently fashionable tendency toward exotic interpretation of the theoretical formalism is recognized not as a mark of ascent for the tower of physics, but rather as an indicator of sway—one that must be dampened rather than encouraged if practical progress is to continue. Indeed, it is the great irony of modern physics that a foundation so stable and reliable as quantum mechanics has proven so vulnerable to the attractor of exotic interpretation. One reason is that after over a century of development, the key conceptual and interpretive problems remain unsettled, even in the wake of evolutionary improvements in technology and experimental methodology. Among these now infamous and interrelated problems are: [1] the problem of measurement—i.e., the dependency of measurement objectivity upon measurement contextuality; [2] quantum nonlocality—e.g., a nonstandard, nonlocal conditional probability such as $P(B|A)$ (the probability of event B given event A) when these events are space-like separated, such that event *A* somehow effects a faster-than-light probability conditionalization of event *B*. (It should be emphasized that classical conditional probabilities are not the same as quantum conditional probabilities, though they are sufficiently analogous to invoke the simpler, classical notation here); [3] the problem of coherently integrating the notions of discreteness and continuity in fundamental physical theories. This is most notoriously exemplified in the problem of integrating 'quantum' and 'classical' physical theories—i.e., 'classical' in the sense of presupposing a physical continuum, with no quantization paradigm for the evaluation of observables. Newtonian mechanics, as well as Einstein's special and general theories of relativity, are both considered classical in this sense.

In the present volume, we demonstrate how all three of these difficulties can be properly understood as interrelated aspects of a single problem: The absence in quantum mechanics of a formal means of depicting local-global relations in an extensive continuum. As mentioned above, while this problem is most popularly instantiated as the incompatibility of quantum mechanics and the general theory of relativity (an aspect of problem [3] above), we argue that its proper solution lies first in recognizing the centrality of local-global relations in *all three* of the aforementioned problems; and second, in recognizing that the overall genesis of difficulty is the presumption of a fundamentally metrical theory of extension grounded in a set-theoretic conceptual framework, the latter being incapable of adequately representing the essential logical and algebraic structural features of quantum mechanics. To be sure, the set theoretic framework has proven extremely fruitful for formal physics throughout the twentieth century; but its adequacy as a conceptual foundation for modern physics is belied not only by the incompatibility of quantum theory and general relativity, but more deeply by the fact that it renders physics vulnerable to paradoxes, singularities, and infinities.

In attending to these incoherent structures, physics has two possible routes forward: It can either incorporate them into its models by proposing their exotic *physical* instantiation—e.g., the paradoxical violation of the principle of non-contradiction in the form of 'physical superpositions' of actual system states in quantum mechanics, or the notion of 'physical singularities' in modern cosmology, etc.; or it can identify and overcome the root source of these vulnerabilities, rendering them epistemic artifacts akin to the multitude of others so explicated throughout the history of science.

In the present volume, we proceed to chart out a pathway for this latter course by first identifying the central deficiency of the conventional metrical, set-theoretic notion of extension: *That it characterizes objects as fundamental to relations*—i.e., such that relations presuppose objects but objects do not presuppose relations. Prior to quantum mechanics, this deficiency went mostly unnoticed; but since it is a signature feature of quantum mechanics that it definitively proscribes specifying the existence of objects *in abstraction* from their relations (viz., system states in abstraction from measurement), the attempt to depict quantum mechanical extensiveness as fundamentally metrical via a set-theoretic formalism—again, such that objects are more fundamental than relations—is, we believe, doomed from the beginning.

The solution we propose begins with the following thesis: Quantum mechanics exemplifies the fact that physical extensiveness is fundamentally topological rather than metrical, with its proper logico-mathematical framework being category theoretic rather than set theoretic. By this thesis, extensiveness fundamentally entails not only relations of objects, but also relations of relations; thus fundamental quanta are properly defined as 'units of logico-physical relation' rather than merely 'units of physical relata.' Objects are, in this way, always understood *as* relata, and likewise relations are always understood objectively. Objects and relations, in other words, are coherently defined as mutually implicative.

In this unique two-part volume, designed to be comprehensible to both specialists and non-specialists, we substantiate this thesis by demonstrating that a revised decoherent histories interpretation of quantum mechanics, structured within a category-theoretic topological formalism, provides a coherent and consistent conceptual framework by which local quantum events can be globally internally related both causally and logically. Further, this framework allows for a quantum mechanical description of spatiotemporal extension that is highly compatible philosophically with the mereotopological model proposed by Alfred North Whitehead, refining and enhancing the latter by elevating it from a set-theoretic basis to a category-theoretic one.

As a philosophical enterprise, we propose this conceptual framework as a speculative ontological program that includes a rigorous mathematical formalism. This provides a uniquely powerful approach to solving the three critical problems of quantum mechanics discussed above, as well as others, by integrating their solution within a coherent and intuitive ontological scheme that is both novel and applicable practically to the physical sciences.

The central thesis of the relational realist speculative philosophical program introduced in this volume is that the classical, conventional conception of the relationship between [a] *'physical object' as ontological extant*, and [b] *'history of facts' as epistemic construct* by which physical objects are characterized, must be reversed if quantum mechanics is to be coherently understood as an ontologically significant theory. That is, the classical conception of a history as essentially contextual and therefore primarily epistemic—a particular story expressing particular knowledge of fundamental physical objects—must be reconceived, such that physical objects are not merely *understood by* their fundamental histories, but rather *understood as* fundamental histories of quantum events. This requires a novel reconceptualization of 'ontological' and 'contextual' as mutually implicative features of every quantum event, wherein the latter is understood as the fundamental, concrete constituent by which the natural world is physically and logically describable.

This primacy of objective quantum events (alternatively, quantum 'facts') is, in the opinion of the authors, the critical starting point for any attempt to provide a viable ontological interpretation of the standard formal framework of quantum theory. In this framework, quantum events are identified as measurement outcomes referent to corresponding physical observables. The theory then provides the means of relating these events. In this respect, the conceptual complexity of any ontological interpretation of quantum theory stems from two factors: [1] The actualization of a measurement outcome event representing the state of a quantum system, though always globally objective, can only be predicted probabilistically and *contextually*—i.e., relative to a particular *local Boolean measurement context* of a selected observable (that is, a context wherein measurement outcomes can be expressed as mutually exclusive and exhaustive true / false propositions). It is always via such local Boolean contextuality that the universe, represented by a global state vector, is decomposed into 'system,' 'measuring apparatus,' and 'environment' with their respective state vectors. Equally important, the probability valuations can be affirmed only retrodictively, *after* a measured result (i.e., a novel fact / unique actualization of a potential outcome state) has been registered by the corresponding measuring apparatus. [2] The totality of events related to the behavior of a quantum system cannot be represented within the *same* local Boolean measurement context due to the property of non-commutativity of quantum observables.

These two factors together necessitate a thorough rethinking of our conceptual and mathematical representation of the notion of a physical continuum suited to the quantum domain of discourse. The first factor illustrates the prominent role that *potential* relations play in the process of quantum measurement: Quantum mechanics always entails the evolution of [a] *potential* outcome states that *cannot* be integrated in terms of classical, Boolean logic (e.g., Schrödinger's Cat is alive *and* dead) to [b] *probable* outcome states that *can* be integrated in terms of classical, Boolean logic (Schrödinger's Cat is alive with probability x, *or* dead with probability y; and since $x + y = 1$, one and only one of these unique outcomes *must* occur). Apart from a local Boolean measurement context, the evolution of potentia to probability cannot ensue; and as we will argue, this evolution is the central engine of quantum mechanics, without which locally contextualized events cannot be integrated into the globally objective, logically consistent histories ubiquitous to experience (empirical or otherwise).

The second factor likewise illustrates the contextual significance of the empirical requirement of preparation procedures for the evaluation of observables via quantum measurement: The determination, by these procedures, of the *initial actual state* of the system and measuring apparatus establishes the local contextualization definitive of their *final actual state* terminal of the measurement. This presupposed contextual correlation is typically understood as an instrumental desideratum rather than a philosophical one—that is, a desideratum satisfied via an ad hoc conceptual intervention arbitrarily imposed upon the standard formalism (viz., von Neumann's projection postulate). By contrast, we will demonstrate that this instrumental desideratum can instead be interpreted both philosophically and logico-mathematically as a *necessarily presupposed feature of quantum mechanics* when these contextual correlations are depicted topologically—viz., a topological localization of quantum observables with respect to local Boolean contexts.

It must be emphasized that the philosophical implications of these two factors cannot easily be separated from their practical implications. This fact has previously been explored extensively by the authors individually—Epperson in his work relating the logical, mathematical, and philosophical features of Alfred North Whitehead's work to the decoherent histories interpretations of quantum mechanics;[1] and Zafiris in his modeling of the topological features of quantum mechanics by way of category-sheaf theory.[2] The deep compatibility underlying these efforts has motivated the authors to seek an integrated philosophical and mathematical framework, which we have termed 'relational realism,'[3] that advances refined conceptions of internal relation and extensive connection in Whitehead's process theory, amenable to a topological reinterpretation, and synthesizes these with the framework of category-sheaf theoretic Boolean localization systems viewed from a process standpoint.

To glimpse the essence of this synthesis, the reader is invited to imagine a conceptual triangle, where on the top node sits quantum theory (formulated in the usual Hilbert space language or in the decoherent histories language) and on the two bottom nodes sit the philosophical structure of Whitehead's process theory on one side, and the mathematical structure of category-sheaf theory on the other. The correlation of the quantum node with the process-theoretic node, as it has been presented by Epperson, finds its conceptual correspondence to the correlation of the quantum node with the category-sheaf theoretic node, as it has been presented by Zafiris—a correspondence both in terms of terminology and interpretation. Thus, the conceptual triangle commutes as we circulate around it, allowing for not only a detailed correlation of the process-theoretic node with the sheaf-theoretic node, but one that also preserves the already established correlations between each of these and the quantum node.

The focus of the present volume is the demonstration of the applicability and ultimate relevance of this synthetic scheme toward a coherent and empirically adequate ontological interpretation of quantum theory, and in this regard, a subtle balance is kept between the philosophical aspects and the corresponding mathematical aspects as these converge. To that end, part I introduces the philosophical foundations of relational realism as instantiated in quantum mechanics, the latter presented in introductory fashion so as to be suitable for non-specialists in physics or mathematics. Part II, in turn, further elaborates this conceptual framework in formal mathematical language appropriate for specialists in physics and mathematics, such that philosophical scheme's technical applicability to modern physics can be rigorously demonstrated. Despite this structure, however, each part is properly understood only within the context of the overall synthesis.

It is important to note that the construction of this synthesis does not follow a strictly linear course. Instead, it unfolds in such a way so as to show, from a convergence of both philosophical process theoretic and mathematical category-sheaf theoretic views, the adequacy and indeed advantage of the relational realist framework in the development of a coherent ontological interpretation of quantum mechanics—one that can be understood as exemplifying, more broadly, a coherent and empirically adequate philosophy of nature.

To be sure, the conceptual breadth of both Whitehead's process theory and categorical sheaf theory far exceeds their restricted application to quantum theory as explored herein, and we consider this to be advantageous for our objectives. By employing these rich conceptual philosophical and mathematical frameworks, taking particular advantage of their systematic coherence, we hope to contribute to the formulation of a realistic ontological interpretation of quantum mechanics that avoids the paradoxes, inconsistencies, and counter-intuitiveness typical of many alternative interpretations.

This research was supported by the John E. Fetzer Memorial Trust (Grants D11C36 and D21C62). The authors are grateful to the following for their invaluable advice and discussion, technical critiques, and collaboration: Jan Walleczek, Tim Eastman, Roland Omnès, Stuart Kauffman, David Finkelstein, Anastasios Mallios, Stephen Selesnick, László Szabo, and Vassilios Karakostas. And thanks especially to our colleague and great friend Karim Bschir, whose insights and contributions to this project have been invaluable. Most of all, we wish to thank Bruce Fetzer and the John E. Fetzer Memorial Trust, without whose sponsorship this work would not have been possible.

Notes

1. See, for example: Michael Epperson, *Quantum Mechanics and the Philosophy of Alfred North Whitehead* (New York: Fordham University Press, 2004); Michael Epperson, "Quantum Mechanics and Relational Realism: Logical Causality and Wave Function Collapse," *Process Studies* 38, no. 2 (2009): 340-66.

2. See, for example: Elias Zafiris, "Quantum Event Structures from the Perspective of Grothendieck Topoi," *Foundations of Physics* 34, no. 7 (2004): 1063-90; Elias Zafiris, "Generalized Topological Covering Systems on Quantum Events' Structures," *Journal of Physics A: Mathematical and General* 39, no. 6 (2006).

3. Relational realism and its topological interpretation of quantum mechanics ought not be confused with the 'relational quantum mechanics' program of Carlo Rovelli. [See, for example: C. Rovelli, "Relational quantum mechanics," *International Journal of Theoretical Physics* 35 (1996): 1637-78.] Many aspects of both programs are sufficiently compatible for fruitful conversation; however the underlying conceptual frameworks and philosophical implications of each are very different.

PART I

Philosophical Foundations of Quantum Relational Realism

Elementary quantum relational events, logical causality, mereotopological extension, and the internal relation of the local to the global in quantum mechanical predication

MICHAEL EPPERSON
A.M., Ph.D. (University of Chicago)

CHAPTER 1

Introduction

Relational Realism: A Program in Speculative Philosophy

> Speculative Philosophy is the endeavour to frame a coherent, logical, necessary system of general ideas in terms of which every element of our experience can be interpreted . . . It is the ideal of speculative philosophy that its fundamental notions shall not seem capable of abstraction from each other. In other words, it is presupposed that no entity can be conceived in complete abstraction from the system of the universe, and that it is the business of speculative philosophy to exhibit this truth.[1]
>
> *Alfred North Whitehead*

The chief mark of progress in the evolution of a philosophical worldview, whether its foundation be scientific, humanistic, theistic, or some integration of these, is the extent to which it is able to coherently accommodate diverse categories of inquiry without either arbitrary dispensation or internal contradiction. The precise manner in which a worldview deals with incommensurable categorical principles and the internal contradictions and paradoxes often borne of them is of first importance, both to the proper understanding of the particular worldview and its implications, and also to its evaluation in contrast to competing worldviews. Indeed, one could argue that the emblematic feature of any philosophical genre in the history of Western philosophy, if not all philosophy, is its method of attending to the conceptual incompatibilities within its scope.

By that general metric, there have been two dominant approaches[2] by which principles incommensurable when treated categorically—e.g., objectivity and subjectivity, necessity and contingency, conceptual and physical, infinite and finite, global and local, continuous and quantum, potential and actual—have been accommodated in Western thought:

[1] Their categorization as fundamentally *mutually implicative* at some deeper level of analysis:

By this method of 'dipolar' relation it is explicitly recognized that the conception of one principle necessarily requires reference to its counterpart principle. Thus, each relatum constitutive of dipolar conceptual pairs is always contextualized by both the other relatum *and* the relation as a whole, such that neither the relata (the parts) nor the relation (the whole) can be adequately or meaningfully defined apart from their mutual reference. It is impossible, therefore, to conceptualize one principle in a dipolar pair in abstraction from its counterpart principle. Neither principle can be conceived as 'more fundamental than,' or 'wholly derivative of' the other.

Mutually implicative fundamental principles always find their exemplification in both the conceptual and physical features of experience. One cannot, for example, define either positive or negative numbers apart from their mutual implication; nor can one characterize either pole of a magnet without necessary reference to both its counterpart and the two poles in relation—i.e., the magnet itself. Without this double reference, neither the *definiendum* nor the *definiens* relative to the definition of either pole can adequately signify its meaning; neither pole can be understood in complete abstraction from the other.

[2] Their categorization as fundamentally *mutually exclusive* at the 'deepest possible' level of analysis:

By this method of 'bipolar' relation it is implied (often erroneously) that the definition of one principle does not necessarily entail reference to its counterpart principle. Therefore it is possible to conceptualize one principle as fundamental to the other. The Platonic dualism separating conceptual and physical objects is perhaps the preeminent example in Western philosophy. Its later rehabilitation in the Cartesian dualism of thought (mind) and extension (matter) is, many would argue, a central feature of the modern Western worldview, given the popular dualistic characterization of mind and brain, or soul and body. Familiar examples of bipolar relation in physics include the classical attempt to characterize basic physical processes as either fundamentally continuous or quantum, fundamentally wavelike or particulate, etc.

1.1 Bipolar Dualism: Reducing Contrast to Mutually Exclusive Relata

When one considers the long lineage by which one might trace the ideas of the ancient Milesians through the ideas of Parmenides and Heraclitus, and onward through Plato and Aristotle, Descartes and the rationalists and Locke and the empiricists, all the way to Bohr and von Neumann, it is clear that the relation of contrasting categorical principles as mutually exclusive relata (Method 2 above) has increasingly dominated Western thought over the centuries. This domination has today become especially acute thanks to its conventional projection onto the hypothetico-deductive and reductive scientific method. Indeed, despite a variety of schematizations throughout the history of philosophy, the general thesis of mutually exclusive fundamental principles, both ontological and epistemic, has arguably become in the twenty-first century the defining characteristic of practically every prevailing scientifically informed worldview. One further sees that there have been two general modes by which mutually exclusive fundamental principles have been brought into relation by this method:

[2a] The qualification of mutually exclusive fundamental principles as *complementary* and *irreducible* characteristics of a more general, unified, and necessarily transcendent ontology that lies beyond the scope of rational systematization:

As exemplified by modern science, for example, Bohr's principle of complementarity describes the fundamental physical properties of reality in terms of conjugate pairs of properties. Epistemically, these pairs are mutually implicative since they are Fourier transform pairs; thus a more precise specification of 'particle position' necessarily entails a less precise specification of 'particle momentum.' But ontologically, the matter becomes more complicated, since the conception and definition of particle position does not require reference to the concept of momentum; whereas the conception and definition of momentum does require reference to the concept of position. As will be seen in the discussion of Method 2b, below, such incongruities often appear to justify the attempt to assimilate one concept to the other—one aspect of nature to another. The present case of position and momentum, for example, is an echo of Parmenides' attempt to assimilate the *appearance of change* to the *fact of permanence*, versus Heraclitus's attempt to assimilate the *appearance of permanence* to the *fact of change*. Similarly, position-momentum complementarity finds its reflection in one of the most profound problems in modern physics—reconciling quantum theory's depiction of nature as fundamentally discrete, and general relativity's depiction of nature as fundamentally continuous. Efforts to quantize the continuum via the quantization of gravity can, in this sense, be seen as an effort to assimilate or reduce the continuous in nature to the discrete.

A cleaner example of complementarity, closely related to that of position and momentum, yet (seemingly) immune to the above difficulty and its solution via assimilation, is wave-particle complementarity, where neither wave nor particle requires direct reference to its counterpart for its definition. Each concept is therefore independent of the other and, in terms of its ontological significance, incommensurable in complementary relation to its counterpart.

In answer to these and other ontological difficulties associated with the concept of complementarity, Bohr's solution was to circumvent them altogether by asserting that physical qualifications are *essentially epistemic*—that is, descriptive of our knowledge of reality rather than reality itself. Any ontological implications of these descriptions therefore lie outside the scope of physics. "In physics," Bohr writes, "our problem consists in the co-ordination of our experience of the external world . . ." such that "in our description of nature the purpose is not to disclose the real essence of phenomena but only to track down as far as possible relations between the multifold aspects of our experience."[3]

Thus the coherence of this method of reconciling mutually exclusive principles via complementary relation rests upon a similarly sharp separation of epistemology and ontology as mutually exclusive concepts themselves, such that a 'fact of knowledge' no longer implies 'knowledge of a fact' as it does when epistemology and ontology are characterized as mutually implicative. Realism, if it is to remain viable in physics and natural philosophy, is by this method recast as 'transcendent realism,' implying an inaccessible level of reality where complementary, mutually exclusive principles would ultimately find their proper unification.

[2b] The qualification of mutually exclusive categorical principles as higher order characteristics or modes of a deeper, fundamental ontology and epistemology that can be further generalized by the reduction or assimilation of one characteristic to the other.

The history of philosophy reveals a clear tendency toward this method; indeed, one could argue that its dominance—seen especially in the philosophy of nature during the early modern period—was the primary fuel for its later dominance in modern science and, via the latter, its increasing popularity in most modern Western worldviews. One can find its origins in much of pre-Socratic philosophy, perhaps most dramatically in the Eleatic School. As an admonishment against the relation of actuality and potentiality as mutually exclusive, fundamental aspects of reality, for example, Parmenides gave a convincing argument for the reduction of potentiality *to* actuality: Anything that "can be," he asserted, "must already be." There can be no "coming into being" because anything that comes into being has only two possible derivations: It either came *from* being, in which case it already exists; or it came from non-being—i.e., nothing—in which case it, too, is nothing, since nothing comes from nothing. Indeed, for Parmenides, one cannot even *imagine* a 'potential' actuality that was not already actual in some way, since imagining something that did not actually exist would literally amount to imagining 'no thing,' and thinking about nothing, Parmenides argues, is the same as not thinking.

Since by this reasoning potentiality is merely a sensory-epistemic abstraction from fundamental actuality, existence is therefore both eternal and unchanging—the Parmenidean 'One'; any perceived differentiation is thus only apparently real as disclosed via sensation, not actually real as disclosed by reason. Though this laid the groundwork for Platonic idealism as another type of philosophical reduction from mutually exclusive fundamental principles—ideal / thought / form vs. material / sensation / extension—it is important to note that Parmenides was not an idealist, but rather a monistic materialist. Thus one can find in many modern physical theories a number of Parmenidean reflections: Everett's 'relative state' interpretation of quantum mechanics[4] (better known as the 'many worlds' interpretation, or MWI, which is an extrapolation of Everett's original concept) makes a similar assimilation of potentiality to actuality, such that every possibility is understood as an actuality in some particular universe. Further, the division of reality into multiple, mutually exclusive universes is understood by the theory as a higher order abstraction beneath which lies a first order, unified 'multiverse'—in many ways akin to the Parmenidean 'One.'

Similarly, the Bohm-Hiley non-local hidden variables interpretation of quantum mechanics[5] describes the universe as epistemically divisible but ontologically undivided. Quantum physics as an epistemic enterprise glimpses this ontological unity via its fundamental characterization of the universe as a unified, 'implicate order' of actualities—which would at first glance seem to disagree with the Parmenidean worldview, since order implies division. But in the Bohm-Hiley interpretation, past and future are symmetrically related, and therefore ontologically indivisible; any potential 'coming to be' in the future is thus already contained in what already exists and existed. Past, present, and future, though epistemically distinct within the context of our finite observational structures, are nevertheless ontologically unified in a quasi-Parmenidean 'One.'

Bohm writes:

> If it were possible for consciousness somehow to reach a very deep level, for example, that of pre-space or beyond, then all "nows" would not only be similar—they would all be one and essentially the same. One could say that in its inward depths now is eternity . . . (But eternity means the depths of the implicate order, not the whole of the successive moments of time.)[6]

Further, for Parmenides, the reduction of potentiality to actuality implied related reductions of other mutually exclusive principle-pairs, such as creativity vs. discovery. As noted above, in Parmenidean philosophy every 'imagining' is ontologically reducible to thinking about existence. And similarly, within the depths of the veiled implicate order underlying the Bohm-Hiley undivided universe, all potentia are ontologically reducible to actuality; novel creativity is ontologically reducible to discovery of the already extant. The ground of any experience interpreted as creativity (instead of understood as discovery) is the epistemic restriction or contextualization of all experience within our necessarily finite observational structures.

Bohm writes:

> As long as we restrict ourselves to some finite structures of this kind, however extended and deep they may be, then there is no question of complete determinism. Each context has a certain ambiguity, which may, in part, be removed by combination with and inclusion within other contexts . . . If we were to remove all ambiguity and uncertainty, however, creativity would no longer be possible.[7]

1.2 Bipolar Dualism and the Scientific Method: The Example of Physical and Philosophical Cosmology

The success of the method of reduction and assimilation in science, at least as measured by the apparent ability to predict and control nature yielded by this method, has inevitably led to a host of physics-inspired philosophical and cosmological reductions such as those discussed above. The philosophical cosmologies borne of the 'many worlds' and 'implicate order' interpretations of quantum mechanics are just two of many modern examples of reductive philosophical cosmologies inspired by and grounded in scientific reductionism. Indeed, for science-dominant worldviews in general, the thesis of fundamental mutually exclusive principles, and their attempted interrelation via either of the two modes, 2A and 2B described above, has today become their defining characteristic. The following pairings, for example, can be considered respectively: classical mechanics vs. quantum mechanics; fundamentally continuous spatiotemporal extension in general relativity vs. fundamentally discrete spatiotemporal extension in quantum mechanics; wave-particle dualism; local-efficient causality vs. nonlocal-quantum causality; classical determinism vs. quantum indeterminism; objective vs. contextual; infinite vs. finite; potentiality vs. actuality; form vs. fact . . .

Even when one turns attention to worldviews whose foundations either exceed or altogether ignore the philosophical foundations of modern science—e.g., various theistic-dominant worldviews—the thesis of mutually exclusive fundamental principles is exemplified in related pairings, including: determinism vs. free will; necessity vs. contingency; perfection vs. imperfection; permanence vs. change; heteronomy vs. autonomy . . . Perhaps most infamously, within the rhetorical spectacle of the modern 'science and religion debate,' the opposition of cosmogonic creation and cosmological evolution finds breathless celebration in popular portrayals of quantum cosmology. The primary fuel by which this debate is sparked into conflict is the stubborn characterization of creation and evolution, as well as necessity and contingency, as mutually exclusive concepts whose only proper mode of coherent relation is the reduction or assimilation of one concept to the other.

Many quantum cosmologists, for example, argue that the universe contingently evolved into existence quantum mechanically as a primordial, initial extant (whether as a singularity or some other form), *ex nihilo*, via a quantum probability function—viz., a 'universal wavefunction' yielding an unconditional probability that 'our' universe would actualize amid the primordially extant distribution of 'alternative' potential universes constitutive of this wavefunction. Some of these theorists have gone still further, promulgating quantum cosmology as a scientific validation of the ideological claim that a primordial, necessary extant such as 'God' is inherently irrational. What is neglected in this argument, however, is that when quantum mechanics is characterized in this way, as a cosmogonic engine operative primordially, *in nihilo*, then the logical structure presupposed by quantum mechanics must *itself* be understood as a primordial, necessary extant—a presupposition which, of course, belies the qualification '*in nihilo*.' In other words, when quantum mechanics, a logico-relational structure expressible in the language of mathematics, is depicted as somehow operative in abstraction from a universe of relata, then this structure itself is necessarily ontological and primordial. One cannot, of course, coherently embrace both this ontological notion of logic and the notion that logic is a purely epistemic epiphenomenon of the human mind. Yet this is precisely what many quantum cosmologists propose to do in their efforts to reduce all necessity to contingency.

Finally, one might also consider those worldviews in which neither science nor religion is seen as foundational. These might simply be termed 'humanistic,' where fundamental mutually exclusive principles are housed, in the most general sense, within the dualism of subjectivity and objectivity. This pairing has its more specific application within a variety of disciplines in the humanities: In literature, deconstruction vs. intentionalism; in philosophy, conceptual vs. physical; mind vs. matter; thought vs. extension; the order of logical implication vs. the order of causal relation; necessity vs. contingency, permanence vs. change, unity vs. diversity, and so on.

In all of these examples, one can find within the history of Western thought a wide variety of systematic attempts to bridge fundamental mutually exclusive conceptual relata by way of either the method of complementarity in the context of transcendent realism (Method 2A above) or by reduction and assimilation (Method 2B). In the early modern period, the overly-general division of philosophy into the rationalist vs. empiricist traditions, each with deep roots in classical philosophy, well reflects the popularity of these methods. Though a careful survey of the philosophy of this period falls outside the scope of this chapter, readers familiar with the history of philosophy will be able to experiment with the above two classifications—Methods 2A and 2B; for example, Spinoza attempted to assimilate the causal order to the logical order, the contingent to the necessary; Locke attempted to assimilate the logical order to the causal order—both examples of Method 2B. Kant's Transcendental Philosophy can likewise be seen as an example of Method 2A, the method of transcendent realism (i.e., in the sense of his transcendental idealism, by which 'things in themselves' transcend the understanding). Fundamental complementarity, within this context, is evinced throughout his system—his discussion of 'conditioned' and 'unconditioned' knowledge in the Transcendental Dialectic, for example.

1.3 The Fallacy of Misplaced Concreteness

As is especially clear in the example of Kant, the chief deficiency of the categorization of incommensurable principles as fundamentally mutually exclusive at the 'deepest possible' level of analysis is its presumption that the bounds of reason have been reached. Fundamental complementarity marks this boundary in Method 2A, where the problem of mutually exclusive foundational principles is relieved only by reference to some ineffable, transcendent unification or implicate order that lies on the other side of the boundary—either beneath the veil of finite observational contexts or, even worse, beyond the supposed scope of reason itself. But this merely exchanges one incoherence for another: the incoherence of incommensurable categorical principles is traded for the incoherence of qualifying (i.e., knowing) the 'unknowable' as unknowable, via the reasoning of the unreasonable—or, in modern science, the misuse of the hypothetico-deductive method to either posit or validate a theory that is, in principle, unfalsifiable.[8]

Method 2B is equally problematic—the attempted reduction or assimilation of mutually exclusive fundamental relata, one to the other, so that one is re-defined as concrete and the other abstract—one ontologically significant, the other an epistemic derivative or artifact. Again, the method of reduction and assimilation rightly recognizes the need to bring into coherence mutually exclusive, incommensurable categorical principles; but it wrongly grasps for that coherence by arbitrarily restricting the speculative schematization of the experience of nature to certain preferred categories of thought, in exclusion of other categories that could just as reasonably be characterized as fundamental. By this method, nature is always *either* fundamentally physical *or* fundamentally conceptual; either fundamentally continuous or fundamentally quantum; either fundamentally finite or infinite; either fundamentally deterministic or indeterministic. When one considers the increasingly profuse inflations of physical cosmological models into metaphysical cosmologies, their stipulated significance is belied by the fact that one can casually assemble practically any combination of the above qualifications and find a correlate interpretation of quantum theory or string theory or some other physical cosmology that can accommodate it.

At the level of principles deemed fundamental to any given scientific theory, then, the method of purely reductive science breaks down. The reason is because theorists enamored of the notion of an 'ultimate unifying reduction' fail to recognize the ineluctable possibility of deeper levels of abstraction underlying any 'ultimately reduced' principle or 'ultimately unified' theory qualified as 'fundamentally concrete' by this method. It is a conceptual hazard that Whitehead famously termed the 'fallacy of misplaced concreteness,' and its pertinence to modern science has never been more important—especially for those worldviews which, in their ongoing construction, use science as bedrock, framework, and scaffold. For these worldviews, science is tasked not only with constructing a fundamental *description* of the universe; the target height of construction is nothing less than a fundamental *explanation* of the universe via sheer deductive reduction. This bounding leap from fundamental scientific description to fundamental scientific explanation is one that over the past several decades has perhaps been less careful than such a leap would warrant. And likewise, these increasingly routine efforts typically receive little critical attention despite the obvious logical obstacles that belie the reasonableness of even attempting such a feat.

Again, one need only consider the growing number of philosophically loaded discussions of various physical cosmologies, the most esoteric of which have been embraced by the popular media for an entertainment-driven education of the public. The tacit stipulation of these presentations is that the demonstrable ability of the scientific method to construct sound reductionist descriptions of nature *in itself* warrants its broader application to the task of explaining or 'accounting for' nature's very existence. What is neglected in this misapplication of the scientific method is the fact that a 'fundamental explanation' of nature constructed via the method of deduction, such as that instantiated in the hypothetico-deductive scientific method, is impossible. This is because the categorical first principles at the base of any deductive scheme are always necessarily presupposed. Science, for example, cannot 'explain' the logical order—i.e., account for its existence—since the language and methodology of science necessarily presuppose this order.

To clarify this point, let us return briefly to the topic of quantum cosmology introduced earlier: In their 1983 paper "The Wave Function of the Universe,"[9] Stephen Hawking and James Hartle propose a quantum mechanical explanation of the origin of the universe whereby the universe spontaneously creates itself *ex nihilo* as a quantum mechanical probability function. (Hawking later asserts the theological implications of this model by stating that it allows for the elimination of the notion of God as a primordial extant in any rational explanation of the origin of the universe.[10] It is a claim that has since been repeatedly echoed by other cosmologists, primarily in mass-market books and media, and as such is arguably one of the driving reasons quantum cosmology has captured such popular notoriety today.)

Elaborating upon the previous discussion, what is neglected by both this model and its characterization by Hawking and other advocates is the fact that a quantum mechanical wavefunction is more than just a 'generic' integration of undefined potential physical states that, when applied universally, yields an unconditional probability distribution of possible universes; the wavefunction, rather, is a fundamentally *relational* structure depicting *conditional* probabilities for the evolution of an actual final physical state—e.g., that of the nascent universe—not from 'nothing,' but from some actual *initial* physical state. The wavefunction, in other words, is an integration of potential final physical states that is always conditionally *contextualized* by some actual, initial physical state. Further, it is a *logically conditioned* integration—specifically a Boolean logical conditionalization—presupposing final states that are always valuated as probabilities, and thus always mutually exclusive and exhaustive. Quantum mechanics, then, whether applied to the description of microscopic systems in a laboratory, or to the description of the universe itself, always presupposes a logical order by which potential outcome states are ultimately reduced into mutually exclusive and exhaustive (i.e., logically coherent) probability outcomes. Indeed, it is only via a presupposition of Boolean logic that probability theory is possible at all; and likewise, it is only via probability theory that quantum mechanics is possible.

But again, more than just presupposing a logical order, quantum mechanics presupposes a correlation of this logical order with *some* causal order of actualities from which potential outcome states derive. Quantum mechanics always begins with actualities, in other words—an initial actual system state—not simply bare, uncontextualized, 'generic potentialities' *in nihilo*. Indeed, it is impossible to define a potential outcome state in quantum mechanics without contextual reference to both an initial actual system state and an anticipated actual outcome state, just as it is impossible to define *any* particular potentiality without contextual reference to some actuality. Assigning the name 'quantum vacuum' to an initial primordial actual state of the universe, thus attempting to characterize it as 'nothing' as is the convention in most quantum cosmologies, neglects the fact that this state is both actual and logically ordered; that is, the causal relations of this primordial state are defined quantum mechanically as an evolution from initial actual state to final actual state, and this definition presupposes the structure of logical implication (as well as other logical structures to be discussed later in this volume). Therefore, quantum mechanical *cosmologies* cannot be properly described as *ex nihilo cosmogonies*, productive of a 'randomly' generated universe from nothing; indeed, even randomness mathematically presupposes an underlying logical order for its definition.

The particular example of quantum cosmology can be seen as exemplifying a broader truth: that any attempt to construct a philosophical cosmology simply by clothing it as a 'purely scientific' physical cosmology, borne of and supposedly validated solely by the scientific method and thus purified of any philosophical presuppositions, is doomed to incoherence. This is because any deductive scheme of reasoning, including the hypothetico-deductive method of modern science, must begin with first principles that are necessarily presupposed, either implicitly or explicitly. The first principles at the foundation of any deductive scheme cannot themselves be deduced, since there is no more general principle by which such a deduction might proceed. Thus, again, science cannot 'explain' the logical order—i.e., account for its origins—since the language and methodology of science necessarily presuppose this order. In the above example, quantum mechanical cosmological models might offer valuable fundamental descriptions of the earliest stages of the evolution of the universe; but when extended to the task of accounting for the origin of the universe *ex nihilo*, the model's reliance on a primordial logical order precludes its success.

Whitehead's fallacy of misplaced concreteness, when applied to these increasingly popular inflations of physical cosmology to philosophical cosmology, is crucial because it reminds us that science can never 'explain away' that which it necessarily presupposes. The logical order underlying mathematics can never be deductively explained or reductively accounted for by a scientific description of the causal order because the logical order is necessarily presupposed *by* the method of scientific description; thus, fundamental reductionist scientific descriptions of nature can never attain the status of fundamental explanation, because there is always a deeper level of abstraction underlying any deductive or reductive scheme, scientific or otherwise.

One of the earliest and clearest illuminations of the deficiency of the reductive-deductive method when applied to the task of 'fundamental' or 'complete' explanation can be found in Plato's *Theaetetus*. There, Plato challenges the idea that a sufficiently deep reduction can serve as a sturdy bridge across the chasm separating description and explanation—appearance and reality. In the dialogue, Socrates tells Theaetetus of a dream he once had wherein he had learned of a theory of explanation by which all things are described as complexes of simpler elements, themselves complexes of still simpler elements. This reduction continues until the simplest elements are apprehended, at which point a complete and true explanation of the initial object is achieved. It is only then, proposes Socrates, that one can be said to possess true knowledge—i.e., explanation—of the object. Theaetetus eagerly accepts this epistemology, but Socrates advises caution; he explains that in his dream, the most fundamental elements are incapable of description by this epistemology, given that they contain no simpler parts. Therefore, the most fundamental elements are themselves unknowable. How, asks Socrates, can the unknowable be the foundation and ultimate justification of knowledge?

Though Plato's admonition on the limits of a purely reductive-deductive epistemology would ultimately be drowned out by the celebration of the manifold profound achievements of the modern scientific method over two millennia later, the historic advances in physics and mathematics emblematic of the twentieth century would also bring with them a number of sharp reminders of these limits. The misapplication of quantum mechanics to the construction of a 'purely physical' scientific cosmogony, discussed previously, is one example; but Plato's admonition would also find its rehabilitation in the key contributions of Gödel and Russell to the philosophy of mathematics—contributions crucial to the construction of a coherent ontological interpretation of quantum theory. One sees this well reflected, for example, in Russell's paradox, which attends to the logical problem of predicating totalities. This difficulty, as will be discussed in detail in chapter 3, finds its relevance in the quantum mechanical predication of system states defined as totalities—e.g., the universe itself when considered quantum mechanically—where the separation of 'measured system,' 'measuring apparatus,' and 'external environment' is properly recognized as arbitrary. This recognition is conventionally accepted in quantum theory, and indeed is explicitly embraced in the case of quantum cosmology, but its underlying implication that all wavefunctions are therefore ultimately 'universal' requires careful exploration.

Another exemplification of Plato's admonition in the *Theaetetus* can be seen in Gödel's two incompleteness theorems, which together establish that no effectively methodical deductive theory capable of expression in the language of arithmetic can ever be both internally consistent *and* complete. The reason, evocative of Socrates' dream, is that any formal theory that is both internally consistent and allows for the deduction of arithmetical proofs will always presuppose some arithmetical statement that is both true and incapable of proof by the theory. The specific relevance of Gödel's theorems to quantum mechanics will be taken up in greater detail in chapters 2 and 3; but in the context of this introductory discussion, it is useful to consider their pertinence to the philosophical problem of relating the *order of causal relation* and the *order of logical implication*, as well as the problem of mutually exclusive categorical principles discussed earlier.

Consider, for example, the various self-referential paradoxes often associated with Gödel's theorem, such as the Epimenides paradox, given in the ancient Cretan philosopher's infamous utterance: Κρῆτες ἀεί ψεῦσται ("Cretans, always liars"). There are two features of this paradox by which its relevance to quantum mechanics will be demonstrated throughout this volume: First, the statement implies both a causal order and a logical order; second, it entails the predication of a totality. As we will see, in the case of both the Epimenides paradox and quantum mechanics, the root cause of the incoherence is the presumption of a *closed totality*, within which causal relation and logical implication—and more broadly, physical relation and conceptual relation—are treated as mutually exclusive categories, as are 'truth' and 'falsity.' The solution we propose in this volume is to instead recognize these as mutually implicative categories within an open totality defined as a history-in-process. The 'decoherent histories' approach to quantum mechanics, as we will see, is a fundamental exemplification of the latter. And likewise, the philosophy of Whitehead, in the context of this understanding of totality, will provide a solid framework for the coherent relation of the physical and the conceptual, as well as the causal and the logical, as mutually implicative features of nature.

To call Whitehead's philosophy revolutionary in this regard would be an understatement; for by the end of the early modern period, during which the foundations of the modern scientific method were forged, contingency via the order of causal relation and necessity via the order of logical implication, as well as other associated bipolar conceptual pairings, were generally treated by natural philosophers as incommensurable, mutually exclusive concepts. This tendency maintained its dominance throughout Whitehead's time, and indeed to the present day. The various proposals for their coherent relation as mutually exclusive typically entailed, as discussed thus far, either their treatment as complementary epistemic features and co-ordinations of our experiences of an otherwise transcendent reality; or as second-order epistemic concepts that can be distilled into a single first-order principle via the reduction or assimilation of one concept to the other. Whether continental rationalist or British empiricist, the common goal was to bridge the two sides of the Platonic chasm separating 'what appears to be' from 'what is reasoned to be.' Despite the richness of detail to be found in the various philosophical systems proposed during this period, the tendency toward the method of mutually exclusive relation of the causal and logical orders—and more broadly, again, the physical and the conceptual, as well as the contingent and the necessary, respectively—resulted in two alternative mappings across Plato's chasm:

[1]	appearance:	sensory perception of contingency via causal relation
	reality:	rational conception of necessity via logical implication
[2]	reality:	sensory perception of contingency via causal relation
	appearance:	rational conception of necessity via logical implication

It is clear that both of these mappings *together*, without any secure bridging, have been inherited by most modern, scientifically informed worldviews, and the unavoidable incoherence generated by the attempted embrace of both mappings at once has proven to be a major impediment to the evolution of philosophically informed science in general. One sees, for example, the epistemic separation of logical implication and causal relation well reflected, respectively, in the increasingly dubious distinction between 'theoretical physics' (a.k.a. 'mathematical physics') and 'experimental physics.' For many, these are seen not just as subdisciplines, but rather as 'branches' of physics that are somehow epistemically separable. Of course, experiments cannot be published without the appropriate theoretical work by which data are analyzed. It is often the case, however, that experimentalists are unable to assess the significance of their own data until a theorist is brought into the team, often long after the experiment has been completed.

Likewise, the straying of theoretical physics from its grounding in falsifiable hypotheses and experiment has indeed led to tantalizing imaginative constructs like the various quantum, multiverse, and string theory cosmologies mentioned earlier. And while it is true that these have captured the popular imagination and inspired increased interest in science, it is equally true that when elegance of formalism too often trumps empirical applicability, the measure of scientific progress begins to derive from the measure of its appeal, rather than its appeal deriving from its progress.

Ironically, for many, the greatest appeal and elegance of theoretical physics comes from its application to the 'external' foundational questions that are, in principle, empirically inaccessible—i.e., those that exceed the universe itself. It is ironic because 'elegance' here implies a brute Pythagorean reduction of the world to bare mathematics—the reduction of all contingency to calculable necessity. But in the same way, one can argue that experimental physics entails its own, complementary reduction; for when applied via disciplines such as molecular biology and neurophysiology to 'internal' foundational questions—e.g., the relationship of mind to brain, of conception to perception, of 'personality' to the physical structure of DNA—it implies a purely physical causal closure—a reduction of the world to 'fundamental' particles and forces.

In summary of the current discussion, one can trace throughout the histories of both philosophy and modern science a tendency to characterize the causal and logical orders (as well as contingency / necessity, and physical / conceptual, respectively) as mutually exclusive foundational concepts that are relatable by either [1] complementarity, or [2] reductive assimilation of one concept to its counterpart. The method of complementary relation is, by its own admitted limitations, incapable of yielding anything more than an arbitrary explanation of either the logical or causal orders as complementary epistemic co-ordinations of our experiences of an ontologically transcendent reality; and the method of ontological reduction and assimilation, when it reaches the level of fundamental explanation, is doomed by the fallacy of misplaced concreteness.

1.4 Dipolar Duality: Contrasts of Mutually Implicative Relata

The paradoxes borne of relating incommensurable categorical principles like causal relation / logical implication, and physical / conceptual as mutually exclusive can be relieved by instead treating such principles as mutually implicative within some presupposed unifying context. The paradoxes are relieved, in other words, when one ceases to consider each principle individually and instead considers it in definitive relation to its counterpart within the context of a presupposed, underlying unity. This underlying context can be *rationally* presupposed, since these principles are, in fact, definitively mutually-referential; the question is simply whether one desires that this mutual reference take one of two possible forms: [1] reduction to incoherence by the arbitrary definition of one principle as substantial and the other as epistemic abstraction; or [2] elevation to coherence via mutual implication. By the first approach, Parmenides might exclaim, "Diversity is illusory. There is only unity! There is no change, only permanence!" while at the same time Heraclitus exclaims, "The only permanence *is* change!" and Protagoras, attempting to silence them both, exclaims, "There is no objectively true statement!" The incoherence of this method is evinced not only by the paradoxical character of these exclamations (the last one closely analogous to the Epimenides paradox discussed earlier); it is also evinced by the fact that unity cannot be defined without reference to diversity, nor diversity without reference to unity, and likewise for the concepts of permanence and change. This clearly belies the assertion of exclusive ontological primacy for either concept over its counterpart.

The second approach—elevation to coherence via mutual implication within some presupposed, speculative, unifying context—is, by contrast, far more promising. Since coherence is a definitive feature of reason itself and an explicit desideratum of the enterprise of logic, the presupposition of a context that provides such coherence is entirely reasonable. While this might seem to warrant a Baconian objection to the intrusion of a priori supposition into the methodology of science, the presupposition of unifying context is no more problematic than the scientific method's equally speculative presupposition of a unifying logical order by which locally testable theories can be presumed to hold universally. It is only when science attempts to 'account for' this order, while at the same time presupposing it, that scientific reason confronts itself with incoherence and paradox.

Thus, the method of relating fundamental cosmological principles, both physical and metaphysical, by mutual implication, requires a philosophical method that transcends the typical restrictions of philosophical deduction and induction, as well as their typical coupling within the hypothetico-deductive method of modern science. A clear example of such a method can be found in Whitehead's concept of 'speculative philosophy,' referenced in the epigraph of this chapter and explicated at length in *Process and Reality*,[11] his epic essay on systematic cosmology. By this method, the presupposed context by which fundamental principles find their mutual implication is not derivable by simple deduction or reduction. Rather, it is the product of imaginative generalization, continuously tested and refined against experience according to the following desiderata: [1] *coherence*, in that the categorical concepts of the speculative scheme are mutually implicative; [2] *logical structure*, such that all constitutive concepts are consistent, free from contradiction, and always exemplify, in every specific instance, the general logical framework. As will be discussed later in this volume, this desideratum when applied to quantum mechanics can be more concretely defined as the system of Boolean logic; [3] *empirical applicability*, such that the speculative scheme can be empirically exemplified in nature; [4] *empirical adequacy*, such that the universality of empirical applicability is reasonably presupposed. Taken together, the two empirical desiderata are, of course, the same desiderata by which scientific theories (e.g., quantum theory, the Standard Model, general relativity, etc.) are likewise presumed to hold universally even though they are testable only in restricted domains. Thus, as has been well evinced throughout the evolution of modern science, empirical adequacy might be lost by one theory because of its failure of applicability in some newly discovered domain, and then regained anew by the theory's refinement or replacement.

In the enterprise of speculative philosophy, as in the enterprise of science, these four desiderata, the rational pair and the empirical pair, are the only metrics by which the success of a particular scheme can be measured. There is never any final arrival at, or ultimate reductive-deductive proof of, the categorical principles constitutive of the scheme's presupposed unifying context; nor is there an arbitrary claim that this context is wholly transcendent and thus exempt from deeper levels of analysis and understanding. There is instead an asymptotic approach toward this context's explication, such that progress is always both measurable via its empirical exemplification and, at the same time, unending.

Process and Reality itself is a clear (and for many, unsurpassed) example of a robust, thoroughly systematized cosmology built upon a foundation of mutually implicative categorical principles via the method of speculative philosophy. The philosophical program of relational realism introduced in the present volume takes as its conceptual foundation the Whiteheadian speculative cosmological scheme, refining and expanding it toward an empirically adequate accommodation of modern quantum theory. To this end, the program of relational realism also entails a refined mathematical formalism, presented in detail in part II. This category-sheaf theoretic formalism not only solves many key conceptual difficulties in Whitehead's attempted formalism; it also addresses many infamous conceptual difficulties in the interpretation of quantum theory, including the latter's integration with relativity theory and, more broadly, clarification of the ontological function of logic in quantum causality.[12]

For Whitehead, the underlying context by which the causal and logical orders are united in dipolar fashion, via mutual implication, is the dipolar quantum 'actual occasion' or 'final real thing'—the imaginatively generalized fundamental unit of reality. But rather than being solely a unit of being, which would expose the actual occasion to the argument of Parmenides, it is also, and more fundamentally, a unit of becoming—a unit of process, or actualization of potentiality, that entails both a causal-physical pole and a logical-conceptual pole. The Platonic chasm separating conceptual and physical (as well as logical and causal) is thus easily bridged in this dipolar scheme because each pole is implicative of the other *within the contextual unity of the actual occasion as a whole*. When actualized, the occasion has its causal efficacy upon subsequent actualizations; and likewise, its process of actualization is conditioned by both [1] its causal physical relations with the actualities 'physically' antecedent to it (prior in time) in terms of spatiotemporal extensiveness—i.e., within its backward lightcone, per the restrictions of relativity theory; and [2] its logical relations with those actualities 'logically' antecedent to it (i.e., prior in implicative order), per the restrictions of the Principle of Non-Contradiction (PNC), among other logical restrictions.

Each actual occasion is thus fundamentally a quantum *unit of relation*. In the process of actualization, the becoming occasion is *internally related* to its entire dative world of antecedently actualized occasions, such that the potentia contextualized by these relations are both causally and logically ordered. By this mode of internal relation, then, every actual occasion, though it is a discrete, quantum unit of relation, entails a vast integration of potential relations between that occasion-in-process and the actual occasions of its dative world. (Thus, another philosophically problematic conceptual pair, 'unity' and 'diversity,' is also brought into coherence via mutual implication. "The many," writes Whitehead, "become one, and are increased by one. In their natures, entities are disjunctively 'many' in process of passage into conjunctive unity."[13]) These integrations are always logically conditioned, such that upon actualization, the novel occasion is internally related to the world in a manner free of violations of PNC, or any other such violation of the logical order.

Writ large, the universe is thus described as an integrated system of serial routes (histories) of actual occasions—that is, a global history of coherently integrated local histories. This is precisely analogous to the concept of local and global system histories in the decoherent histories interpretations of quantum mechanics introduced later in this volume.[14] Coherence at all scales derives from the fact that the internal relations of every actualization constitutive of each local history are logically conditioned, such that the orders of causal relation and logical implication are always correlated. Relations *among* histories therefore also evince this correlation—even those that are spacelike separated and thus incapable of causal physical ordering—i.e., ordering within a 4D spatiotemporal extensive framework. Indeed, it is by the physical-logical dipolarity of the actual occasion in relation to its dative world that the inability to *physically* or *causally* order spacelike separated occasions per the restrictions of relativistic spacetime (i.e., the impossibility of defining a total order in relativity theory) does not in any way imply that these spacelike separated occasions similarly lack a logical order. This is clearly evinced by the fact that partial ordering—asymmetry and transitivity—*does* hold in relativity theory, as it must, lest the theory collapse into incoherence. In other words, the relativistic proscription against specifying a *total* physical-causal order among spacelike separated occasions does not negate the *necessary presupposition* of a unifying, logical ordering of these occasions if the theory is to maintain its coherence. In the relational realist philosophical program, this coherence is grounded in the speculative presupposition that the causal and logical orders are mutually implicative via the unifying context of the dipolar actual occasion.

In the conventional mechanistic-materialist worldview, a history is understood to be a story about the objective world—an abstract scheme of epistemic relations among facts descriptive of concretely 'material' objects. In the relational realist worldview, a history is understood to be a concrete scheme of ontological relations among facts constitutive of abstractly 'material' objects. In both classical and relational realist histories, coherence is maintained when every novel fact is asymmetrically internally related to the totality of its antecedent facts; but in the case of classical histories, this is merely a conceptual, epistemic desideratum, and one impossible to achieve categorically. Contradictions always abound, indicating error and inspiring revision. By contrast, in the case of the ontological histories of relational realism, facts are not merely conceptual, logical objects; they are dipolar—objective both causally / physically and logically / conceptually. Coherence is categorical, and (if the speculative scheme is successful) impossible to avoid—at least by the metrics of empirical science, which thus far has given no evidence whatsoever of incoherent relations among the constituents of nature—no physical exemplification of conceptual paradox.

To be sure, such exemplifications have been conceived theoretically and in the exacting language of mathematics, and these attempts will be a topic of discussion in chapter 2—traveling 'backwards' through time and 'undoing' the facts of the past via so-called 'retro-causality,' physical 'bilocation,' and other such physical violations of PNC. To date, however, there has not been a single experiment in the sciences decisively demonstrating a *physical* instantiation of these or any other concept that would obviate the presupposed structural logical coherence of the world. Indeed, apart from this categorical logical structure, the scientific method would not only lose its applicability to the task of coherently describing nature; it would literally lose itself—and thereby, paradoxically, the very means by which it would claim success in these dubious efforts.

In this regard, the utility of the relational realist speculative philosophical scheme is perhaps best evinced by its application to the task of coherently interpreting scientific theories that are conventionally accepted as fundamental, yet are nevertheless profoundly resistant to coherent integration. The incompatibility of quantum theory and general relativity, as mentioned earlier, is a particularly problematic example. Its solution has proven unusually elusive, with little definitive progress made after decades of effort. If the various string theoretic and quantum cosmologies have fallen short, it is arguably because of their shared reliance upon the centuries-old physico-reductive, mechanistic-materialistic approach to relating the causal and logical orders as mutually exclusive features of reality. And likewise, a speculative philosophical relation of the causal and logical orders as mutually implicative features of reality lays the groundwork for an alternative and arguably improved set of physical theories—theories framed within a systematic ontology as robustly informed and confirmed by modern physics as the old mechanistic-materialistic approach, but without the underlying conceptual incoherence.

Notes

1. Alfred North Whitehead, *Process and Reality: An Essay in Cosmology, Corrected Edition*, ed. D. Griffin and D. Sherburne (New York: Free Press, 1978), 3.

2. Michael Epperson, "Relational Realism: The Evolution of Ontology to Praxiology in the Philosophy of Nature," *World Futures* 65, no. 1 (2009), 19-41. The present chapter is a revision and expansion of the ideas presented in this article.

3. Niels Bohr, *Atomic Physics and Human Knowledge* (New York: Wiley, 1958), 18.

4. Hugh Everett, "'Relative State' Formulation of Quantum Mechanics," *Reviews of Modern Physics* 29 (1957): 454-62.

5. David Bohm, "A Suggested Interpretation of the Quantum Theory in Terms of 'Hidden' Variables." *Physical Review* 85 (1952):166-93; David Bohm, "Time, the Implicate Order, and Pre-space," in *Physics and the Ultimate Significance of Time*, ed. David R. Griffin (New York: State University of New York Press, 1986), 177-208; David Bohm and B. J. Hiley, *The Undivided Universe: An Ontological Interpretation of Quantum Theory* (London: Routledge, 1993).

6. Bohm, "Time, the Implicate Order," 199.

7. Bohm, "Time, the Implicate Order," 198.

8. Again, the 'many worlds' interpretation of quantum mechanics is but one of the more infamous examples.

9. James Hartle and Stephen Hawking, "The Wave Function of the Universe," *Physical Review D* 28 (1983): 2960-75.

10. R. Weber, *Dialogues with Scientists and Sages: The Search for Unity* (London: Routledge, 1986), 212.

11. Whitehead, *Process and Reality*, 3-17.

12. Evinced, for example, by recent experiments exploring the phenomenon of phase decoherence in condensed matter physics. This topic will be explored in more detail in subsequent chapters.

13. Whitehead, *Process and Reality*, 21.

14. See also, Michael Epperson, *Quantum Mechanics and the Philosophy of Alfred North Whitehead* (New York: Fordham University Press, 2004).

CHAPTER 2

Substance and Logic in Quantum Mechanics

This chapter is intended to provide a general overview of the conceptual framework presented over the course of part I. As such, it gives an introductory discussion of concepts that will be revisited in greater detail throughout the remainder of the book.

The notion of an ontologically significant scientific theory—that is, one that aims toward an understanding of nature rather than just the nature of understanding—has undergone an evolution in the history of Western philosophy that has been punctuated by repeated cycles of approbation-as-profound and dismissal-as-vacuous, with each trend taking as its justification the same issue: The necessary presupposition, in *any* ontological interpretation of a scientific theory, of some particular framework of categorical principles. Whether generated by deduction, induction, imagination, or some combination of these, history is replete with examples whereby the profundity of first principles evolves first to dogmatic acceptance, then inevitably to rejection on the grounds of experimental disconfirmation. The first principles and their associated methodologies of inquiry are then modified or replaced, again via some combination of deduction, induction, and imagination, and the cycle repeats. The shift from Aristotelian hylomorphic physics in the medieval period to classical mechanistic-materialistic physics in the early-modern period is but one example, and the grounding of this shift in the specific issue of first principles is well represented in key writings of this period; the introduction to Book I of Bacon's *New Organon* and Boyle's "Of the Excellency and Grounds of the Corpuscular or Mechanical Philosophy" are but two examples. The most recent iteration of this cycle is the shift from classical mechanistic-materialism to quantum physics, the aftershocks of which have yet to subside.

After centuries of these tectonic shifts, one could argue that it is for lack of sure footing that in today's academy, the very notion of the systematic development of a coherent ontological interpretation of fundamental physical theories is little more than a cottage industry in most university philosophy departments. But the mere existence of presupposed foundational categorical principles and their recurrent displacement, abandonment, and replacement throughout the history of philosophy does not in itself warrant the wholesale repudiation of scientific realism by way of sheer reduction of ontological concepts to merely epistemic concepts as has been fashionable over the past century. One can trace the roots of such reduction back to Protagoras, and likewise Plato's riposte to the Protagorean doctrine maintains its relevance: The statement "it is always the case that there is no objectively true statement, only subjectively true statements" is self-refuting and therefore paradoxical, as is the statement "there exists no ontologically significant statement, only epistemologically significant statements." The fundamental incoherence of such claims is grounded in the fact that epistemic assertions of this kind—and indeed of any kind—themselves presuppose and exemplify an underlying ontological commitment. As suggested in the previous chapter, Plato's argument is ultimately grounded in, and can be extended further to include, the recognition that the mutually implicative relationship between epistemology and ontology wholly obviates the possibility of simply reducing or assimilating one mode of inquiry to the other.

One could perhaps counter that the Platonic refutation lacks self-sufficiency, since it relies upon a presupposed scheme of first principles themselves taken to be objectively true—namely, those of classical propositional logic which include, among others, the Principle of Identity

(PI), the Principle of Non-Contradiction (PNC), and the principle of the excluded middle (PEM). But of course the Protagorean refutation of philosophical realism is equally reliant upon this same scheme, even as it is undone by it; and the same can be said of the great variety of analogous present-day attempted refutations of scientific realism. Indeed, not only does the modern scientific method itself presuppose these first principles of classical logic, but so does any philosophical evaluation of the *meaningfulness* of scientific theories. Put most broadly, the possibility of any epistemically meaningful claim and any ontological claim of meaning presupposes an underlying logical structure of first principles including PI, PNC, and PEM. Indeed, this structure is a necessary presupposition of any coherent and consistent framework of conceptual relations of any kind. As discussed in the previous chapter, even in the case of mathematics, the most objective and universal scheme of conceptual relations yet conceived, Gödel demonstrated with his two incompleteness theorems the requirement of presupposed first principles of relation—namely, relations among the natural numbers that are necessarily true yet unprovable via any consistent mathematical axiomatic scheme. And likewise, the consistency of any such scheme itself requires reference to the same necessarily presupposed principles of relation. The Epimenides Paradox, as discussed earlier, provides a heuristically useful analog: The Cretan's claim, "Cretans, always liars," cannot be true, but nor can it be false. Thus it is clear that PNC [$\neg$ (P $\land$ $\neg$P)] is not merely a feature of mathematical conceptual relations; it is, rather, a necessary presupposition for the possibility of *any* scheme of conceptual relations that is coherent and consistent. Apart from such a scheme, the conception and evolution of the modern scientific method would have been impossible.

2.1 'Liberating' Science from Boolean Logic: A Sisyphean Adventure

Despite the centrality of a necessarily presupposed logical framework to both the scientific method and, more broadly, to the conventional Western worldview of scientific realism, one sees two puzzling trends that have been gaining traction over the past several decades: First, the attempt by some physicists to apply the scientific method to the task of 'accounting for' the Boolean logical structure *presupposed by* the scientific method. As discussed in the previous chapter, it is an ambition well exemplified by quantum cosmological models purporting to account scientifically for the origin of the universe and its logico-mathematical structure *ex nihilo*.

Second, and more puzzling, is the attempt by some quantum theorists to apply the scientific method to the task of invalidating the Boolean logical structure presupposed by the scientific method. It has become increasingly fashionable, for example, to claim that PNC and Boolean logic in general are routinely violated in quantum physics and that actual observable physical instantiations of these violations—i.e., 'physical paradoxes'—have been experimentally confirmed even at the macroscopic level. One such experiment recently published in the journal *Nature*,[1] for example, involved coupling a macroscopic mechanical oscillator (a 60 micron long silicon 'paddle') to a quantum system such that superpositions of *potential* system states violating PNC in the quantum system were interpreted by the paper's authors, as well as the journal's editors, as producing *actual* 'macroscopic superpositions' violating PNC in the mechanical oscillator. In a companion article describing the experiment, *Nature* reported that the experimenters

> used the weird rules of quantum mechanics to simultaneously set the paddle moving while leaving it standing still . . . Through a series of careful measurements, they were able to show that the paddle was both vibrating and not vibrating simultaneously . . . The experiment shows

> that the principles of quantum mechanics can apply to everyday objects as well as atomic-scale particles.[2]

The exotic notion of quantum measurements generating macroscopic, 'physical superpositions' of observable, alternative *actual* states rather than merely alternative *potential* states (one cannot, of course, 'observe' potentia, despite the fact that one can represent them mathematically) is as old as the Schrödinger's Cat thought experiment. The latter, it is often forgotten, was conceived by Schrödinger not as a celebration of the violation of PNC in quantum theory, or an endorsement of the idea that such violations are physically instantiated; it was intended, rather, as an admonishment that any interpretation of quantum theory implying 'observable' superpositions—viz., observable physical violations of PNC—must be considered deficient. "It is typical of these cases," he wrote, "that an indeterminacy originally restricted to the atomic domain becomes transformed into macroscopic indeterminacy." But this indeterminacy, Schrödinger argued, "can then be resolved by direct observation. That prevents us from so naively accepting as valid a 'blurred model' for representing reality."[3]

The key phrase for Schrödinger, here, is "by direct observation" and it provides the fundamental principle by which any experiment involving quantum mechanical measurement should be interpreted. One could, for example, improperly interpret the interference fringe of the familiar double-slit experiment[4] as a kind of 'macroscopic superposition' such that a single photon is envisioned as 'actually going through' both slits at the same time. By such an interpretation, the interference fringe itself is thus depicted as evincing a fundamental violation of PNC. But this interpretation is easily confuted by the fact that in observing an interference fringe, one is not 'directly observing' a macroscopic superposition; the interference fringe in a double-slit experiment is generated as a sequence of individual, discrete measurement events, wherein each photon impinges upon the detector at a particular point. After a sufficiently large succession of these discrete events, the interference fringe begins to manifest as a spatial relationship among these points.

Thus the form of the interference fringe, and by analogy, the interference effect described in the quantum oscillator experiment discussed above, is not properly understood as a macroscopic 'object-superposition' violating PNC but rather as a statistical relation among discrete quantum measurement outcome events, each of which satisfies PNC. In this regard, the quantum superposition associated with each individual photon is properly understood as a superposition of *potential* outcome states prior to the unique actualization of one of these states. And though such superpositions of potential outcome states (e.g., the 'pure state' in quantum theory) do violate PNC (potential passage through Slit 1 *and* potential passage through Slit 2), it is a fact of quantum mechanics that these superpositions of potential outcome states *always* evolve to become a discrete, actual measurement outcome, registered as a discrete impingement of the photon at the detector. Only after an ensemble of such detections has been recorded can an interference fringe be defined—again, evincing the fact that the latter is only statistically and retrodictively representative of the superpositions of potentia ingredient in each discrete actualized photon detection. This fact is especially crucial when interpreting so-called 'double-slit quantum eraser' experiments,[5] by which many theorists purport to 'erase,' via a kind of 'retro-causality,' these discrete actualizations of potentia, and with them the path information contained in the outcome states recorded by these measurements. These experiments will be discussed further in chapter 4.1.

With respect to the oscillator experiment discussed above, one should likewise interpret the superposition of oscillator states not as a superposition of 'contradictory actual states' that is somehow directly observable and persistent over a period of time as was implied in the *Nature* article, but rather as a superposition of potential outcome states by which actual initial states and actual final states of the device are statistically related. These potential relations are calculable predictively, but 'measurable' only retrodictively—that is, *after* an ensemble of discrete, actual outcome states has been registered.

The crucial point here is that *apart from the statistical analysis of a succession of actual, discrete outcome states, the superposition of states cannot be measured.* More crucially: Even though a time interval associated with a superposition is definable by calculation—i.e., a 'coherence time' during which the coherent superposition of potential states is interpreted by many theorists as 'persisting' (in the oscillator experiment, for example, this time was calculated to be 17 *ns*)—it is nevertheless impossible to directly observe a superposition of potential states *during* this defined time interval. Again, the very notion of 'observing potentia' is in itself problematic conceptually—a problem that is easily resolved by acknowledging that both theoretically and in scientific practice, as well as at the level of common intuition, *potential* states are always defined and measured via analysis of the *relation between actual states*—that is, the initial and final actual states that bound this relation. In quantum experiments such as the one discussed here, these relations take the form of oscillations ('Rabi oscillations', or 'Rabi cycles' named after physicist Isidor Isaac Rabi), such that a sequence of these oscillations is understood as a sequence of these relations. Thus, the 'coherence time' pertaining to the intervals 'between' the actual states constitutive of such a sequence, though it can be analytically *defined by calculation*, is always *observed by measurement* only retrodictively and statistically—that is, always after a sequence or history of discrete actual states has been registered and recorded.

This admonition is often neglected, in large part because of a failure to explicitly acknowledge that time is not itself an 'observable' in standard quantum theory—i.e., it is not properly expressible as a self-adjoint operator.[6] The technical meaning of this statement will be explored later in this volume, but for the current discussion its significance can be understood thus: While one can use quantum theory to compute the probability that an observable like an electron's spin polarization will be 'up' or 'down' upon arrival at a particular detector, one cannot use standard quantum theory to determine *when* an electron will arrive at the detector (this is commonly referred to as the quantum 'time of arrival' problem). In this way, and reflective of the literal sense of 'observable,' time is no more an observable in quantum physics than it is in classical physics, being neither an object itself, nor definable as a 'quality' intrinsic to the object measured. It is, rather, a *relational inference*, quantified via an *external parameter* by which an initial actual state of a measured system is metrically related to a subsequent actual state.

In summary, it is a fundamental feature of quantum mechanics that the object of observation is always a system in an actual state, and never a superposition of potential states. One cannot 'directly observe' potentiality, but rather only infer it as a calculable relation between actual initial and final relata—an assertion that is both intuitively comprehensible and experimentally demonstrable; yet the increasingly popular conflation of 'calculation of the potential' and 'measurement of the actual' in experimental reporting persists, even to the point of becoming the standard interpretation of these experiments as disseminated by the popular media.

Still more puzzling, this popularized interpretation directly contradicts fundamental concepts in quantum mechanics that are otherwise unquestionably regnant in the more familiar scenarios such as the double-slit interference experiment. Likewise, even in the most basic college physics course, it is well understood that quantum mechanics proscribes the possibility of evaluating any observable associated with an electron 'in between' its quantum states, or even conceptualizing the electron 'moving as an actual object through the space between' the spatial configurations associated with these states. Relative to this interstitial space, it is understood that the electron is only definable as a superposition of potential states (one of which will be actualized upon measurement), and thus not directly observable; it is not, as depicted in the analogous quantum oscillator experiment, a superposition of actual states that is somehow directly observable.

Put most simply, the depiction of a potential system state, or a superposition of potential states, as an observable physical extant, rather than as a relational conjunction of actual states, is fundamentally incoherent. It is akin to depicting 'causality' as an observable physical extant in abstraction from the actual states qualified as 'cause' and 'effect.' In both cases, the incoherence derives from the attempt to assimilate the idea of

[a] using the quantum *formalism* to *calculate* the probabilities of various potential states becoming actualized, into the idea of [b] using quantum *mechanics* to *measure* these potential states via observation. Philosophically, this attempted assimilation of 'observation of what is' to 'calculation of what is reasoned to be' is the modern scientific epitome of the attempted assimilation of ontology to epistemology discussed earlier in this chapter. It is a misplaced concretizing of calculation that neglects the fact that even in quantum mechanics, calculated superpositions of potential outcome states necessarily presuppose discrete, observable, actual initial and final system states and their logical relation, and it is only via the latter that predictive calculations are confirmed retrodictively. The electron, in other words, is always *observed as actualized*, in either one state or another, in satisfaction of PNC, and never observed as potentialized—i.e., as a superposition of potential states in violation of PNC. In this way, superpositions are properly understood as relations of successive actual states, initial and final, via an appropriate measurement interaction.

In quantum mechanics, this relation is fundamentally describable as an evolution of: [1] a pure state of *potential* outcome states (these are not mutually exclusive and can violate PNC), to [2] a mixed state of *probable* outcome states that are [a] mutually exclusive (satisfying PNC, i.e., '*at most* one outcome state will be actual upon measurement') *and* [b] exhaustive, in that the probability valuations, by definition, must sum to unity (satisfying PEM, i.e., '*at least* one outcome state will be actual upon measurement). Though this logically conditioned evolution of potential to probable outcome states is a fundamental feature of quantum mechanics (Von Neumann formalized the mathematics of this evolution as his 'Process 1'[7]), it is important to emphasize that quantum theory *presupposes* this evolution. The theory does not, in other words, entail any physical dynamical mechanism that 'generates' the evolution of potentiality to probability. Likewise, the theory also presupposes the evolution of probable outcome states to unique actual outcome state, in satisfaction of PEM. These are both aspects of the 'problem of state reduction' in quantum mechanics, which many theorists have cited as evidence of the incompleteness of quantum theory—that it presupposes [1] the existence of the actualities it measures, and [2] that its measurements will satisfy PNC and PEM.

As has been argued thus far, and will continue to be argued throughout the remainder of this volume, neither of these presuppositions can be reasonably assessed as a theoretical deficiency, since both are necessarily presupposed by the scientific method itself—the method by which quantum theory was initially conceived and validated, and by which it continues to be developed and implemented. In the case of the quantum oscillator experiment discussed above, it is the failure to explicitly recognize these presuppositions, coupled with the recently fashionable bent toward seeking profundity and mystique over intuitive coherence in the experimental application of quantum theory, that leads to unfortunate interpretive claims such as "the experimenters used the weird rules of quantum mechanics to simultaneously set the paddle moving while leaving it standing still . . . vibrating and not vibrating simultaneously . . ."[8]

Beyond being a necessary presupposition of both the scientific method and the mathematical language by which the method is formalized and applied to experiments, and by which the data of experiments are evaluated and interpreted, the presupposition of PNC is foundational to the possibility of coherence and consistency within any conceivable domain of experience. Thus, interpretive claims such as the one cited above seem to advocate a fundamental schism of critical reasoning, where PNC and Boolean logic in general are depicted as epistemic constructs borne of our classically restricted experiences of the world; and the world itself—the actual object of science disclosed most fundamentally via quantum theory—is at the same time depicted as routinely and categorically violating these constructs, even though it is these very constructs that allow science to proclaim this schism in the first place through experiments such as the one cited above.

Quantum theorist Roland Omnès describes the problem thus:

> There is indeed a crisis, for unlike the flourishing situation in the history of knowledge, the philosophical reflection about science has lost its

way—or stagnates. The fashionable authors see only uncertainties, paradigms without enduring principles, an absence of method, and a presence of erratic revolutions, precisely when we should be trumpeting the success of a science whose extent and consistency are unprecedented ... Beyond the shadow of a doubt, the origin of this crisis is to be found in an event that no one has fully recognized in all its significance: the irresistible irruption of the formal approach in some fundamental sciences such as logic, mathematics, and physics. As a consequence, these disciplines have become practically impenetrable, which explains the capitulation or the adventurousness of so many commentators, not to mention the disarray of the honest man or woman who wonders what those who should understand these subjects are talking about.[9]

2.2 Abandoning the Logic of One World for the Logic of Many Worlds: An Unlikely Liberation

There are three central principles of quantum theory that have contributed heavily to the 'impenetrability' and 'adventurousness' Omnès forewarns against in the above quotation, and which taken together have made it increasingly fashionable to dismiss as vacuous the entire enterprise of constructing a persuasive ontological interpretation:

[1] Objective Indeterminacy: Quantum indeterminacy displaces the classical physical first principle of 'objective determinism as a necessary implication of mathematical objectivity'—that mathematical necessity at the conceptual level implies deterministic contingency at the physical level. Instead, in quantum theory, mathematical probability implies indeterminacy at the physical level.

[2] Objective Local Contextuality: The local context dependence of quantum measurement is evinced by the fact that [a] probable outcome states of measured systems are definable only according to the Boolean-logical contextual measurement basis of a particular chosen detector. In the conventional Hilbert space formalism, this is exemplified by the requirement that the measurement basis be orthonormal, such that potential

outcome states are always represented as self-adjoint operators whose projections upon this orthonormal basis define these potential states as Boolean-logical, mutually exclusive 'True-False' propositions;[10] and [b] actual outcome states of measured systems always conform to this contextual measurement basis. This is typically interpreted as a problem because in the pure state, superpositions of potential outcome states *can* be defined without reference to this Boolean contextual measurement basis (per Principle 3, below); but this reasoning is undermined by the fact that the evolution of the pure state to the mixed state always entails such contextualization, such that the probability-valuated outcome states constitutive of the latter *always* conform to the Boolean contextual measurement basis of the particular quantum measurement, as does the unique actual outcome state terminal of the measurement.

[3] Objective Superposition of States: The nature of the pure state in quantum measurement—a coherent superposition of potential measurement outcome states—violates the first principles of classical logic, namely PNC (*at most* one outcome state is true) but also the Principle of the Excluded Middle (PEM) (*at least* one outcome state is true). PNC [$\neg$ (P $\wedge$ $\neg$P)] is violated because the potential outcome states integrated in the pure state are not necessarily mutually exclusive; and because they are related only by conjunction without complement in the pure state ('Schrödinger's cat is alive AND dead'), it is impossible to schematize these potential outcome states in terms of Boolean logic, which also requires relation by disjunction ('Schrödinger's cat is alive OR dead'). Likewise, PEM [P $\vee$ $\neg$P] is violated because there is no mechanism in quantum theory by which one of the potential outcome states *must* be actualized.

Together, these three features have inspired a variety of ontological interpretations of quantum theory, many of which advocate some signature radical conceptual innovation grounded in one or more of these features. The problem of objective local contextuality, for example, combined with the problem of objective superposition of states—specifically, reducing the pure state to a unique actual outcome state—inspired the following innovative idea: Because [a] the object of direct experimental observation in a quantum measurement interaction is always a system in a unique actual state, and [b] unique outcome states are always consequent of experimental observation in quantum mechanics, it is thus reasonable to suppose that the human mind, as the final observational mechanism in any chain of instrumental observation, somehow participates in the creation of the objective facts yielded by quantum measurement. Early explorations of this idea by Wigner and von Neumann, for example, have been refined into formalized ontological interpretations by theorists such as Henry Stapp.[11]

Likewise, a similarly radical and increasingly popularized interpretation of quantum mechanics is the aforementioned 'Many Worlds Interpretation' (MWI) introduced in chapter 1. MWI begins with the conceptual innovation of treating the pure state as a superposition of alternative *actual* outcome states rather than a superposition of alternative *potential* outcome states, such that the concept of potentiality is wholly assimilated to the concept of actuality. In other words, 'everything that can happen does happen'—a presupposed first principle of MWI which, despite its frequent characterization as a 'scientific' deduction, is consistently defied experimental demonstration (viz., as evidenced in every recorded experimental application of the scientific method to date). MWI advocates counter this criticism by appealing to an additional presupposed principle—that of the 'multiverse'—by which it is stipulated that each alternative outcome state is actualized in its own universe, with each universe spatiotemporally isolated from the others, and thus experimentally inaccessible to the scientific method. But this, of course, only reinforces the original criticism: that the popular portrayal of MWI and its categorical principles as scientifically deduced and scientifically meaningful, is wholly inappropriate.

Nevertheless, it continues to be argued by advocates of the MWI interpretation that because the pure state entails violations of both PNC and PEM, quantum theory therefore entails the scientific invalidation of PNC and PEM—that is, it invalidates the presupposition of exclusive disjunction for contradictory actual outcome states ('It is necessarily true that Schrödinger's cat is *either* alive *or* dead') and therefore entails a comprehensive, categorical invalidation of Boolean logic in fundamental physics. By this argument, physics requires the replacement of Boolean logic by some particular formalization of modal logic which, for an ontologically significant interpretation of quantum theory, implies at least some commitment to modal realism. Again, MWI entails that every potential quantum measurement outcome is actualized in its own relative world, and that there exists a 'multiverse' of countless alternative worlds, each spatiotemporally isolated from the others. While the conceptual framework of MWI was originally conceived by Hugh Everett (his 'relative state interpretation') as a novel conceptual solution to the measurement problem rather than an ontological implication of modal realism, recent emphasis of the non-Boolean, modal logical nature of quantum mechanics naturally implies an MWI-type ontological scheme. In his *On the Plurality of Worlds*, for example, where modal realism is formalized, philosopher David Lewis writes,

> I advocate a thesis of plurality of worlds, or modal realism, which holds that our world is one world among others. There are countless other worlds . . . They are isolated: there are no spatiotemporal relations at all between different things that belong to different worlds. Nor does anything that happens at one world cause anything to happen at another. . . The other worlds are of a kind with this world of ours . . . Nor does this world differ from others in its manner of existing.[12]

It should be emphasized here that despite the obvious conceptual compatibilities, for technical reasons that exceed the purpose of the current discussion, Lewis did not believe MWI to be a precise ontological exemplification of modal realism.[13] The point of the comparison here is simply that MWI, as an ontological interpretation of quantum mechanics, requires that 'meaningful statements' generated by the interpretation must resort to modal logical relations because they exceed the restrictions of Boolean logical relations. This, in turn, allows for interpretive modal logical statements such as 'Schrödinger's cat *actually* exists as a superposition of both live and dead states,' rather than the Boolean logical interpretive statement, 'Schrödinger's cat *potentially* exists as a superposition of both live and dead states, such that *at most* and *at least* one of these states will be actual upon measurement.'

Of course, the more intuitive Boolean logical interpretive statement is far less exotic than the modal logical statement; but more important, it is *only* the Boolean logical interpretation that has seen universally consistent empirical validation, given that observation of alternative outcome states across multiple universes is not possible even in principle, according to the theory's own parameters. MWI, in other words, is not falsifiable by the scientific method. Nevertheless, recent experiments in quantum mechanics such as the quantum oscillator experiment discussed earlier have been reported with a commitment, either explicit or implicit, to the more exotic modal logical interpretation. It is not surprising, then, that the idea of an ontologically significant interpretation of fundamental physical theories is once again trending from approbation-as-profound to dismissal-as-vacuous.

2.3 Quantum Mechanics and Boolean Logic

It is the Boolean logical interpretive statement, 'Schrödinger's cat *potentially* exists as a superposition of both live and dead states, such that *at most* and *at least* one of these states will become actual upon measurement,' that has been consistently exemplified via a history of clear empirical demonstration (even in the quantum oscillator experiment discussed earlier), where the pure state superposition of *potential* outcome states always evolves to a reduced matrix of *probable* outcome states. This is crucial because as probabilities, the alternatives must be [a] mutually exclusive, in satisfaction of PNC (i.e., 'at most one actual outcome state'), and [b] exhaustive, in satisfaction of PEM, in that when added together the probabilities sum to unity (i.e., 'at least one actual outcome state'). This categorical presupposition of PEM, like PNC, is thus arguably a first principle of any ontological interpretation of quantum mechanics, and it is represented in quantum theory by Born's rule, which likewise is accepted as an underivable presupposition of quantum theory.

The probabilistic character of quantum mechanics formalized via Born's rule is of first importance in this regard because it makes explicit the theory's necessary presupposition of not just PNC and PEM, but also, more broadly, the first principles of Boolean logic (conjunction, disjunction, and complement), and the derivative principle of material implication (the latter of which will be introduced later in this section). For it is only via a presupposition of Boolean logic that probability theory is possible,[14] in particular the sum rule for probabilities and the quantum theoretical exemplification of the latter in Born's rule.

It is clear, then, that the categorical first principles of PNC and PEM, and their function within a Boolean logical framework of additional presuppositions, are central to each of the three problematic issues discussed above: objective indeterminacy, objective local contextuality, and objective superposition of states. Indeed, there are only two aspects of quantum mechanics that, when considered in isolation (i.e., in abstraction from the theory as a whole), violate these Boolean first principles. The first, as has already been discussed, is the pure state superposition of alternative potential outcome states. The second is closely related to the first, and can be understood as its connection to the problem of objective local contextuality: Although individual local measurement contexts are Boolean, the attempt to relate *multiple* local measurement contexts together in a single measurement—a normative function of quantum mechanics—yields non-Boolean potential relations *across* these contexts, because the variously contextualized observables do not commute. In other words, in a composite quantum system, when considering the joint probability of outcome a_1 generated by measurement of observable a via its local Boolean contextual measurement basis A *and* the probability of outcome b_1 by measurement of observable b via *its* local Boolean contextual measurement basis B, one finds that $(a_1 \cdot b_1) \neq (b_1 \cdot a_1)$. There is thus a *fundamental asymmetry of relation* in quantum mechanics that is not present in classical mechanics where observables do commute.[15] Of course, relational asymmetries of this kind also exist in classical physics;

conditional probabilities, for example, are perfectly expressible via classical probability theory—e.g., in cases where $P(A|B)$, the probability of '*A given B*,' is not equivalent to $P(A)$ alone. But whereas classical probability theory allows for cases where $P(A|B)$ *is* equivalent to $P(A)$, such that A and B can be qualified as mutually independent, this is not possible in quantum mechanics because of the non-commutativity of quantum observables (the phenomenon of 'quantum entanglement' is a related physical indication of this.) Thus, quantum mechanics requires a different kind of probability conditionalization rule than its classical analog—namely, *one that depicts the evaluation of quantum observables as a fundamentally asymmetrical relational process.*

To further clarify this point, consider that the conventional ontological implication of commutativity in classical mechanics is that the order of observation (i.e., the order of predication via measurement) is irrelevant because all observables are thought to possess well-defined values at all times, regardless of whether or not they are measured. But in quantum mechanics, this is not the case; asymmetrical probability conditionalization evinces that the act of observation is generative of novel facts, not merely revelatory of pre-existing facts.[16] That is, outcome states yielded by quantum mechanical measurement are not merely *revealed subsequent to measurement*, but rather *generated consequent of measurement*, as evinced by the Heisenberg uncertainty relations as well as other equally fundamental principles of quantum theory. The outcome states generated by quantum mechanics, then, are properly understood as ontological events, not merely epistemic qualifications, and as such, their asymmetrical relational dependencies are likewise ontologically significant. For example, if a = 'opening a screen door' and b = 'walking through a screen door,' it is quite obvious why $(a \cdot b) \neq (b \cdot a)$. Each conjunction represents a possible physical state, but these states and their ingredient conditional probabilities are quite different (as anyone fortunate enough to witness this particular demonstration will verify).

Thus the order of generation of predicative facts in quantum mechanics yields asymmetrical potential relational dependencies among these facts and among their measurement contexts. These potential dependencies are represented in the conventional Hilbert space formalism by using the tensor product to relate potential outcome states subsuming multiple observables and their respective measurement contexts. If, for example, observable a has two potential values, a_1 and a_2 in Boolean measurement context A, and observable b has two potential values, b_1 and b_2 in Boolean measurement context B, then $a_1 \cdot b_1$ (the statement, 'observable a is in state a_1 and observable b in state b_1') is expressed as the tensor product $a_1 \otimes b_1$. This means that the specification of a is impossible without reference to the specification of b, and vice versa. This is the essence of quantum entanglement—the impossibility of defining the combined states of a and b as $a \cdot b$, as though a and b were mutually independent. Thus the pure state Ψ of this composite system must represent *all* potential propositional relations between a and b—but more than that, it must represent all potential relations among *all* observables constitutive of the system, in abstraction from any particular contextualization. Further complicating matters, it is an essential and infamously challenging feature of quantum mechanics that these manifold potential relations constitutive of the pure state are formally equivalent to a reduced, contextualized selection of these potential relations. In the present example, this is expressed as:

$$\Psi = \frac{1}{\sqrt{2}} (a_1 \otimes b_1 + a_2 \otimes b_2)$$

$1/\sqrt{2}$, as will be explained presently, represents the probability 1/2 for either the composite state $a_1 \otimes b_1$ or $a_2 \otimes b_2$, and probability 0 for either the composite state $a_1 \otimes b_2$ or the composite state $a_2 \otimes b_1$. The reason that the possible dependencies do not include $a_1 \otimes b_2$ and $a_2 \otimes b_1$ is that the individual local contexts A and B are Boolean, such that a_1 and a_2 are mutually exclusive (i.e., $a_1 \veebar a_2$) in context A, and b_1 and b_2 are mutually exclusive ($b_1 \veebar b_2$) in context B, *and* these Boolean relations are correlated from A to B via a 'compatibility condition' (discussed further in section 2.5). In other words, the Boolean structure of context A 'overlaps' with the Boolean structure of context B, such that if contexts A and

B are each represented as Boolean subalgebras, or equivalence classes of Boolean subalgebras,[17] the Boolean structure of each context can be extended to that of the other where these subalgebras overlap. It must be emphasized here that Boolean measurement contextuality in this sense is not merely an epistemic coordination, but rather an ontologically significant, logical conditioning of *potential* outcome states such that they can evolve, through coherent integration, into *probable* outcome states[18]—a conditioning that *always* occurs in quantum mechanics. Indeed, this is arguably its most essential feature, if not its signature feature—that quantum measurement always terminates in probability outcomes, and therefore always presupposes a Boolean logical relational structure. The fact of this presupposed structure, then, confutes the often repeated claim that quantum mechanics is somehow *essentially* barren of the principles of classical logic, just because it contravenes certain principles of classical, deterministic physics.

As discussed earlier, the presupposition of a Boolean-logical measurement context is represented in the conventional Hilbert space formalism of quantum mechanics via [1] the requirement of an orthonormal measurement basis (representing the Boolean measurement context), and [2] the requirement that observables in quantum mechanics (in this case, *a* and *b*) are representable only by self-adjoint operators projected onto this measurement basis.

As a heuristic model evincing these requirements, consider, for example, a 2-dimensional Hilbert space—an abstract vector space named after mathematician David Hilbert. In quantum mechanics, a Hilbert space can be thought of as a 'potentiality and probability space' where logically conditioned propositions about a measured system state can be mathematically represented in terms of vectors and their relations. In a Hilbert space of two dimensions, representable as a simple Cartesian *x*-*y* coordinate system, there can be sets of, at most, two mutually orthogonal vectors. In a space of three dimensions, *x*-*y*-*z*, there can be sets of, at most, three mutually orthogonal vectors, and so on. This correspondence serves as the foundation for the concept of an orthonormal measurement basis, where vectors representing actual system states are of unit length and mutually orthogonal.

It is easy to see how this measurement basis (i.e., measurement context) might be used to represent a Boolean-contextualized quantum observable *a* that has only two possible outcome values, a_1 and a_2: One first recognizes that as a Boolean-contextualized observable, these possible outcome values *must* be relatable as mutually exclusive and exhaustive when actualized as a measurement outcome. To this end, a_1 and a_2 are represented as mutually orthogonal vectors in our 2-dimensional Hilbert space. (As noted above, however, if there were three possible outcome values, a_1, a_2, and a_3, we would require the addition of a third dimension—i.e., a *z*-axis—to accommodate sets of three mutually orthogonal vectors, and so on.) Thus, for observable *a*, the orthonormal measurement basis $a_1 \perp a_2$ represents Boolean-contextualized possible outcome values $a_1 \vee a_2$ (that is, a_1 'or' a_2). In quantum mechanics, however, the simple disjunction $a_1 \vee a_2$ always evolves to become an *exclusive* disjunction $a_1 \veebar a_2$ ('*either* a_1 or a_2') with the actualization of a unique outcome state; that is, the simple disjunction 'a_1 or a_2' is additionally conditioned by PNC ('if a_1 then *not* a_2' and vice versa) thus becoming an exclusive disjunction of contradictories, '*either* a_1 *or* a_2'.

Given this Boolean framework of quantum mechanical *contextualization* of observable *a*, the quantum mechanical *evaluation* of the observable as either a_1 or a_2 upon measurement can be formalized by simply adding a third vector, ψ, representing the objectively indeterminate state of the system *in abstraction from* the contextualized measurement, and projecting it, via a projection operator, onto the vectors a_1 and a_2. Since a vector's unit length represents 'probability = 1,' one can simply invoke the Pythagorean theorem to show that the length of each projection, ψ onto a_1 , and ψ onto a_2 , when squared, represents the probability valuation of outcome states a_1 and a_2 , respectively.

The real power of this formalism as a means of representing Booelan-contextualized quantum observables becomes apparent when one considers less idealized systems—i.e., those comprising observables with more than two possible outcome values. As noted above, for example, if a quantum observable has three possible outcome values, a_1, a_2, and a_3, adding a third dimension to our Cartesian coordinate system (*x*-*y*-*z*) allows for sets of—at most—three mutually orthogonal vectors. Thus $a_1 \perp a_2 \perp a_3$ represents $a_1 \veebar a_2 \veebar a_3$. In this case, $a_1^{\perp}$ is a subspace that includes a_2 and a_3. But in less idealized systems, where observables can have countless values, the expression $a_n \veebar a_n^{\perp}$ is far from trivial; indeed, its equivalence in quantum theory to $a_n \veebar \neg a_n$ clearly evinces the centrality of quantum theory's presupposition of Boolean-contextualized observables, and more generally, its presupposition of PNC and PEM. Thus, for any Boolean measurement context *A*, the following truth-functional, material implication holds:

Context *A*		
a_n	$\neg a_n^{\perp}$	$\rightarrow$
T	T	T
T	F	F
F	T	T
F	F	T

It is important to emphasize here that this truth table represents the Boolean *contextualization* of a quantum state in terms of a particular observable, not the *evaluation* of this observable. Recall that the evaluation of a quantum observable entails the *actualization* of one of its potential values—i.e., it is the unique, actual outcome state consequent of measurement. In the above example, we can signify this actualized outcome state as α_1 or α_2 which is the actualization of *either* potential outcome state a_1 or a_2, respectively, according to the probability valuation of each. Again, the presupposition of this actualization, in satisfaction of PNC and PEM, is reflected in the formalism by the fact that together, the probability valuations of a_1 and a_2 sum to unity.

More broadly, then, the concept of 'evaluation of observables via quantum measurement' entails the totality of the process by which an initially indeterminate, uncontextualized actual state ψ, defined as an abstract integration of countless *potential* outcome states (where neither PNC nor PEM hold) evolves, *by way of* its local Boolean contextualization in terms of some particular observable a, to become a reduced integration of *probable* outcome states a_1 or a_2 , thus satisfying PNC and PEM, and terminating in the actualization α_1 or α_2 according to the probability valuations of a_1 and a_2 respectively. Thus in quantum mechanics probability presupposes actuality (as it does in classical mechanics) *and* actuality presupposes probability (which it does not in classical mechanics). Likewise, evaluation presupposes Boolean contextualization *and* Boolean contextualization presupposes evaluation. Evaluation and contextualization, in other words, as ontologically relational features of actuality and potentiality, are thus mutually implicative in quantum mechanics.

Indeed, it is in light of this mutual implication that quantum indeterminacy can be more comprehensively understood—that is, beyond just the probabilistic character of quantum measurement; for while it is true that the evaluation of observables presupposes Boolean contextualization, it is never the case that the latter *determines* the former. Rather, evaluation is *conditioned* by Boolean contextualization, and this conditioning is indeterminate. Row 3 of the above truth table, for example, describes a state evaluable as *neither* a_n *nor* $a_n^\perp$ where the contextual implication $a_n \rightarrow \neg a_n^\perp$ is nevertheless true. Likewise, this contextual implication itself, as a truth-functional, is logically equivalent to the expression $\neg a_n \vee \neg a_n^\perp$ indicating, again, that the state *is* evaluable as *neither* a_n nor $a_n^\perp$. The state can, in other words, be defined *objectively* in a way that *exceeds the particular Boolean context* given by $a_n \rightarrow \neg a_n^\perp$. In quantum mechanics, as noted above, this objectively indeterminate state of the system *in abstraction from* any particular local Boolean contextualization is represented by the pure state vector ψ.

Formally, the evaluation of an observable a via the Boolean contextualization of ψ is given in the expression:

$$|\psi\rangle = \alpha\,|\,a_n\rangle + \beta\,|\,a_n{}^{\perp}\rangle$$

with $|\alpha|^2 + |\beta|^2 = 1$. (For present purposes, the bracketing of terms with the symbols $|\ \rangle$ is simply understood as signifying, in the conventional Dirac notation, that these terms are quantum state vectors.) One can easily interpret this formal expression by revisiting the earlier discussion: $|\psi\rangle$ is a vector of unit length, thus representing an objective, though indeterminate, uncontextualized state. The right side of the equation represents the contextualization of $|\psi\rangle$ by Context *A*, as described above. The *evaluation* of the observable *a* defined by that contextualization entails the projection of the state vector $|\psi\rangle$ upon the contextualized, mutually exclusive potential outcome states (eigenstates) $|\,a_n\rangle$ and $|\,a_n{}^{\perp}\rangle$. $|\alpha|^2 + |\beta|^2 = 1$ is reflective of the presupposition of the Principle of the Excluded Middle (PEM), such that the coefficients α and β,[19] when squared, give the probabilities of $|\,a_n\rangle$ and $|\,a_n{}^{\perp}\rangle$, respectively, as actual outcome states terminal of the measurement interaction. That is, since these probabilities sum to unity (i.e., they are exhaustive), it is presupposed that *one* of them will become actualized as a measurement outcome state.

In summary, the requirements of [1] a local Boolean measurement context, represented in the standard formalism as an orthonormal measurement basis, and [2] observables represented as self-adjoint projection operators associated with this basis, together guarantee that any quantum mechanical observable is always definable in terms of mutually exclusive and exhaustive outcome states—that is, outcome states relatable as an exclusive disjunction of contradictories, in satisfaction of both PNC and PEM. As a result, local Boolean measurement contexts are *themselves* always relatable coherently and consistently in quantum mechanics; but as discussed earlier, and as will be explored further presently, relations among contexts are always *internal relations*, as evinced by the non-commutativity of quantum observables.

2.4 Internal Relations in Quantum Mechanics

An internal relation is one in which the objective properties of a relatum are modified *by* the relation. They are relations, in other words, that are *constitutive* of a given relatum, rather than external to it.[20] For example, a child's genetic history is understood to be *internally* related to that of his grandfather, since the grandfather's history is internally constitutive of the child's. Conversely, the grandfather's genetic history is *externally* related to that of the child, since it is objectively independent of any relationship with the latter. Likewise, with respect to the concept of a history in general, including the concept of quantum mechanical histories,[21] time itself is asymmetrical in this way, such that facts of 'the present' are understood as internally related to the facts of the past, while the facts of the past are externally related to those of the present.

In precisely the same way, in quantum mechanics, the outcome state of a measured system is internally related to its initial state, with its particular contextualization. Further, since relata are qualified as probabilities in quantum mechanics, probability conditionalization in composite quantum systems is another example of asymmetrical internal relation. While classical conditional probabilities are not the same as quantum conditional probabilities (it can be argued, however, that the latter are properly understood as a generalization of the former), they are sufficiently analogous to employ the simpler classical notation here: $P(A|B)$—the probability of A given B—entails that A is internally related to B. When it is also true that B is externally related to A (i.e., A and B are not *mutually* internally related) this asymmetry is reflective of the non-commutativity of quantum observables, such that $P(A|B) \neq P(B|A)$.

At the most fundamental level, there are only three possible modes of relation among local Boolean measurement contexts in quantum mechanics:

[a] disjoint: $A \cap B = \emptyset$
[b] overlapping: $A \cap B = A \otimes B$
[c] implicative: $A \Rightarrow B$ or $B \Rightarrow A$

Though these modes will be explored in more detail later in the chapter, for the purposes of the present discussion, only [b] and [c] involve internal relations, so we will begin with them. With respect to [c], the mode of implicative internal relation, note that in distinction from the logical connective $\rightarrow$ signifying material implication (a purely syntactic, truth-functional implication) employed earlier in the relation of contextualized observables (where $a \rightarrow b$ is read 'if *a* then *b*'), asymmetrical internal relation *between measurement contexts A and B*, as well as *between an actualized outcome state* α_n *and its particular measurement context A*, is both syntactic *and* semantic, and thus requires a different connective. To this end, the statement 'Context *A* is internally related to Context *B*, and Context *B* is externally related to Context *A*' is signified herein via the expression $A \Rightarrow B$, alternatively read '*A* only if *B*.' Likewise, the statement 'actualized outcome state α_n is internally related to Context *A*' is signified via the expression $\alpha_n \Rightarrow A$.

While this is analogous to the concept of 'logical entailment' ('*A* entails *B*,' or '*B* is deducible from *A*'), as well as to C. I. Lewis's notion of 'strict implication' (both notions being semantic rather than syntactic), internal relation cannot be simply assimilated to these. G. E. Moore defines internal relation thus: For any internal relational property *F* belonging to *x*, the following must be true:[22]

$$[1] \qquad (\forall x)\,(\forall y)\,((Fx \rightarrow (\neg\, Fy \vDash y \neq x))$$

In other words, "if *x* has the relational property *F*, then from *y*'s lack of this property it can be *deduced* that *y* is not identical with *x*." If this proposition is true for a particular value of *F*, then *F* is an internal relational property. If false, *F* is an external relational property.

It is interesting to note that in comparison with [1], there is a similar proposition which, as Moore argued,[23] can be said to hold true for *all* relational properties *F*, whether internal or external:

$$[2] \qquad (\forall x)\,(\forall y)\,((Fx \vDash (\neg\, Fy \rightarrow y \neq x))$$

This expression is identical to [1], except that the connectives → (material implication) and ⊨ (entailment) are transposed. Expression [2] thus reads, "given that x has the relational property F, one can *deduce*, as a matter of *fact*—i.e., semantic content, not just syntactic form—that if y lacks this property, then it cannot be identical with x." Recall that in quantum mechanics, this distinction between form and fact is reflected in the distinction between contextualized *potential* outcome states and *actual* outcome states, respectively. 'Facts' are evaluated observables—i.e., *actualized* potentia; they are actualizations α_n of potential measurement outcomes a_n whose Boolean contextualization A allows these potential outcomes to be valuated as probabilities. Since, for internal relations, *both* [1] and [2] are true (again, such that the connectives → and ⊨ can be transposed) it would seem that the statement 'α_n is internally related to A' exhibits a mutually implicative relationship between the concept of material implication and the concept of logical entailment in quantum mechanics—i.e., the mutual implication of syntax and semantics, of form and fact. In this way, internal relation in the relational realist interpretation of quantum mechanics can be thought of as connoting a kind of 'ontological implication' that subsumes but exceeds logical implication at the level of fundamental physics.

In quantum mechanics, there are two basic categories of internal relation, both of which are asymmetrical, and both of which are operative in every quantum measurement interaction: [1] *global outcome state* internally related to locally contextualized measurement outcomes (i.e., *extension of the local to the global*); [2] locally contextualized measurement outcomes internally related to the *global initial state* (i.e., *restriction of the local by the global*).

By category [1], locally contextualized measurement outcomes—'quantum facts'—are understood as constitutive of a novel global totality consequent of measurement—i.e., a novel augmentation of the initial global state. By category [2], locally contextualized measurement outcomes are internally related to the global totality of facts subsuming the measured system's initial actual (though indeterminate) state. More specifically, both the potential and actual outcome states of a measured system are internally related to the system's local Boolean contextualization of the global totality of facts constitutive of the measured system's initial state (cf., von Neumann's projection postulate and Lüders rule). In quantum mechanics, in other words, a measurement's own particular local contextualization of the initial global state is always understood to be internally constitutive of the measurement outcome. For example, in the expression introduced earlier,

$$|\psi\rangle = \alpha\,|\,a_n\rangle + \beta\,|\,a_n{}^{\perp}\rangle$$

it is always the case that $\alpha\,|\,a_n\rangle + \beta\,|\,a_n{}^{\perp}\rangle$ is internally related to the objective but indeterminate totality of facts constitutive of $|\psi\rangle$.

As we will see in the following section, nonlocal probability conditionalization is a well-known exemplification of categories [1] and [2] together, where locally contextualized *potential* measurement outcomes of one component of a composite quantum system are internally related to (and thus logically conditioned by) locally contextualized *actual* outcomes within a different component (cf. von Neumann's projection postulate and Lüders rule applied to nonlocal EPR correlations, chapters 2.5, 2.8, and 9.4).

As the conceptual framework presented in this volume unfolds, we will see that [1] and [2] above are, in fact, dipolar aspects of a single, unified relational process inherent in every quantum measurement event. A key principle of this framework, as is evident above, and as will be explored further in the next section, is that locally Boolean-contextualized quantum mechanical internal relations always require reference to a totality of facts whose *global* relations are non-Boolean. This key principle is most easily understood as an analog to the concept of the measured system's state vector $|\psi\rangle$ introduced in the previous section: Recall that $|\psi\rangle$ represents an objective though indeterminate state of a measured system *in abstraction from any particular local contextualization*, and is thus non-Boolean. Likewise, the global state vector $|\Psi\rangle$, also introduced in the previous section, represents an objective though indeterminate state of a composite quantum system, where 'global' is defined as the totality of this composite system. As we saw earlier, $|\Psi\rangle$, as an uncontextualized representation of the global state, is, like $|\psi\rangle$, non-Boolean; the individual local systems constitutive of the global composite system, one will recall, are not mutually independent and therefore relations between them must be expressed as a tensor product. Writ large, of course, $|\Psi\rangle$ could just as easily represent the totality of the universe itself considered quantum mechanically—i.e., as an uncontextualized, objective though indeterminate state of the universe. In that light, it becomes intuitively clear why there is no sense in which a 'global Boolean context' for a global state vector $|\Psi\rangle$ can be specified quantum mechanically in the same way that local Boolean contexts can be specified for $|\psi\rangle$: it would be akin to measuring the universe *itself* as either 'here' or 'there' in the same way that any system *within* the universe can be so measured. (This is an exemplification of the logical and philosophical problem of predicating totalities, discussed in chapter 3.) Further still, as seen in the previous discussion of composite quantum systems, quantum mechanics does not allow for the analytic depiction of a 'global Boolean contextualized state' as simply an exhaustive concatenation of all local Boolean contextualized states, as though these local states were mutually independent.[24] As we will see, this limitation is equally intuitive, even though at first glance it would seem less so. (Both of these related proscriptions are features of the Kochen-Specker theorem, and will be explored in detail in chapter 3.)

What can be specified quantum mechanically is always a mutually implicative relationship between local and global, such that neither can be abstracted from the other in any quantum measurement event. This relationship has only two possible modes, and these correspond, respectively, with the two categories of asymmetrical internal relation discussed above; and like the latter, these two modes are mutually implicative and always jointly operative in every quantum measurement event: [1] *extension of the local to the global*, wherein locally contextualized facts (i.e., measurement outcomes) condition global potentia nonlocally. This is evinced, for example, by nonlocal probability conditionalization in quantum mechanics, to be further discussed in section 2.8; [2] *re-*

striction of the local by the global, wherein global facts condition local potentia and their local contextualization. This is evinced, for example via the phenomenon of environmental decoherence, which will be introduced in section 2.7.

The central conceptual challenge, then, for any ontological interpretation of quantum mechanics, is not only the problem of measurement, nor is it the problem of actualization of potentia (i.e., the problem of the existence of facts); the central challenge underlying both of these problems, rather, is properly understanding, via a coherent and empirically adequate conceptual scheme, the mutually implicative relationship between local and global in quantum mechanics. This necessarily entails the construction of a formal philosophical and mathematical framework that adequately depicts how the logical features of this relationship can be shown to condition the causal features.

In summary of the discussion so far, the fundamental dipolar process in which this framework is anchored—the essential process of quantum mechanics—is [1] the asymmetrical internal relation of a *global outcome state* to locally contextualized measurement outcomes (i.e., *extension of the local to the global*); [2] the asymmetrical internal relation of locally contextualized measurement outcomes to the *global initial state* (i.e., *restriction of the local by the global*). Again, it is a fundamental principle of quantum mechanics per the Kochen-Specker theorem that this dipolar process excludes the possibility of global Boolean contextualization, either synthetically via extension, or analytically via restriction. The constituent facts of this totality, in other words, are not determinate; they cannot *all* be assigned definite bivalent truth values (i.e., *either* true or false) such that PNC and PEM are satisfied among *all* possible relations of these facts. This totality, then, is not only epistemically indeterminate globally; it is ontologically indeterminate locally—that is, as locally *constitutive* of the particular quantum measurement event internally related to it. However, as we will see, it is via this dipolar process of asymmetrical internal relation that manifold local Boolean contextualized facts *can* be coherently and objectively integrated nonlocally—even if not comprehensively as a global Boolean totality—such that PNC and PEM condition causal relations not only *within* local Boolean contexts, but also *across* these contexts, even when measured systems are spatially well-separated. It is via asymmetrical internal relation, in other words, that a global totality of facts can be *coherently* internally constitutive of a local quantum process.

2.5 The Compatibility Condition for Logical Causality in Quantum Mechanics

Asymmetrical internal relations among local quantum measurement contexts, combined with the presupposition that individual local measurement contexts are structurally Boolean, together constitute the conceptual foundation of the *compatibility condition for logical causality* in quantum mechanics, which will be introduced here, and explored in greater detail in chapter 4, as well as in part II, chapters 9 and 10. Here, 'logical causality' ought not be interpreted as meaning 'logic that is causal'; it means, rather, 'causality that is logical.' More accurately, it means 'causality that is logically conditioned.' The compatibility condition is exemplified in physics most generally as the universal, categorical correspondence of [a] the asymmetrical order of material implication and logical consequence (for the purposes of the present discussion, these together can be referred to simply as 'logical implication') with [b] the asymmetrical order of causal relation.

The correlation of the asymmetrical relationship depicted by 'cause-effect' statements with the asymmetrical relationship depicted by 'if-then' statements has been an infamous philosophical problem since the time of Plato because while it is easily demonstrated, it cannot be proven by derivation or deduction from some more fundamental principle. Even classical mechanics provided philosophers of the early modern period with no readily discernible indication of *why* the physical order of causal relation correlated with the conceptual order of logical implication. As Hume famously argued, one cannot ever 'directly observe' causal connection, any more than one can directly observe logical implication; rather, we infer causal connection only after directly observing system states in conjunctive relation—e.g., before and after a measurement interaction. This is well exemplified in quantum mechanical measurement, as discussed earlier, by the fact that one cannot directly observe superpositions of potential outcome states; rather, we infer these superpositions only after directly observing a conjunction relating [a] an actual initial system state prior to a measurement interaction, and [b] an actual final system state consequent of the measurement interaction.

The resulting philosophical proposals of the early modern period either attempted to assimilate the conceptual logical order to the physical causal order (e.g., the philosophy of Locke), to assimilate the causal order to the logical order (e.g., the philosophy of Spinoza), to depict the correlation of the logical and causal orders as a schematization by which we experience the otherwise transcendent 'things in themselves' (e.g., the philosophy of Kant), or to admonish against the philosophical presupposition of either order (e.g., the philosophy of Hume). Quantum mechanics, however, provides a strong indication that the asymmetry of the order of causal relation *presupposes* the asymmetry of the order of logical implication, yet cannot in any way be *reduced* to the latter; for its fundamentally indeterministic nature, both at the level of theory and empirical practice, utterly precludes any such sheer assimilation of contingency to necessity—e.g., in the spirit of Spinoza or Leibniz. Rather, quantum causality's necessary presupposition of the logical order entails simply that the causal and logical orders are properly understood as mutually implicative at the level of fundamental physics. Again, it is an essential presupposition of quantum mechanics that universally, every local measurement context is structurally Boolean—that is, every local context can be represented mathematically as a Boolean subalgebra, or alternatively as an equivalence class of Boolean subalgebras. It is by this presupposition that all local contexts are globally relatable coherently and consistently (as is presupposed by every physical theory whose laws are presumed to hold universally).

The compatibility condition is built upon two foundational concepts introduced previously, which will be further elaborated here:

[1] *locally*, every measurement context must be Boolean, such that in the mixed state,[25] Boolean material implication holds—e.g., for any measurement context A it will always be the case that for potential outcomes a_1 and a_2 , $a_1 \rightarrow \neg a_2$ ('if a_1, then not a_2') and $a_2 \rightarrow \neg a_1$. This, of course, is just PNC for an observable a with only two possible eigenstates (i.e., potential outcome states) a_1 and a_2. This number, however, is potentially infinite in quantum mechanics; again, because the contextual measurement basis is orthonormal (where n mutually exclusive eigenstates—e.g., representing n possible pointer positions on a particular detector—are depicted in the formalism as n mutually orthogonal vectors in a Hilbert space of n dimensions) one can represent the Boolean complement $\neg a_1$ by simply grouping all of the alternative mutually exclusive eigenstates into a subspace $S^{\perp}$ of A, such that $a_1 \rightarrow \neg a_1^{\perp}$. As noted earlier, this presupposition of local Boolean contextuality, yielding mutually exclusive outcome states *regardless of the number of possible outcome states*, is a necessary prerequisite for the probability valuation of these alternative outcome states (the Born rule), which is a categorical presupposition of quantum mechanics.

[2] *globally* (i.e., when local contexts are brought into nonlocal relation), intra-contextual Boolean material implication (that is, *within* individual local measurement contexts) must be relatable inter-contextually *across* these local contexts (i.e., 'globally'). In quantum mechanics, this is expressed as a tensor product relationship of potential outcome states.

Thus, continuing with the above example, in a composite quantum system, for local measurement contexts A and B (and their associated detectors), if potential outcome a_i is internally related to Boolean context A, and potential outcome b_i is internally related to Boolean context B, as given in the expression:

$$(a_i \Rightarrow A) \wedge (b_i \Rightarrow B)$$

then the state of the composite quantum system, as we saw in section 2.3, is expressed as:

$$\Psi = \frac{1}{\sqrt{2}} (a_1 \otimes b_1 + a_2 \otimes b_2)$$

which yields the probabilities:

$$\begin{aligned} P(a_1 \cap b_1) &= 0.5 \\ P(a_2 \cap b_2) &= 0.5 \\ P(a_1 \cap b_2) &= 0 \\ P(a_2 \cap b_1) &= 0 \end{aligned}$$

Once a measurement outcome has been registered by one of the detectors—say *A*—the integration of potential outcome states at *B* is *revised* via its internal relation to the outcome at *A*, and this revision is manifest as a probability conditionalization. Thus, it is via the asymmetrical internal relation of *B* to *A* that the asymmetrical structure of material implication intra-contextually—within each local Boolean measurement context individually—can be extended inter-contextually, *across* these local contexts via the tensor product relationship of potential outcome states at *A* and *B* together.

As a special case, depending upon the individual contexts, the internal relation of *B* to *A* can yield probability valuations that take the form of logical entailment. For example, if an electron-positron pair is emitted during a particle decay, measurement of the electron's quantum spin as 'up' on a particular axis / measurement context entails that the positron's spin will be 'down' if the same axis is chosen for the positron's measurement context. (This logical entailment is reflected in the conservation of momentum law.) However, even when internal relations across local contexts take the form of logical entailment, these relations are always probability valuations yielded via the tensor product relationship of potential outcomes across the component local contexts—again, such that any 'global' state specification can only be *inductively* evaluated in terms of the local contexts it subsumes. In other words, because neither local observables nor their measurement contexts are related deterministically, the 'global' state can never be analytically evaluated as simply a classical sum of its local constituent states.

This is evinced most clearly in experiments on composite quantum systems that are often depicted as exhibiting 'nonlocal causal' relations, where, per the current discussion, local measurement contexts *A* and *B*—i.e., detectors *A* and *B*—are spatially well-separated, each measuring a different component of the composite system. Because of their spacelike separation, it is assumed that given the relativistic speed of light limitation, the order of component measurement (*A* then *B*, or *B* then *A*) should be irrelevant, since any physical causal correlations between components would require a faster-than-light propagation of energy between *A* and *B*. The most well-known of these experiments are the EPR[26]-type experimental arrangements, which will be discussed further in section 2.8 and chapter 3, and formally depicted via the relational realist sheaf-theoretic framework in chapter 9.4. What these experiments reveal is that while there is, indeed, no measurable nonlocal, efficient causal influence between *A* and *B*, there *is* a measurable, nonlocal probability conditionalization between *A* and *B* that always takes the form of an asymmetrical internal relation. For example, as discussed above, if *A* registers first, the outcome at *B* is internally related to the outcome at *A*. This, again, is manifest as a probability conditionalization of the *potential* outcomes at *B* by the *actual* outcome at *A*; specifically, the integration of *B*'s contextualized potential outcomes, represented as an equivalence class of Boolean subalgebras, is 'revised'[27] by the actual outcome at *A*. This revision, indicative of the asymmetrical internal relation of *B*'s outcome to *A*'s outcome, has been well demonstrated in numerous EPR-type experimental investigations of quantum nonlocality.

While some interpretations resort to exotic explanations (e.g., superluminal propagations of hidden energy, or other efficient causal, physical-dynamical mechanisms) the phenomenon of quantum nonlocality can instead be intuitively and coherently understood simply as a logical conditioning of causal relations—a conditioning implicit in the logical relational structure presupposed by all scientific theories. This logical conditioning of causal relations, defined above as the compatibility condition for logical causality in quantum mechanics, renders explicit the implicit presupposition of: [1] the Boolean relational structure of each local context *A* and *B* (again, as a necessary presupposition of the scientific method in general) and [2] the coherent global relation of these local contexts via this structure.[28]

From this perspective, in summary of the present example, the equivalence class of Boolean subalgebras representing the integration of potential outcomes at *B* is 'logically affected' (or better, ontologically revised) by the measurement outcome at *A*, thus exhibiting *B*'s internal relationship to *A*, even when *A* and *B* are spacelike separated. This nonlocal revision entails no propagation of energy of any kind from *A* to *B* and is thus not properly understood as an efficient causal influence of the actualized outcome at *B* by that of *A*; rather, it is a logical conditioning (viz., a nonlocal probability conditionalization) of the contextualized *potential* outcomes at *B* via the internal relation of these outcomes to the *actualized* outcome at *A*.

The advantage of the category-sheaf theoretic formalism introduced in the present volume is that it explicitly reveals the formal mereotopological structure by which these nonlocal internal relations are integrated via the compatibility condition in quantum mechanics; whereas in the more conventional Hilbert space framework, even when extended to include concepts such as overlapping equivalence classes of Boolean subalgebras, this structure is only implicit and not formally defined. Nevertheless, for the present introductory discussion, this more conventional depiction will serve as a useful propaedeutic to the category-sheaf theoretic formalism presented later in the book.

2.6 Inducing the Global from the Local

It is via the compatibility condition for logically conditioned causal relations that local Boolean measurement contexts are representable as Boolean subalgebras, or equivalence classes of Boolean subalgebras, relatable globally via their overlaps. Such global relations in composite quantum systems are always asymmetrical internal relations of [a] *potential* measurement outcomes contextualized by one local context, to [b] an *actualized* measurement outcome contextualized by another local context. The mathematical representation of this conceptual framework is presented formally in part II of this volume by Elias Zafiris in terms of Grothendieck topology and category-sheaf theory.[29] As groundwork for this later discussion, it will be helpful to explore in greater detail the concept of overlapping equivalence classes of local Boolean contextual algebras as an inductive representation of a 'global' state (Ψ).

As discussed previously, the manifold locally contextualized potential measurement outcomes constitutive of a global composite quantum system are related inter-contextually (i.e., across the component local contexts) via the tensor product. When a local context is represented as a Boolean subalgebra, or as an equivalence class of Boolean subalgebras, the tensor product relationship can thus be thought of as including an 'overlap' of these Boolean subalgebras, and this overlap is representative of compatible potential measurement outcomes (i.e., those satisfying PNC and PEM inter-contextually). In classical mechanics, by contrast, the global state is characterized as a straight-forward conjunction of all local Boolean-contextualized states, such that the global totality of all contexts is itself fully Boolean (i.e., the whole is simply the sum of its parts). In quantum mechanics, however, the global can only be depicted via *induction*, not conjunction; and it is via the overlap of Boolean subalgebras representing the various local measurement contexts constitutive of the global system that such induction is possible.

As we have already seen, one reason the global can never be deduced from the local quantum mechanically, but only induced, is that observables contextualized by multiple local measurement contexts do not commute; and as Kochen and Specker[30] have shown, it is impossible even in principle to embed local contextual Boolean subalgebras into a 'global' Boolean algebra. (As will be argued further in chapter 3, the ontological significance of the Kochen-Specker theorem with respect to relating local and global states is intimately related to the logical problem of predicating totalities.)

Quantum theory's signature replacement of classical local-global mereology with the concept of 'inducing the global' via overlapping local contextual Boolean subalgebras, where *potential* measurement outcomes in one context are always internally related to *actual* measurement outcomes in others, is central to the argument that topology is the proper means by which to formalize quantum mechanical relations, given their underlying asymmetrical logical order. This is because quantum mechanical relations, as we have seen thus far, are not properly understood as relations of independently extant *objects*; rather, they are *asymmetrical internal relations of relations considered objectively*. Thus, as object relata, quantum mechanical internal relations are *always structure preserving*, as can be depicted topologically (i.e., mereotopologically via category theory) allowing for inductions like overlapping Boolean subalgebras to have ontological and not merely epistemic significance. By contrast, the conventional metrical approach to quantum mechanics, with its adherence to classical mereology, is incapable of depicting structure preserving 'relations of relations' in this way. In the conventional set theoretic framework, for example, 'elements'—the framework's fundamental object relata—are only relatable externally, never internally; there is no formal means by which to depict an element's 'internal relational structure'—i.e., its internal relations to other elements and their relations. (The category theoretic framework, unlike the set theoretic framework, does precisely this, as will be further elaborated throughout the course of this volume.)

Indeed, it is arguably because of the conventional tacit presupposition of classical mereology, where the whole is simply the totality of its parts, that it is often considered a theoretical deficiency of quantum mechanics that it cannot define the global as a totality of this kind—i.e., via a single, comprehensive Boolean algebra. As discussed earlier in this chapter, this has inspired many theorists to conclude that Boolean logic must therefore be invalid at the level of fundamental physics. However, the fashionable solution of replacing classical Boolean logic with a non-

Boolean quantum logic is, again, a dubious one, given that Boolean logic is a necessary presupposition of the scientific method itself—the very conceptual framework by which quantum mechanics, and every other scientific theory, is formalized, implemented, and evaluated. Indeed, it can be argued that Boolean logic is necessarily presupposed in the coherent conception and expression of any theory, model, or rational construction of any kind—including the non-Boolean logics intended to replace it. But 'presupposition of' here ought not be misread as 'reducible to'; for the argument is not that all rational relations of *fact* are reducible to Boolean, bivalent *forms* of relation. The mutual implication of fact and form, respectively correlate with the mutual implication of actuality and potentiality discussed earlier, precludes any such sheer assimilation. The claim, rather, is simply that any scheme of coherent and consistent conceptual relations, particularly any scheme that is sufficiently coherent and consistent to accommodate the modern scientific method, necessarily presupposes Boolean logic. Again, this is true even for schemes constructed to invalidate Boolean logic at some fundamental level such as that depicted via quantum physics; for regardless of their breadth of formal complexity, the depth of invalidation achieved by such schemes is nevertheless measurable only via a still deeper level of Boolean presupposition, without which the purported invalidation could be neither coherently formulated nor coherently expressed.

Thus, it would seem that a proper assessment of the inability of quantum mechanics to depict a fully Boolean global state cannot simply be that Boolean logic is invalid at the level of fundamental physics; it seems more likely, rather, that it is the conventional, classically mereological conception of 'global state' via the purely analytical relation of local and global that requires revision—specifically, one amenable quantum mechanics' double assertion of [1] the universality of local Boolean measurement contextualization, and [2] the impossibility of defining a universal global Boolean measurement context.

As is clear from the discussion thus far, the relational realist framework makes both of these assertions via a depiction of the relationship between local and global that is both analytic *and* synthetic. Instead of recasting local-global quantum mechanical relations as fundamentally non-Boolean rather than Boolean, it recasts local-global relations as fundamentally mereotopological rather than metrical and classically mereological. In this way, the global is always coherently (i.e., Boolean logically) relatable to its locally contextualized constituents, while at the same time never wholly reducible to an analytic description of these constituents.

As a heuristic example of a topological induction of the global from the local, consider the familiar Möbius strip. Taken as representing a global form, we can depict multiple local Boolean quantum measurement contexts on this form as fibers spanning the width of the strip (i.e., perpendicular to its length), bundled together and running lengthwise along its entirety. The total bundle of these 'local' fibers thus constitutes an inductive representation of the entire strip. But topologically, this inductive 'global' fiber bundle takes the form of a cylinder, not a Möbius strip. This is because locally, a cylinder and a Möbius strip are topologically identical, such that the topological relations among the 'local' fibers do not yield an inductive approximation of the global sufficient to capture the distinctive twist characteristic of the actual global form of the Möbius strip. In the same way, the topological relations among local Boolean measurement contexts / Boolean subalgebras in quantum mechanics do not allow one to completely define a 'global' state simply as a totality of 'local' states.

Nevertheless, the fact that local quantum measurement contexts are coherently relatable *at all* necessarily presupposes the compatibility condition introduced in the previous section, without which it is impossible to coherently depict quantum mechanics as a universal, and thus ontologically significant, theory. It is via the compatibility condition that all local measurement contexts in quantum mechanics are characterized as fundamentally Boolean relational structures whose relational asymmetry locally is preserved when extended globally.[31]

As has previously been demonstrated by Zafiris in other writings,[32] and further discussed by him in part II, this presupposition of local Boolean measurement contexts relatable globally via induction (i.e., 'gluing' in sheaf theory) is represented topologically as a presupposed uniform fibration of the global topological space into a structure of individual, local Boolean subalgebras. In the relational realist philosophical cosmology this is a formal expression of the fact that the universe is a logically structured totality of quantum facts, but one that subsumes manifold diverse (though always logically relatable) local contextualizations. Further, since quantum measurement events are always predicative—i.e., always generative of predicative facts—this totality is synthetic, such that it increases, in a structure-preserving way, with each novel, locally contextualized, quantum fact; it is not, in other words, a closed, predefined totality, analytically or deterministically reducible to the individual states of its constitutive parts.

This presupposition is most clearly evinced in quantum mechanics in three ways, as discussed previously: [1] quantum indeterminacy; [2] the presupposition of orthonormal measurement bases (i.e., Boolean measurement contexts) in quantum measurement; and [3] the possibility of relating, as overlapping equivalence classes of Boolean subalgebras, multiple local Boolean measurement contexts in a logically coherent and consistent way that preserves both the asymmetrical relational structures (e.g., material implication and logical consequence) within these local contexts, and the asymmetrical internal relations among them globally.

2.7 Decoherence: Causal Relation or Logical Integration?

With respect to the above discussion, the logical integration of [a] a matrix of *potential* outcome states in the pure state to [b] a reduced matrix of *probable* outcome states in the mixed state, is a logical integration of potential relations conventionally expressed as a tensor product, by which equivalence classes of local Boolean-contextualized potential outcomes are relatable inter-contextually among the various local measurement contexts constitutive of a composite quantum system. The reduction of pure state potentia to mixed state probabilities via equivalence classes of relations is conventionally attributed to the process of quantum decoherence,[33] of which there are a number of interpretative models. It should be emphasized, however, that despite this diversity of interpretation, all of these models presuppose the principles of Boolean logic in both the formation of equivalence classes of locally contextualized observables, and the elimination of logically incompatible relations (i.e., 'quantum interference' violating PNC) from within these classes.[34]

With respect to the discussion thus far, decoherence occurs when the Boolean subalgebra representative of a particular, locally contextualized *measured system observable* is integrated with equivalence classes of Boolean subalgebras representing all *unmeasured environmental observables* per their own individual local contextualizations. As unmeasured observables, the latter are undefined; nevertheless, they are presumed to be locally Boolean, and thus representable as equivalence classes of Boolean subalgebras whose overlap with the Boolean subalgebra representing the measured system observable *can* be defined. These unmeasured environmental observables are, in other words, the countless environmental degrees of freedom operative when the relations between a quantum mechanical measured system *and* its environment are included in the measurement formalism.

This inclusion might seem unjustified instrumentally, given that the totality of these unmeasured degrees of freedom is, in fact, immeasurable practically. However, the inclusion of environmental relations in the quantum measurement formalism is not only justified; it is arguably prescriptive if quantum mechanics is to be considered a fundamental physical theory rather than merely a methodology. This is because 'system,' 'detector,' and 'environment' are always formally entangled in quantum mechanics, such that their partitioning is purely arbitrary. Thus, the immeasurability of environmental relations as an epistemic issue cannot be taken to imply the irrelevance of these relations ontologically, and this distinction is the hallmark principle of those interpretations of quantum mechanics in which decoherence is recognized as an essential feature.

In certain of these interpretations, such as the dynamical environmental decoherence models,[35] system-environment relations are depicted as purely efficient causal relations via energy transfer between the environment and the measured system. (Experimentally this is achieved, for example, by modeling environmental degrees of freedom as an oscillator bath.)[36] A popularly presumed implication is that if this energy transfer can be prevented or dampened (e.g., by bringing the system to a sufficiently low temperature), decoherence might be prevented—or at least delayed—leaving the system in an 'actual superposition' for an extended period of time.

However, it is now understood that there are quantum system-environment relations that entail no energy transfer whatsoever, yet nevertheless generate decoherence. Nuclear spins, when treated as environmental degrees of freedom, are one example.[37] Since there is no energy transfer in nuclear spin bath decoherence, quantum mechanical relations between measured system and environment cannot be characterized as purely efficient casual relations.[38] Thus, experiments on nuclear spin related decoherence indicate that phase exchange between system and environment is more accurately understood as a Boolean logical conditioning of potentia via 'if-then' relations, rather than a purely causal influence through energy exchange via 'cause-effect' relations. Recall that this is precisely the same understanding depicted in the EPR-type investigations of 'causal quantum nonlocality' introduced in section 2.5, wherein the physical efficacy of logical relations was likewise evinced in absence of energy transfer—i.e., in absence of 'purely' efficient causal relations. (We will revisit EPR in greater detail in the following section.) Thus, logical causality in quantum decoherence well exemplifies the central thesis explored thus far in this chapter: In quantum mechanics, all causal relationships presuppose logical relationships, and logical relationships can have physical significance.

2.8 EPR and Quantum Nonlocality

In addition to the phenomenon of quantum decoherence briefly introduced above, perhaps the clearest exemplification of the thesis of logical causality is the familiar thought-experiment of Einstein, Podolsky and Rosen[39] (EPR) introduced in section 2.5 in the form of its more recent experimental incarnations[40]: A singlet (total spin zero) system decays into an electron-positron pair, with the two particles propagating in opposite directions (*z* and –*z* on an *x*-*y*-*z* coordinate system). The spin of each particle (i.e., its intrinsic angular momentum) is measured by a detector at the end of each path (detector *A* and detector *B*). Each detector can be oriented to measure the particle's spin direction on perpendicular axes (e.g., either the *x*-direction or the *y*-direction relative to the *z* axis).

There are three central principles operative in this experiment as it pertains to the conceptual framework presented thus far:

[1] Once the particle's *x*-spin component in measured, it is impossible to measure its *y*-spin component, and vice versa (*x* and *y* spin components are, in other words, complementary observables like position and momentum, which were the observables of the original EPR thought experiment).

[2] Because the initial singlet has zero angular momentum, the total angular momentum of the two particles must also be zero. Many fundamental particles such as electrons and quarks have only two possible spin directions, 'up' and 'down' (↑ or ↓) on any particular axis.[41] This means that if detector *A* measures particle *A* with spin up on the *x*-axis, (a_x↑), then detector *B*, if it is also set to measure spin on the *x*-axis, will always show (i.e., with probability = 1) spin down (b_x↓).

[3] Detectors *A* and *B* are spatially well-separated such that a measurement outcome at detector *A* could only *causally* affect a subsequent measurement at detector *B* via a faster-than-light energy

> transfer propagating from *A* to *B*. Since relativity theory proscribes the possibility of such a transfer, *A* and *B* are considered to be causally externally related—i.e., nonlocally independent with respect to efficient causality.

As discussed earlier, the classical ontological understanding of the evaluation of an observable like *x* and *y* spin direction via a measurement interaction is that the latter merely reveals pre-existing and always well-defined values. Thus the state of a system is understood classically as a *complete* specification by which the values of all its observables are, in principle, capable of revelation by an appropriate measurement. This allows the direct evaluation of some restricted set of observables to indirectly reveal, by logical implication, the values of any other observables. This is the classical ontological understanding of Principle 2 above: Particles *A* and *B* have their *x* and *y* spin components at all times, such that direct measurement of Particle *A* as $a_x\uparrow$ reveals, by logical entailment, that Particle *B* is $b_x\downarrow$ even if Particle *B* is not itself directly measured. Entailment, here, is thus symmetric and purely epistemic.

When interpreted according to the classical realist ontology, then, the complementarity of *x* and *y* spin measurements (Principle 1) should be avertible with an appropriate sequence of measurements. Evaluation of *both* *x* and *y* spin components for Particle *A*, for example, could be achieved via a combination of [1] *direct* evaluation of Particle *A*'s *x*-spin, and [2] *implicative* evaluation of Particle *A*'s *y*-spin via direct evaluation of Particle *B*'s *y*-spin. For example, if Detector *A* measures the spin direction of Particle *A* as $a_x\uparrow$ and Detector *B* measures Particle *B* as $b_y\downarrow$, then by logical implication per Principle 2 above, Particle *A*'s *y*-spin can be inferred with certainty (probability = 1) to be $b_y\uparrow$.

Yet when the experiment is actually conducted, the indirect, logically implicative measurement of Particle *A*'s *y*-spin component fails. When the *x*-spin component of Particle *A* is measured by Detector *A* (rendering its *y*-spin component undefinable by Detector *A*), the attempted measurement of the *y*-spin component of Particle *B* at Detector *B* *likewise yields undefined values*. It is only when the *y*-spin of Particle *A* is measured that Particle *B*'s *y*-spin is well-defined (such that $b_y\downarrow \Rightarrow a_y\uparrow$ with probability = 1). But if Particle *A*'s *x*-spin is measured first, it is always the case that Particle *B*'s *y*-spin component is $b_y\downarrow$ with probability = 0.5, and $b_y\uparrow$ with probability = 0.5.

Thus a classical ontological interpretation of these results would lead to the bizarre question: "Is Particle *B* *causally influenced* by the choice of locally contextualized measurement (*x*-spin or *y*-spin) at Particle *A*? And if so, what is the faster-than-light mechanism by which this causal influence can be 'transmitted' nonlocally from Detector *A* to Particle *B*?" For it seems that the measurement outcome at Detector *A* (i.e., the evaluation of either a_x or a_y), a function of the Boolean measurement context chosen for Detector *A* (measurement context A_x or A_y), somehow 'non-locally affects' the possible measurement outcomes at Detector *B* such that:

If Detector *A* measures $a_x\uparrow$ or $\downarrow$
then at Detector *B*, $P(b_y\downarrow) = 0.5$ and $P(b_y\uparrow) = 0.5$

If Detector *A* measures $b_y\uparrow$
then at Detector *B*, $P(b_y\downarrow) = 1$

If Detector *A* measures $b_y\downarrow$
then at Detector *B*, $P(b_y\uparrow) = 1$

By contrast, the quantum ontological interpretation of this experiment, per the relational realist framework presented thus far, depicts these results as illustrative of an intuitively simple idea: Potential outcomes are always logically conditioned by extant actual outcomes, regardless of any spatial separation of the regions associated with these outcomes. In absence of any superluminal, physical propagation of energy from the spacelike separated regions characteristic of EPR experiments, it is only via such logical conditioning that outcomes in both regions can be coherently integrated—i.e., as coherently related *facts* satisfying both PNC and PEM, such that despite their individual local contextualizations and mutual efficient causal independence, these facts are coherently integrable within a shared global history. Thus, when searching for Socrates ('measuring his position') the probability of discovering him in Athens is nonlocally, logically conditioned by his sighting in Crete. Classically, of course, this intuition is unremarkable since by that worldview, still regnant in our common sensibility, all physical observables are thought to possess precise values at all times. But quantum mechanically, as we have seen, measurement outcomes can only be understood as generated consequent of measurement, not merely revealed subsequent to measurement—an understanding that has been consistently confirmed empirically, against every attempted imposition of classical intuition upon quantum theory.[42] The actualization of a potential measurement outcome, in other words, is always understood as a *novel* fact in quantum mechanics.

But this need not be considered counterintuitive to common sense in general (as EPR-type experiments are typically portrayed), just because it defies a particular intuition concerning classical observables. Indeed, an equally intuitive understanding of quantum mechanical logical-causal correlation among diverse locally contextualized facts can be found in the common conception of a 'history of facts'; for this concept always entails a natural presupposition that a global history will coherently subsume (i.e., without violations of PNC and PEM) the locally contextualized histories constitutive of it. Whether or not the objects depicted via these local histories are spacelike separated is irrelevant to this presupposed correlation.

Indeed, to the extent that quantum mechanics is accepted as the fundamental theory by which we properly understand the notion of 'physical object,' it is not the history that depicts the object, but rather the object that depicts the history. The classical understanding of a history as an epistemic representation of an ontologically fundamental 'material object' is thus reversed—a quantum Copernican turn wherein it is the classical notion of material object that is understood to be the epistemic representation of ontologically fundamental histories of quantum facts. Though the latter are always locally contextualized, they are always coherently relatable globally. Note that in the present example, it is only via the presupposed Boolean structure of each local context—'either in Athens or not' and 'either in Crete or not'—and the overlap of these local Boolean contexts, that their associated locally contextualized facts are coherently relatable globally—i.e., within a wider shared history that integrates, without violations of PNC or PEM, the local histories of Athens and Crete individually.

Likewise, in the case of the EPR experimental arrangement, this logical integration is expressed as a non-local ('global') inductive relationship among local Boolean measurement contexts. In this case, there are four possible contexts (A_x, A_y, B_x, B_y), each of which is locally Boolean such that, for example, in Context A_x, $a_x\uparrow \rightarrow \neg a_x\downarrow$. As discussed in sections 2.3 and 2.5, it is via this local Boolean contextuality and the associated compatibility condition that relations of potential outcome states *across* contexts is possible. Recall that these relations are expressible in the form of a tensor product:

$$\Psi = \frac{1}{\sqrt{2}}\left(a_x \otimes b_x + a_y \otimes b_y\right)$$

with each Boolean measurement context *A* and *B* represented as a Boolean subalgebra. The tensor product relationship in the above equation defines the overlap of these Boolean subalgebras, by which compatible measurement outcomes at *A* and *B* are relatable. An *actualized* measurement outcome within any one of the four contexts, A_x, A_y, B_x, B_y , then, has the effect of altering this overlap, producing what Jeffrey Bub[43] describes as a 'revision' of the overlap, which includes within it the *potential* measurement outcomes in the counterpart context. In this way, every potential measurement outcome in the composite system is internally related to every logically antecedent actual measurement outcome.

It is crucial, however, to distinguish between 'logical antecedence' and 'temporal antecedence' here, for these are often casually assimilated. Temporal antecedence refers to an asymmetrical *metrical* relation of events—that is, a 'distance' relation of events as *objects* according to the parameter of time (or more accurately, spacetime). Logical antecedence, by contrast, refers to an asymmetrical logical supersession of events which are *themselves* internal relational structures—that is, an internal relation of relations, such as that implied by the notion of conditional probability or more broadly, propositional logic, operative quantum mechanically. For example, in a classical conditional probability P(*B*|*A*), 'the probability of *B* given *A*,' there is no requirement that *A*'s logical conditioning of *B* be understood as a supersession of events *A*-*B* since classical conditional probabilities are purely epistemic; that is, as classically conceived, conditional probabilities presuppose that all observables have precise values at all times, such that any logical dependency of *B* upon *A* is reflective only of one's knowledge of *A* and *B* as already extant facts. The historical significance of the EPR argument and its modern experimental incarnations is the definitive demonstration that this classical conception of conditional probabilities is entirely invalid in quantum mechanics,[44] where measurement must be understood as generative of novel facts (measurement outcome events) and not merely revelatory of already extant facts. Thus in quantum mechanics, logical dependence depicted by a conditional probability like P(*B*|*A*) can only be understood as a Boolean logical conditioning of potential measurement outcomes at *B* by an actual measurement outcome *A*, such that P(*B*|*A*) depicts a supersession of ontological events and their asymmetrical logical relation.

In this way, the traditional conflation of temporal and logical antecedence, and more generally, the causal and logical orders—a conflation that has dominated the philosophy of nature since the Enlightenment—has been definitively invalidated by quantum mechanics. There remain, however, a number of fundamental physical theories that maintain this conflation—most notoriously the general theory of relativity, which represents our current understanding of gravity. The latter's tacit embrace of the conflation of the causal and logical orders, in sharp contrast to quantum mechanics' elementary differentiation of these orders, is the central conceptual root of the notorious incompatibility of these two fundamental theories. But this root has thus far gone largely unrecognized because until now the quantum mechanical distinction between temporal and logical antecedence had not been formalized in a manner practically applicable to the general theory of relativity. The sheaf-theoretic topological formalism of the relational realist philosophical cosmology, however, does precisely this, and its successful application to the reconciliation of quantum theory and the general theory of relativity has already been demonstrated in two recently published papers.[45] This will be discussed further in chapter 11.6.

For the purposes of this introductory discussion, however, the central concept by which the relational realist formalism can be understood within the context EPR is that whereas temporal antecedence in causal relations is formally expressed *metrically and set theoretically*, in the relational realist framework logical antecedence in casual relations is formally expressed *topologically and category theoretically* as an antecedence of internal relations. Further, in the same way that metrical relations presuppose topological relations, asymmetrical temporal antecedence presupposes asymmetrical logical antecedence. For example, metrical comparisons of diverse geometrical forms, such as the area of a circle compared to that of a square, presuppose a topological scheme of congruence relations by which these distinct metrical forms can be meaningfully correlated. To this end, topological relational structures are 'elastic' in a way that is structure-preserving across compatible geometrical forms. Thus a circle and a square as distinct geometrical forms are topologically homeomorphic, each morphable to the other in a manner that preserves the internal structure of each. Analogously, with respect to physical causality, metrical comparisons of diverse spatiotemporal causal contexts—e.g., the spacelike separated measurement contexts of the EPR arrangement—presuppose a logical, topologically formalizable scheme of congruence relations by which these local contexts, with their respective Boolean intra-contextual asymmetrical logical relations (e.g., material implication *within* a given local context), can be meaningfully compared and coherently integrated inter-contextually, across Boolean contexts, via structure-preserving asymmetrical internal relation.[46]

In section 2.4, it was noted that there are three possible modes of relation among local Boolean measurement contexts:

[a] disjoint: $A \cap B = \emptyset$
[b] overlapping: $A \cap B = A \otimes B$
[c] implicative: $A \Rightarrow B$ or $B \Rightarrow A$

Mode [a] depicts mutually external relations among local measurement contexts whose Boolean intra-contextual structures are *not* compatible for integration inter-contextually (e.g., no correlation of material implication in Boolean context *A* with material implication in context *B*, or vice versa). As disjoint contexts, in other words, there is no overlap of the Boolean subalgebras representative of each local context *A* and *B*. With respect to EPR, this is illustrated by the case where Detector *A* measures $a_x\uparrow$ or $\downarrow$, such that at Detector *B*, $P(b_y\downarrow) = 0.5$ and $P(b_y\uparrow) = 0.5$. As a purely random conjunction, there is nothing logically inducible or deducible globally—i.e., inter-contextually across *A* and *B*—from locally contextualized measurements at *A* and *B* individually.

Mode [c], implicative relation, was previously introduced in sections 2.4 and 2.5. With respect to the present discussion, logical antecedence is manifest via this internal relational mode as nonlocal, inter-contextual logical entailment. Thus, if Particle *A* is measured first, then either:

$b_x\uparrow \Rightarrow a_x\downarrow$ or $b_x\downarrow \Rightarrow a_x\uparrow$ with probability = 1 for contexts A_x and B_x

or

$b_y\uparrow \Rightarrow a_y\downarrow$ or $b_y\downarrow \Rightarrow a_y\uparrow$ with probability = 1 for contexts A_y and B_y

Where this quantum internal relatedness takes the form of logical entailment, in other words, the global (i.e., relations across local measurement contexts) can be *deduced* from the local.

Mode [b], by contrast, depicts those cases where quantum internal relatedness remains at the level of the tensor product—i.e., where relations among potential outcome states remain coarse grained and thus veiled within the overlapping equivalence classes of local Boolean subalgebras. By this mode of internal relation, the global can only be *induced* from the local.[47]

It is this mode in particular that makes EPR experiments even more fascinating: When Detector *A* is set to either context A_x or A_y, and Detector *B* is set to an axis that is *neither* B_x nor B_y, but rather one rotated to some intermediate angle θ, the experiment demonstrates the following: The relations between Context *A* and Context *B* are *neither* implicative (e.g., as they are in the case of A_x and B_x), *nor* disjoint (e.g., as they are in the case of A_x and B_y). The contexts instead exhibit the second mode of relation, [b], given above: They remain overlapping, such that the tensor product relationship yields neither strictly random conjunction as exhibited when the contexts are disjoint, nor conjunction by logical entailment as exhibited when the contexts are implicative, but rather something in between—a global state subsuming both *A* and *B* that is definable via internal relations between *A* and *B*, but only inductively.

In summary, it is a central claim of the relational realist interpretation of quantum mechanics that the calculable and empirically verifiable nonlocal revision of relations among potential outcome states is not properly understood as a mechanism of efficient causality—e.g., where a measurement outcome at Detector *A* is understood to 'causally affect' Particle *B nonlocally* via some presumed, but as yet undiscovered, superluminal, physical-dynamical mechanism.[48] Rather, our interpretation simply presupposes, as a categorical feature of logical causality, that once a measurement outcome occurs at Detector *A* (an outcome that is locally Boolean-contextualized by Detector *A*), the integration of potential measurement outcomes for Particle *B* (locally Boolean-contextualized by Detector *B*), as well as the unique actualization of one of these potential outcomes, will be logically internally related to that outcome at Detector *A*.

This presupposition of internal relatedness among quantum measurement outcomes would seem to be crucial for any ontological interpretation of quantum mechanics, given the clear experimental disconfirmation of the classical ontological notion that measurement is merely revelatory of already extant well-defined values for all observables, and the associated notion that logical relations are purely epistemic. Again, quantum mechanics evinces that measurement outcomes are not merely revelatory of extant facts, but generative of novel facts. Thus, the coherent interpretation of quantum mechanics necessarily presupposes (either implicitly or explicitly) an internal relational structure by which novel facts constitutive of a measured system's outcome state can be logically integrated with those constitutive of its initial state, and by which the diverse local measurement contexts associated with these states can be logically integrated globally.

This presupposed internal relational structure is thus a *mereotopological* structure—i.e., one wherein part-whole and local-global relations are fundamentally topological rather than metrical, such that locally contextualized facts, as quantum 'units of relation' themselves, can be logically related globally in a *structure-preserving* way. In the relational realist interpretation, this presupposition is rendered explicit, and thus requires a proper mathematical formalism by which this mereotopological internal relational structure can be practically bridged to the conventional, metrical formalism; for until now, it is only by this latter formalism that the extensive features of physical relations have been depicted in modern science.

2.9 Topological Relations vs. Metrical Relations: Quantum Mechanics and Spacetime

That local measurement outcomes are logically relatable globally is not only a presupposition of the relational realist ontology; it is a presupposition of the scientific method itself. Thus a careful philosophical exploration of the mathematical form of these relations via a coherent speculative ontological framework could prove extremely important to the solution of fundamental conceptual problems in modern physics. The phenomenon of quantum nonlocality, introduced in the previous section, is one example, as is the general incompatibility of quantum theory and relativity theory. In both cases, by representing local quantum measurement contexts as Boolean subalgebras, such that locally contextualized measurement outcomes can be internally related globally, one discovers a crucial point—one central to the relational realist program: These internal relations are not fundamentally metrical and continuous, pertaining to *elementary objects*, as conventionally understood (e.g., requiring a point-set structure, where points are related by continuous spatiotemporal intervals); they are, rather, fundamentally topological in that they pertain to *elementary units of relation*—i.e., elementary relations that are fundamentally discrete, not continuous.

As will be demonstrated throughout the remainder of this volume, the concept of a topological structure of internal relations is ideal for mathematically formalizing quantum mechanical, nonlocal logical relations such as those depicted by EPR-type experiments, and those which lie at the heart of the incompatibility of quantum theory and the general theory of relativity. But just as important, the concept of a topological structure of internal relations is equally well-suited to the task of formalizing coherent and intuitive *interpretations* of these fundamental physical theories.

With respect to the issue of mathematical formalism, as mentioned before, the work of my colleague Elias Zafiris has generated significant advances in this exploration, using Grothendieck topology to construct a category-sheaf theoretic interpretation of quantum mechanics.[49] Ontologically, this implies that the scientific evaluation of fundamental physical relations does not, as is conventionally accepted and prescribed by general relativity, require the presupposition of a metrical continuum as a fundamental background or foundation for the description of physical processes. Quantum mechanics, we believe, exemplifies the idea that the fundamental background by which physical relations are evaluated is topological, not metrical. This ought not be taken to imply, however, that metrical relations are fundamentally irrelevant to physics, or that they are to be understood as ill-defined shadows of relations whose actual essence is sheerly topological. Metrical relations, rather, are simply higher-order types of relation that presuppose fundamentally topological relations.

(Analogously, the relation of classical and quantum mechanics is properly understood in this same way, such that the latter is presupposed by the former.)

Since the mereological structure of the topology proposed in the relational realist philosophy accommodates the concept of internal relations, it is more accurately described as *mereo*topological—a topology of whole-to-part, global-to-local—whereby internal relations among quantum actualizations (e.g., measurement outcomes) are both [1] structure-preserving in terms of their asymmetrical logical order; and [2] generative of novel measurement outcomes, in that quantum mechanical relations are synthetically predicative and not merely analytically revelatory of pre-existing facts. Outcome states in quantum measurement, as emphasized earlier, are always properly understood as *generated consequent* of relations among system-detector-environment, and therefore ontologically significant; they cannot, per classical convention, be accurately characterized as merely *revealed subsequent* to these relations and therefore merely epistemically significant.

With respect to its applicability to the solution of problems like the incompatibility of quantum mechanics and general relativity, the relational realist approach begins with the assertion that continuous metrical relationships presuppose discrete topological relationships. Whereas metrical relations derive from set-theoretic points on a continuum as their fundamental elements, topological relations derive from discrete regions as their fundamental elements. Further, the metrical concept of 'point' presupposes the topological concept of 'region,' or more broadly, 'relation,' in that one can define a topological region / relation without reference to 'point';[50] but one cannot define point in abstraction from relation—e.g., relation to other points, or relation to some presupposed coordinate domain definable via boundary conditions.

The relational realist conception of an elementary unit of relation as fundamental actual constituent of nature should not, however, be interpreted as a sheer reduction or assimilation of 'object relatum' to relation, as has been prescribed by advocates of so-called 'radical ontic structural realism.'[51] Though the lure of this kind of conceptual reduction is well precedented in the history of philosophy, as discussed at length in chapter 1, equally precedented is the reduction to incoherence that invariably ensues when the concepts to be assimilated are mutually implicative, like relations and relata (and likewise their respective counterpart pairs, potentiality and actuality, continuous and quantum, subjective and objective.) The ability to coherently and consistently define relata *as* relations—i.e., in the sense of 'relations of relations'—via a formalism such as category theory does not, as is sometimes claimed,[52] obviate the presupposition of 'object relata' inherent in the concept of relation—or, more ambitiously, obviate the notion of 'objects' in fundamental physical theories; it merely objectifies relations so that they can function *as* object relata. Indeed, when objectified relations are not properly recognized as object relata, one is left with the incoherent notion of 'relations without relata'—again, an incoherence that is easily avoided by recognizing that relations and relata are mutually implicative concepts.

Since quantum relations, as we have seen, formally entail both objective *and* subjective (i.e., contextual) features, the notion of an elementary 'unit of relation' capable of serving as an object relatum in quantum mechanics would therefore require a dipolar character capable of accommodating both of these features. To that end, relational realism as a speculative ontological program begins with the Whiteheadian categorical principle that fundamentally, all elementary object relata are quantum as 'objective' data—the individual brute facts of the world—*and* quantum as unique, 'subjective' (i.e., contextualized) integrations of internal relations with *other* object data. (This is the 'subject-superject' character of Whitehead's elementary 'actual occasions' as dipolar units of relation, which will be discussed further in chapter 4.) In this way, the classical notion of 'continuous relations among objects' defined metrically and set theoretically is more fundamentally definable topologically and category theoretically as 'discrete relations among relations.'

As applied to the interpretation of quantum mechanics discussed thus far, in the same way that a fact cannot be meaningfully defined without reference to or relation with other facts, a quantum measurement *outcome* (a 'quantum fact') cannot be defined apart from its relation to the quantum facts constitutive of an *initial*, locally contextualized global state. This is the initial system-detector-environment state—i.e., the universe itself relative to that particular, locally contextualized measurement outcome. But likewise, ontologically, a quantum measurement outcome is generated *by* these relations as a synthetic predication of the initial system-detector-environment state, such that with each quantum measurement outcome, the universe has been objectively augmented by a novel predicative fact. (This is discussed further in chapter 3.)

With respect to the relativistic spatiotemporal extensive aspects of these quantum mechanical relations, rather than beginning with a metrical continuum, like the background of spacetime in relativity theory, and attempting to quantize it to accommodate quantum theory, the relational realist approach instead begins with a topological space as the fundamental formal background. It then defines continuous metrical relations (e.g., spatiotemporally extensive relations) as higher-order *potential* relations of initial and final *discrete*, *actual* quantum states—such that continuous relations among potentia, in other words, presuppose discrete topological relations among actualities.

As a mathematical analogy, consider the relationship between $\mathbb{R}$ (the set of real numbers—a continuum and thus uncountable), $\mathbb{Q}$ (the set of rational numbers—discrete and thus countable), and $\mathbb{Z}$ (the set of integers—also discrete and countable). The convention by which we intuitively relate these sets is embodied by the very first representation of mathematical relations typically presented to children in their early education: the real number line. Thus, we intuitively take the 'foundational' set to be the continuum $\mathbb{R}$ and then define $\mathbb{Q}$ as a discretized subset of $\mathbb{R}$ (i.e., we quantize the continuum $\mathbb{R}$) and likewise $\mathbb{Z}$ as a subset of $\mathbb{Q}$. In this way, the conceptual structure of one's mathematical intuition corresponds comfortably to that of one's classical physical intuition, by which discrete objects are understood as existing and moving within a fundamental spatiotemporal continuum, as both of these intuitions are developed together throughout the course of one's education. The result is a conventional understanding of the world as a set of fundamentally material objects whose qualifications are fundamentally continuous; that is, the existence, motion, and every other qualitative feature of these objects is ultimately expressible as a metrical quantity within the spacetime continuum, wherein the latter is simply a formal extension of the real number line.

The problem, of course, is that quantum mechanics has long since invalidated this intuition, recasting it as a practically useful but entirely superficial and fundamentally incorrect understanding of the physical world. Fortunately, however, the revised understanding given in quantum mechanics is equally intuitive; for the inversion of the classical priority of 'fundamental continuum' over 'derivative quantum' is likewise reflected in a mathematical inversion that is just as intuitive as the original priority of continuum over quantum, embodied by the real line: We begin instead with $\mathbb{Z}$, the integers, as the 'fundamental background' of discrete units and derive $\mathbb{Q}$, the rational numbers, as a set of *relations* (in this case, ratios) among these discrete units of $\mathbb{Z}$, and likewise derive $\mathbb{R}$, the real numbers, as a set of relations among the units of $\mathbb{Q}$,[53]—that is, relations of relations.

The relational realist, category-sheaf theoretic interpretation of quantum mechanics, in its depiction of the quantum and continuous features of nature, is precisely reflective of this mathematical analogy. To the extent that the relational realist philosophical cosmology provides a coherent and empirically adequate ontological interpretation of quantum mechanics that is compatible with relativity theory,[54] the exploration of the interpretation's broader and deeper philosophical implications will, if the scheme is successful, likewise find its relevance to still broader and deeper domains of experience. In this regard, the underlying conceptual correlations of the relational realist scheme and the philosophical cosmology of Whitehead will prove to be crucial.

2.10 Substance in Quantum Mechanics: Relational Realism and the Philosophy of Alfred North Whitehead

Any assertion as to the ontological significance of quantum mechanics—especially the relationship between a quantum measurement outcome and 'reality' (observable or otherwise)—presupposes a commitment to the general Aristotelian concept of substance. Substance in this regard ought not be conflated with the concept of physical matter, despite the material aspect of Aristotelian hylomorphism; for in the latter, matter is properly understood not as 'physical substance,' but more broadly as a generic medium by which potential forms of definiteness can exist—a 'bearer of potentia.' One could similarly characterize the pure state in quantum theory as a substantive bearer of potential relations / potential forms of definiteness, one of which becomes actualized as a measurement outcome; and likewise, the mixed state in quantum mechanics, mediating potentiality and actuality, can be thought of as a bearer of *probable* relations / probable forms of definiteness. In the ontological interpretation of quantum mechanics, then, potential and probable forms of definiteness are substantively 'real' in the Aristotelian sense, with some particular form becoming 'actual' as a unique measurement outcome. In this way, both potentiality and actuality are properly understood as mutually implicative features of reality—of 'quantum substance'—analogous to matter and form in Aristotelian substance.

Fundamentally, substance in this Aristotelian sense is "the essence of a thing . . . what it is said to be *in respect of itself*" (*Metaphysics* Z.4. 1029b14) such that the thing defined is further reducible neither physically nor conceptually. The dipolar 'actual occasion' of Alfred North Whitehead, introduced briefly in chapter 1.4, is perhaps the most revolutionary modern rehabilitation of the Aristotelian notion of substance for two reasons: First, the physical and conceptual features of the quantum actual occasion are mutually implicative poles of a single substance rather than mutually exclusive essences of separate substances (e.g., Cartesian *res extensa* and *res cogitans*); and second, an actual occasion 'in respect of itself' can never be abstracted from its internal relations with other actual occasions.

With regard to the ontological interpretation of quantum mechanics, many of the more infamous conceptual difficulties discussed thus far—the measurement problem, the problem of state reduction, etc.—can be traced to an uncritical acceptance of the classical mechanistic-materialistic conception of substance, further speciated via the dualism of Descartes, into mutually exclusive physical and conceptual realities. Likewise, it can be argued that the Whiteheadian rehabilitation of the Aristotelian concept of substance, as a repudiation of Cartesian dualism, paves a sound pathway around these difficulties—a pathway that is, by Whitehead's own desiderata for speculative philosophy, both coherent and empirically adequate.

As I have previously argued,[55] the systematic ontology Whitehead presented in *Process and Reality* is well exemplified in quantum mechanics, and the conceptual correlations are easily mappable. Subsequent refinements of this correlation, constitutive of the relational realist interpretation of quantum mechanics introduced in this volume, are all entirely compatible with the Whiteheadian philosophical cosmology such that familiarity with the latter will be heuristically useful to the understanding of the quantum mechanical concepts presented herein. The central conceptual correlation upon which the broader framework of relational realism is grounded is that by which substance is properly understood: Both Whiteheadian actual occasions and quantum mechanical measurement outcomes are fundamentally discrete, dipolar units of contextualized internal relation; they are quanta of substance.

A 'becoming' actual occasion / quantum measurement-in-process begins as a matrix of potential internal relations with (Whiteheadian 'prehensions' of) a dative world of antecedently actualized occasions / measurement outcomes. The becoming occasion / quantum measurement-in-process is logically internally related to these data, and the potential relations / prehensions are initially uncoordinated (e.g., the pure state in quantum mechanics, or equivalently, the Whiteheadian 'conceptual phase' of actualization), such that the vast majority of conjunctions of prehensions / potential relations violate PNC. But in the Whiteheadian 'comparative phase' of concrescence (cf., the mixed state in quantum mechanics), the local Boolean context of the becoming actual occasion / quantum measurement-in-process (its 'subjective standpoint,' in Whiteheadian phraseology) is logically integrated with the Boolean contexts / standpoints of *all* prehended data constitutive of the actual world (i.e., system-detector-environment) to which the becoming actual occasion / quantum measurement-in-process is internally related. Thus, writes, Whitehead,

> to 'function' means to contribute determination to the actual entities in the nexus of some actual world. Thus the determinateness and self-identity of one entity cannot be abstracted from the community of diverse functionings of all entities[56] . . . *The real potentialities relative to all standpoints are coordinated as diverse determinations of one extensive continuum.* This extensive continuum is one relational complex in which all potential objectifications find their niche . . . All actual entities are related according to the determinations of this continuum; and *all possible actual entities in the future must exemplify these determinations in their relations within an already actual world.*[57]

The result of this logically conditioned integration of relations is a negative selection process—a distillation of the quantum mechanical pure state / Whiteheadian 'conceptual phase' of concrescence into the mixed state / 'comparative phase'—whereby prehensions / potential relations that do not satisfy PNC are eliminated from the integration. In the comparative phase / mixed state, the remaining contextualized *potential* measurement outcomes (Whiteheadian 'subjective forms') now form a reduced matrix of *probable* measurement outcomes—i.e., potential outcomes / subjective forms that are mutually exclusive and exhaustive, and valuated such that they sum to unity. That is, *at most and at least one* of these subjective forms / probable outcome states will be satisfied / actualized as a unique outcome, in satisfaction of PNC and PEM.

This brief summary glimpses a number of conceptual refinements of the Whiteheadian process of concrescence, which will be explored in greater detail throughout the remainder of part I. For those already familiar with Whiteheadian philosophy, it should be noted that none of these refinements contradicts his original ideas; they merely give greater specificity to the definition of certain concepts Whitehead had left more general. For example, the valuation of subjective forms in the comparative phase is, for Whitehead, merely 'valuation up' or 'valuation down';[58] whereas in quantum mechanics, these are refined as probability valuations, which not only captures the notion of 'valuation up or valuation down' but also the satisfaction of PEM, in that the probabilities sum to unity.

Similarly, for reasons to be explored presently, Whitehead never specifies the logical criteria for the negative selection process by which prehensions 'incompatible for integration'[59] are eliminated from concrescence as 'negative prehensions.' He does not specify the satisfaction of PNC as a logical condition for integration; nor does he specify that the subjective standpoint of a concrescing occasion must be Boolean, such that PNC can be used as such a condition—i.e., both for the integration of prehensions and the valuation of their subjective forms.

Finally, Whitehead had great difficulty in relating his Theory of Prehension to his Theory of Extension, a difficulty closely related to the problematic integration of quantum theory and relativity theory discussed previously. Whitehead's attempt to relate prehension and extension in part IV of *Process and Reality* is unsuccessful primarily because his conception of 'extensive connection,' when formally expressed, was somewhat handicapped by the nature of the mathematical tools of his time, which were heavily grounded in set theoretic, metrical depictions of relations within a continuum. This became especially problematic when he attempted to formalize a theory of extensive connection that was fundamentally discrete and mereological, not metrical.

> Considered in its full generality, apart from the additional [relativistic] conditions proper only to the cosmic epoch of electrons, protons, mole-

> cules, and star-systems, the properties of this [extensive] continuum are very few and *do not include the relationships of metrical geometry*. An extensive continuum is a complex of entities united by the various allied relationships of whole to part, and of overlapping so as to possess common parts, and of contact, and of other relationships derived from these primary relationships. This extensive continuum expresses the solidarity of all possible standpoints throughout the whole process of the world. It is not a fact prior to the world; it is the first determination of order—that is, of real potentiality arising out of the general character of the world . . .[60]

In precisely the same way, and via the same concepts given in the above quote, the algebraic formalism of Grothendieck topology and category-sheaf theory provides the basis for coherently relating the ontological interpretation of quantum mechanics with the metrical-extensive features of relativistic physics,[61] and likewise, provides the basis for relating Whitehead's Theory of Prehension with his Theory of Extension. Had these mathematical tools been available in Whitehead's time, it is entirely likely that the category-sheaf theoretic formalizations of internal relation would have formed the basis of his work in part IV of *Process and Reality*.[62]

With respect to PNC, PEM, and Boolean logic, the fact that Whitehead says little of logic in *Process and Reality* is not surprising, given that he was wary of the increasingly popular positivist trend toward the dogmatic reduction of all physical experience to logical and mathematical formalism. Indeed, even in philosophy today it is frequently the case that what begins as a clear philosophical argument expressible *via* a representative mathematical formalism quickly devolves into an arcane argument exclusively *from* formalism. Whatever intuitive features the argument may have initially entailed quickly become lost amid ponderously technical (and indeed, often practically hermetic) concepts accessible only to an increasingly small community of specialists. One will recall this very concern expressed by quantum theorist Roland Omnès in the quotation from section 2.1—a concern Whitehead had similarly forewarned against. Thus, given his previous technical work with Russell culminating in their *Principia Mathematica*, it is understandable that he did not want his philosophical work to be either reduced to his mathematical work, or taken as groundwork for a logical positivist, mathematical structuralist cosmology. This wariness is especially evident in the following letter (only recently discovered) written by Whitehead to one of his doctoral students, Henry Leonard, who had indicated an interest in applying Whitehead's work in a positivist direction:

> The bridge which you construct between my own expositions of methodology—e.g., in *Process and Reality*—and Logical Positivism is a topic rarely distant from my thought. Every mathematician and symbolic logician is, in his habit of thought, a logical positivist. Yet to some of the expositions I find myself in violent opposition—especially to the very habit of dismissing questions as unmeaning, i.e., unable to be expressed in existing symbolism. Wittgenstein annoys me intensely. He is a complete example of the saying, "I am Master of this College / What I know not, is not knowledge."
>
> Logical Positivism in this mood—its early mood—will produce a timid, shut in, unenterprising state of mind, engaged in the elaboration of details. I always test these general rules by trying to imagine the sterilizing effect of such a state of mind, if prevalent at any time in the last ten thousand years.
>
> The fact is that thought in the previous two centuries has been engaged in disengaging itself from the shackles of dogmatic divinity. Thus it unconsciously seeks new fetters, viz. anything offensive to the Pope of Rome. But I see no reason to believe that the stretch of Bertrand Russell's mind, or of Wittgenstein's mind, or of Carnap's mind, has attained the limits of insight or expression possible in the evolution of intelligent beings. They are bright boys, good representatives of a stage of rationalism, but nothing more.[63]

In light of this admonition, and in full agreement with it, it must be emphasized that despite the unavoidable fact of technical mathematical formalism in quantum mechanics, and despite the underlying logical structure of this formalism, the ontological interpretation of quantum mechanics given in the relational realist philosophical cosmology is in no way a logico-reductive, 'mathematical structuralist' or 'structural realist' scheme. It does not employ mathematics to assimilate all relations to those definable by the physical sciences, or to determine physical relations via a 'mathematization' of the world and all of its constituent relations; rather, it simply attempts to identify the logical conditions by which coherent and consistent relations are *possible* in a cosmology that—as evinced by quantum mechanics—depicts predication not as merely revelatory of pre-existing facts, but generative of novel facts.

To elaborate: If [1] every novel predicative fact (viz., quantum measurement outcome) is both *contextualized* via internal relations with already extant facts, and potentially *contextualizing of* logically subsequent facts (via subsequent measurements), as posited in both Whiteheadian and relational realist philosophy, and [2] every fact is *generated* via internal relation with all other facts and *their* associated contextualizations, and [3] one presupposes that the asymmetrical order of causal relation and the asymmetrical order of logical implication are always aligned, then it is clear that any ontological interpretation of quantum mechanics necessarily presupposes a framework of logical causality by which facts can be coherently and consistently related, both locally within and globally among, multiple diverse measurement contexts.

The relational realist interpretation, as a speculative philosophical program, attempts to schematize and elucidate precisely such a framework. Beginning with quantum mechanics as an exemplification of each of the three principles above, we posit a formalized mereotopological structure by which Boolean-contextualized predicative facts are internally related, and by which these internal relations are generative of novel predicative facts. The categorical logical principles presupposed by the relational realist speculative philosophy and its proposed mereotopological structure are the same as those necessarily presupposed by the modern scientific method: the Principle of Identity (PI), the Principle of Non-Contradiction (PNC), and the Principle of the Excluded Middle (PEM). Likewise, both presuppose the Boolean relational forms of conjunction, disjunction, implication, complement, and exclusive disjunction by which PI, PNC, and PEM are defined.

As will be explored in chapters 3 and 5, and more technically in part II, relational realism's empirical applicability to quantum mechanics is well demonstrated in that the former's mereotopological relational structure can be precisely applied formally to the latter via the mathematics of category-sheaf theory and Grothendieck topology. Indeed, the successful and practical application of this formalism to quantum mechanics in particular is a key feature of the relational realist philosophical cosmology, given that quantum mechanical relations uniquely constitute a fundamental exemplification of the ontological relational structure we propose.

As a Whiteheadian speculative philosophical enterprise, however, the true test of the scheme's utility will be its meaningful applicability beyond the restricted locus of quantum mechanics. As mentioned earlier, the adequacy of the relational realist ontology to the task of reconciling quantum mechanics and general relativity is currently being explored,[64] with strong indications that the mathematical formalism of relational realism does, in fact, yield a promising means to this important goal. Likewise, the application of the relational realist ontology to the study of complex adaptive systems is also proving to be a fruitful avenue of inquiry and is currently being explored with our colleague Stuart Kauffman.

As for the applicability of the relational realist philosophical cosmology beyond the domain of scientific inquiry, again, its firm grounding in the philosophy of Whitehead will prove to be most beneficial. Among the many deep inheritances from Whitehead, his notion of and commitment to speculative philosophy, especially palpable in his letter to Henry Leonard, is perhaps most important of all. Not only did it open the door to the further development and refinement of his own speculative scheme; it continues to demand that this development and refinement always exhibit an intrinsic resistance to either misconstrual as a dogmatic, mathematical-structural ontology, or to adulteration into such an ontology. Thus although the relational realist philosophy includes both an explicit reference to its presupposed logical first principles, and a rigorous mathematical formalism by which it can be applied to the interpretation of scientific theories, the relations described by the philosophy are neither reducible to its formalism, nor to its categorical principles. Fact is not reducible to form. Explanation is not reducible to description.

The cardinal concept of relational realism is that fact and form are mutually implicative in every quantum of substance—in every actual occasion. Indeed, whatever ones preferred ontology, the mutual implication of fact and form is easily stipulated at least as an epistemic certainty, since it is impossible to give formal definition to a fact without reference to (relation with) other facts. In the relational realist ontology, however, an asymmetry of relation is presupposed, such that a novel potential fact is always internally related to a world of already actualized, dative facts. Again, this internal relation takes the form of a logically conditioned integration of potential relations between [1] the novel contextualized fact-in-process and [2] the dative, antecedently actualized facts of the world, including the contexts by which these data were themselves relationally generated.

This logically conditioned integration of potential internal relations, exemplified in a quantum measurement interaction as the reduction of the pure to the mixed state, presupposes the first principles of Boolean logical relation, and likewise presupposes that all individual contexts are locally Boolean, such that not only the facts themselves but also their contextualizations can be related coherently and consistently globally. The result of this process of integration of potential relations is an actualized predicative fact, coherently formalized via its definite internal relations with the world. Thus in the relational realist ontology, formal relations presuppose the facts related; but at the same time, fact also presupposes form, since facts are generated *by* these formal internal relations—i.e., formal integrations of potential measurement outcomes in quantum mechanics *always yield a unique actual measurement outcome*—a novel fact.

This mutually implicative relationship of fact and form is well instantiated in quantum mechanics in that alternative probable outcome states (alternative probable forms of definiteness) can only be defined by double-reference to [a] a Boolean-contextualized, *antecedently actualized initial state* of the measured system (i.e., an initial fact determined by a preparation measurement), and [b] a Boolean-contextualized *anticipated actual outcome state* whose actualization is presupposed by the mechanics, in satisfaction of PNC and PEM.[65]

Thus form presupposes fact in quantum mechanics; but at the same time fact presupposes form, in that the actualized outcome state is only definable in quantum mechanics as the output of this formal relational process. It is a process wherein potential measurement outcomes (potential forms of definiteness) in the pure quantum state are logically integrated into probable measurement outcomes (probable forms of definiteness) in the mixed quantum state, and finally into a unique actual outcome state (an actualized form of definiteness—i.e., a quantum 'fact'). Thus in quantum mechanics, novel facts are not merely analytical predications revealed subsequent to measurement per the classical ontological worldview; they are synthetic predications, generated consequent of measurement—that is, consequent of the ontological relational process exemplified by quantum measurement.

Notes

1. A. D. O'Connell et al., "Quantum ground state and single-phonon control of a mechanical resonator," *Nature* 464, no. 7289 (2010): 697-703. (See also its companion news article referenced in note 3 below.)

2. Geoff Brumfiel, "Scientists Supersize Quantum Mechanics," *Nature* (2010) 17 March 2010, doi:10.1038/news.2010.130.

3. Erwin Schrödinger, "The Present Situation in Quantum Mechanics" (translated by J. Trimmer), in *Proceedings of the American Philosophical Society* 124 (1980): 323-38.

4. This experiment on the wave-particle nature of light originated with Thomas Young in the early nineteenth century, and variants of it are still undertaken today to explore quantum superposition and entanglement. The basic experiment entails photons propagating through either of two parallel slits prior to impinging upon a detector screen. When either slit is closed, the photons are

well localized at the detector, thus exhibiting their 'classical' particulate character; and when both slits are open, the photons form an interference fringe at the detector, thus exhibiting their 'quantum superpositional' wave-like nature.

5. See, for example: S. Walborn et al., "Double-slit Quantum Eraser," *Physical Review A* 65, no. 3 (2002): 033818.

6. J. J. Halliwell and E. Zafiris, "Decoherent Histories Approach to the Arrival Time Problem," *Physical Review D* 57 (1998): 3351.

7. J. Von Neumann, *Mathematical Foundations of Quantum Mechanics* (Princeton: Princeton University Press, 1955), 352.

8. Brumfiel, "Scientists Supersize Quantum Mechanics."

9. Roland Omnès, *Philosophie de la Science Contemporaine* (Paris: Editions Gallimard, 1994). English translation published as *Quantum Philosophy: Understanding and Interpreting Contemporary Science* (Princeton: Princeton University Press, 2002), xi.

10. The local measurement context can likewise be represented by a Boolean subalgebra. The advantages of this representation will be discussed later in this volume.

11. See, for example: Henry P. Stapp, *Mindful Universe: Quantum Mechanics and the Participating Observer* (Berlin: Springer, 2007).

12. David Lewis, *On the Plurality of Worlds* (Oxford: Basil Blackwell, 1986), 2-3.

13. See, for example: David Lewis, "How Many Lives Has Schrödinger's Cat?" in *Lewisian Themes: The Philosophy of David K. Lewis*, ed. Frank Jackson and Graham Priest (Oxford: Oxford University Press, 2004), 4-23.

14. See, for example: R. T. Cox, "Probability, Frequency and Reasonable Expectation," *American Journal of Physics* 14 (1946): 1-13.

15. In Whitehead's *Treatise on Universal Algebra*, he refers to the inherent 'truism' and 'paradox' of such non-commutativity (discussed further in chapter 3.4).

16. As evidenced by the experimental disconfirmation of local hidden variables theories. This will be discussed further in chapter 2.8 and chapter 3. See also: J. S. Bell, "On the Einstein Podolsky Rosen Paradox," *Physics* 1, no. 3 (1964): 195-200. See also, the Kochen-Specker Theorem: Simon Kochen and Ernst Specker, "The problem of hidden variables in quantum mechanics," *Journal of Mathematics and Mechanics* 17 (1967): 59-87. See also, the famous experiments of Aspect et al.: A. Aspect et al., "Experimental Tests of Realistic

Local Theories via Bell's Theorem," *Physical Review Letters* 47, no. 460 (1981).

17. Jeffrey Bub, "Quantum Logic, Conditional Probability, and Interference," *Philosophy of Science* 49 (1982): 402-21. See also: Jeffrey Bub, "The problem of properties in quantum mechanics," *Topoi* 10, no. 1 (1991): 27-34.

18. This is a conditioning process analogous to von Neumann's 'Process 1.' See, for example: J. Von Neumann, *Mathematical Foundations of Quantum Mechanics* (Princeton: Princeton University Press, 1955), 352.

19. The italicized Greek characters representing these coefficients should not be confused with the non-italicized Greek characters representing actualized outcome states (i.e., where a potential outcome state a_n is actualized as α_n upon measurement.)

20. See, for example: G. E. Moore, "External and Internal Relations," *Proceedings of the Aristotelian Society* 20 (1919-1920): 40-62.

21. See, for example: Roland Omnès, *The Interpretation of Quantum Mechanics* (Princeton: Princeton University Press, 1994), 122-42. See also: Robert Griffiths, "Consistent Histories and the Interpretation of Quantum Mechanics," *Journal of Statistical Physics* 36, no. 1 (1984): 219-72, and J. J. Halliwell, J. Pérez-Mercader, and W. H. Zurek, *Physical Origins of Time Asymmetry* (Cambridge: Cambridge University Press, 1996), 210.

22. Moore, "External and Internal Relations."

23. Moore, "External and Internal Relations."

24. Simon Kochen and Ernst Specker, "The Problem of Hidden Variables in Quantum Mechanics," *Journal of Mathematics and Mechanics* 17 (1967) 59-87.

25. Introduced in section 2.1.

26. A. Einstein, B. Podolsky, and N. Rosen, "Can Quantum-Mechanical Description of Physical Reality Be Considered Complete?" *Physical Review* 47, no. 10 (1935): 777-80.

27. Jeffrey Bub, "Quantum Logic, Conditional Probability, and Interference," *Philosophy of Science* 49 (1982): 402-21. See also: Jeffrey Bub, "The problem of properties in quantum mechanics," *Topoi* 10, no. 1 (1991): 27-34.

28. Cf., for example: Roland Omnès, *The Interpretation of Quantum Mechanics* (Princeton: Princeton University Press, 1994), 163: "Rule 4: Any description of the properties of an isolated physical system must consist of propositions belonging together to a common consistent logic. Any reasoning to be

drawn from the consideration of these properties should be the result of a valid implication or of a chain of implications in this common logic" and his No-Contradiction Theorem (162). Cf., also, in this same volume, chapter 5: "The Logical Framework of Quantum Mechanics," (144-200).

29. See, for example: Elias Zafiris, "Quantum Event Structures from the Perspective of Grothendieck Topoi," *Foundations of Physics* 34, no. 7 (2004). See also: Elias Zafiris, "Generalized Topological Covering Systems on Quantum Events' Structures," *Journal of Physics A: Mathematical and General* 39 (2006).

30. Kochen and Specker, "The Problem of Hidden Variables in Quantum Mechanics," 59-87.

31. This is discussed further in chapter 3.

32. See note 29.

33. Decoherence, introduced briefly here, will be discussed in detail in chapters 3.1 and 4.2.

34. This is given via the mathematical trace function, where the off-diagonal terms of the density matrix (representing relations that violate PNC) are eliminated. What remains in the reduced state density matrix is a set of mutually exclusive and exhaustive probable outcome states.

35. See, for example: Wojciech Zurek, "Decoherence and the Transition from the Quantum to the Classical," *Physics Today* 44, no. 10 (1991): 36-44. See also: Wojciech Zurek, "Decoherence, Einselection, and the Quantum Origins of the Classical," *Reviews of Modern Physics* 75, no. 3 (2003): 715-75.

36. See, for example: A. Caldeira and A. Leggett, "Quantum Tunneling in a Dissipative System," *Annals of Physics* 149 (1983): 374-456.

37. S. Bertaina et al., "Quantum oscillations in a molecular magnet," *Nature* 453 (May 2008): 203-6.

38. That is, reduced entirely to the dynamics of an effective Hamiltonian.

39. A. Einstein, B. Podolsky, and N. Rosen, "Can Quantum-Mechanical Description of Physical Reality Be Considered Complete?" *Physical Review* 47, no. 10 (1935): 777-80.

40. While these experiments differ with respect to the particular observables measured (photon polarization, particle spin, etc.) they all similarly investigate the thesis of nonlocal 'EPR-type' quantum correlations among measurement events in a composite quantum system. The experimental model explored here is one based on: D. Bohm and Y. Aharonov, "Discussion of Experimental

Proof for the Paradox of Einstein, Rosen and Podolski," *Physical Review* 108 (1957): 1070-76.

41. Electrons, as well as all other elementary fermions, have a spin quantum number of 1/2, and are thus referred to as 'spin ½' systems. Along any given axis perpendicular to the direction of propagation (e.g., along the x or y axis if the direction of propagation is either $+z$ or $-z$), there are only two possible spin directions relative to each axis. These spin directions are represented by the spin projection quantum numbers +1/2 and −1/2. One conventionally takes the signs + and − to indicate direction; thus +1/2 and −1/2 are often read as 'spin up' and 'spin down,' respectively. Spin ½ systems will be referred to extensively throughout the remainder of part I.

42. See note 16.

43. Jeffrey Bub, "Quantum Logic, Conditional Probability, and Interference," *Philosophy of Science* 49 (1982): 402-21. See also: Jeffrey Bub, "The problem of properties in quantum mechanics," *Topoi* 10, no. 1 (1991): 27-34.

44. Indeed, the EPR experimental arrangement was initially conceived by Einstein, Podolsky, and Rosen in order to demonstrate the applicability of this classical conception to quantum measurement. It was only in later decades, when technology became sufficient to actually execute EPR-type experiments that their argument was definitively invalidated (see note 16).

45. A. Mallios and E. Zafiris, "The Homological Kähler-De Rham Differential Mechanism—Part I: Application in the General Theory of Relativity," *Advances in Mathematical Physics* 2011 (2011), doi:10.1155/2011/191083, and A. Mallios and E. Zafiris, "The Homological Kähler-de Rham Differential Mechanism—Part II: Sheaf-Theoretic Localization of Quantum Dynamics," *Advances in Mathematical Physics* 2011 (2011), doi:10.1155/2011/189801.

46. A more technical depiction of the EPR experiment using the relational realist formalism of category-sheaf theory is given in chapter 9.4, for which the present discussion is intended to serve as an introduction.

47. In part II, this concept is further developed as the 'inductive limit' in the category-sheaf theoretic interpretation of quantum mechanics.

48. As mentioned earlier, among other problems, this would require a physical, dynamical mechanism of superluminal energy transfer from A to B. While there do exist interpretations of quantum mechanics that propose such mechanisms, there have as yet been no experimental observations or detections of superluminal energy by which the nonlocal *logical* correlations operative in

EPR-type experiments could be exploited as *efficient-causal* correlations. The latter, for example, would allow for faster-than-light communication—a theoretical phenomenon that has been much studied, but never realized experimentally.

49. See note 29.

50. This is a foundational principle of point-free topology, mereotopology, and category-theoretic Grothendieck topology, which will be explored in detail in part II.

51. See, for example: S. French and J. Ladyman, "Remodeling structural realism: Quantum physics and the metaphysics of structure," *Synthese* 136 (2003): 31-56. See also: S. French, "On the Withering Away of Physical Objects," in *Interpreting Bodies: Classical and Quantum Objects in Modern Physics*, ed. E. Castellani (Princeton: Princeton University Press, 1998), 93-113. See also, Ladyman, J. and Ross, D. *Every Thing Must Go: Metaphysics Naturalized* (Oxford: Oxford University Press, 2009).

52. See, for example: Jonathan Bain, "Category-theoretic Structure and Radical Ontic Structural Realism," *Synthese* 0039-7857 (2011): 1-15.

53. For example, the Cauchy sequence.

54. See note 45 and chapter 11.6.

55. Michael Epperson, *Quantum Mechanics and the Philosophy of Alfred North Whitehead* (New York: Fordham University Press, 2004).

56. Alfred North Whitehead, *Process and Reality: An Essay in Cosmology*, Corrected Edition, ed. D. Griffin and D. Sherburne (New York: Free Press, 1978), 25.

57. Whitehead, *Process and Reality*, 66 (italics added).

58. See, for example, Whitehead, *Process and Reality*, 241.

59. See, for example, Whitehead, *Process and Reality*, 148.

60. Whitehead, *Process and Reality*, 66 (italics added).

61. Mallios and Zafiris, "Homological Kähler-De Rham Differential."

62. This thesis is discussed in detail in chapter 3, where Whitehead's work in *A Treatise on Universal Algebra* and *Principia Mathematica* is examined in terms of conceptual compatibilities with both category theory and Whitehead's later philosophical work.

63. Letter from Alfred North Whitehead to Henry Leonard (January 10, 1936). United States Library of Congress, Manuscript Division (http://www.loc.gov/today/pr/2011/11-020.html).

64. See chapter 11.6; see also, Mallios and Zafiris, "Homological Kähler-De Rham Differential."

65. As discussed earlier, the presupposed satisfaction of PNC and PEM is the emblematic feature of the reduced density matrix of mutually exclusive and exhaustive probability-valuated outcome states constitutive of the mixed state in quantum mechanics.

CHAPTER 3

Predication in Quantum Mechanics

If there is a common thread woven through the great variety of ontological interpretations of quantum mechanics, it is the belief that discrete measurement events are the fundamental objects of the theory, such that the possibility of specifying the state of a physical system continuously—i.e., 'between' these events—is, in principle, excluded. A quantum measurement event is the quantum mechanical evaluation of an observable, and this entails the assignment of a truth value *F* or *T*, via projection operators with eigenvalues 0 or 1, respectively, to each potential measurement value (e.g., particle spin 'up' or 'down') within some well-defined range of values. As discussed in chapter 2, these values are always defined in terms of a local Boolean measurement context and its associated orthonormal measurement basis. By the presupposition that system states are defined by eigenvalues,[1] measurement can thus be characterized as the actualization of some potential outcome state of the measured system, with each potential outcome state entailing an alternative maximal scheme of truth values for its constituent observables. Quantum observables, in other words, are always properly understood as potential facts, and an evaluated observable is likewise properly understood as an *actualized* potential fact.

In this way, as discussed previously, a fact yielded via quantum measurement is understood as generated consequent of measurement, such that measurement events are ontologically significant; they are facts constitutive of object systems and not merely constitutive of our knowledge of object systems. Classical measurement, by contrast, is understood as merely revelatory of facts—that is, revelatory of already extant truth values pertaining to the evaluated observables. As a result, classical observables are co-determinate and commutative, such that classical system states can be completely specified and represented by a global Boolean algebra—the essence of a Newtonian block universe model.

But unlike classical states, quantum mechanical system states, because they are specified as synthetically predicative facts generated by measurement rather than analytically predicative facts revealed by measurement, can never be *completely* specified—only maximally specified. This is evinced by the fact that the representative algebra of a quantum system is non-commutative and only partially Boolean globally; for the generation of a novel fact *consequent* of measurement clearly precludes the possibility of defining a measured system as a *complete* specification of a co-determinate set of facts. And likewise, there is no structure-preserving map on the global partial Boolean algebra that distinguishes between [a] facts constitutive of the system undergoing measurement and [b] facts generated *by* measurement. Against the argument that this is merely an epistemic limitation ensuing from the incompleteness of quantum theory, such that the disclosure of hidden variables would render such a mapping and thereby a complete state specification as in the case of classical mechanics, Kochen and Specker[2] have proved that the partial Boolean algebra of projection operators characteristic of quantum mechanics cannot, in general, be embedded into a global Boolean algebra.[3]

Closely associated with this difference between quantum and classical measurement are two others: [1] the relationship between the observer and the observed in quantum mechanics is considered to be ontologically significant, such that the eigenstates of a measured system always correlate with the particular Boolean measurement context (e.g., the preferred orthonormal measurement basis and representative Boolean subalgebra) of the chosen measuring apparatus. Thus the objectivity of facts that are generated by measurement is at least in some sense conditioned by the subjective context of the measurement—i.e., the facts constitutive of the measuring apparatus by which the preferred basis is defined; [2] the relationship between the observer and the *unobserved* in quantum mechanics is considered to be ontologically significant, as evinced by the phenomenon of quantum decoherence discussed in chapter 2.7, such that the probability valuations of the eigenstates of the measured system are logically (and in some cases physically) conditioned by facts 'environmental' to the measured system. This is because, in the words of Heisenberg, the description of the measuring apparatus

> contains all the uncertainties concerning the microscopic structure of the device which we know from thermodynamics, and since the device is connected with the rest of the world, it contains in fact the uncertainties of the microscopic structure of the whole world[4] . . . The transition from the 'possible' to the 'actual' takes place as soon as the interaction of the object with the measuring device, and thereby with the rest of the world, has come into play.[5]

Thus the 'objectivity' of a particular actualized local observable is conditioned by unmeasured global observables that exceed the local context of the measurement. But as noted in the previous chapter, unlike classical local system-environmental relations, the nonlocal relations of quantum physics do not necessarily entail a transfer of energy,[6] and are therefore not fully describable in the conventional sense of classical efficient causality. Thus, the causal conditioning of potential local measurement outcomes by global actualities—i.e., facts environmental to the local measurement context—is not primarily a conditioning via efficient causal relation, but rather a conditioning via logical correlation. In this way, unlike the classical conception of the latter as a purely epistemic epiphenomenon—that is, logic as a function of mentality—the ontological interpretation of quantum mechanics depicts logical correlations as a physically significant feature of nature itself.

As introduced in chapter 2, this notion of non-efficient causal conditioning by logical correlation can be expressed in terms of relations among the various maximal and nonmaximal Boolean subalgebras by which composite quantum systems can be defined. Jeffrey Bub, for example, has shown[7] that prior to a measurement interaction, a composite system $S_1 + S_2$ can be defined as a global equivalence class $\mathcal{B}$ of maximal Boolean subalgebras in the partial Boolean algebra of properties of $S_1 + S_2$. A local measurement M_2 at S_2 selects a local nonmaximal Boolean subalgebra $\mathcal{B}_2$ (that is, it locally contextualizes S_2 in terms of the preferred orthonormal measurement basis of M_2). This *local* selection of $\mathcal{B}_2$ produces a non-efficient-causal conditioning, or "revision," to use Bub's term, of the *global* equivalence class $\mathcal{B}$ of maximal Boolean subalgebras in the partial Boolean algebra of properties of $S_1 + S_2$, to a new global equivalence class $\mathcal{B}'$. This logical conditioning of $\mathcal{B}$, that is, its revision to $\mathcal{B}'$, thus entails a revision of all constituent subalgebras of $\mathcal{B}$, including both $\mathcal{B}_1$ (the local measurement context of S_1) *and* $\mathcal{B}_2$ (the local measurement context of S_2) even if S_1 and S_2 are spatially well separated, as in an EPR-type arrangement. In other words, any local contextualization of S_1 by a measurement M_1 (i.e., any possible nonmaximal Boolean subalgebra $\mathcal{B}_1$) must now be logically compatible with $\mathcal{B}_2$—that is, it must be a Boolean subalgebra yielded by the union of $\mathcal{B}_2$ and the set of elements in $\mathcal{B}_1$ that are compatible with $\mathcal{B}_2$.[8]

This logical conditioning, in concert with the efficient causal mechanisms of physics, together exemplify a fundamental *logical causality* in nature. As clearly demonstrated by quantum mechanics, the unspecified actual facts of the world, even if spatially well-separated from a particular local measurement context, always have a non-negligible, logical conditioning influence upon what is physically possible within that local context.

3.1 EPR and Decoherence

As introduced in chapter 2, there have been, over the past several decades, two key types of experimental exploration of logical causality in quantum measurement: experiments investigating quantum decoherence and experiments investigating non-local correlations among subsystems of a composite quantum system. The latter type, already introduced in the previous section, is well represented by the EPR-type experimental arrangement[9] discussed in chapter 2.8, where two measured local subsystems S_1 and S_2 of a composite system $S_1 + S_2$ are spatially well-separated—i.e., mutually environmental—such that any transfer of energy between the subsystems (energy intended to account for the causal conditioning of one subsystem's measurement outcome by that of the other) would need to be superluminal, in violation of relativity theory. To date, no such energy transfer has been detected, though several non-local, superluminal efficient causal mechanisms have been proposed over the decades.[10]

With respect to logical causality in quantum decoherence, the latter can be generally understood as a sub-process of every quantum measurement interaction, entailing [1] the individuation of potential measurement outcomes that are initially coherently superposed in the pure state (e.g., Schrödinger's cat as alive AND dead), and [2] the logically conditioned evolution of those individual *potential* outcome states that satisfy both PNC and PEM, to *probable* outcome states that are mutually exclusive and exhaustive (e.g., such that Schrödinger's cat is *either* alive XOR[11] dead). Together, these now individuated / decoherent states comprise the probable measurement outcomes of the mixed state. This evolution from the pure state to the mixed state is achieved via a *negative selection* of those potential outcome states whose conjunction violates the logical principle of non-contradiction (PNC)—i.e., those potential outcome states that 'logically interfere' with each other. These are formally represented by an 'interference term' corresponding to the off-diagonal terms of the density matrix; and likewise, the negative selection mechanism is represented as a mathematical self-cancellation of these off-diagonal terms, leaving only the diagonal terms, which do satisfy PNC.[12]

Thus, as discussed in chapter 2, the mutually exclusive nature of the probable outcomes of the mixed state reflects the satisfaction of PNC—i.e., 'at most one outcome state'; and likewise the exhaustive nature of the mixed state reflects the satisfaction of the logical principle of the excluded middle (PEM)—i.e., 'at least one outcome state.' The intrinsic presupposition of PNC and PEM together in any ontological interpretation of quantum mechanics is thus reflected in the fact that each outcome state is valuated as a probability and not merely a potentiality once decoherence occurs; that is to say, all probable outcome states are mutually exclusive per PNC, and exhaustive per PEM, such that together they must sum to unity. And as will be further explored presently, the function of probabilities in quantum mechanics presupposes not just PNC and PEM, but also the laws of Boolean logic[13] by which the Boolean algebras discussed in the previous section are constructed; for the mixed state is always defined according to the local Boolean measurement context corresponding to a particular observable.

The logical conditioning of potential outcome states into probable outcome states via local Boolean measurement contextualization is the central import of decoherence—and arguably of quantum mechanics in general. The result is the enabling of the distributive law and the probability sum rules that follow from it, which, because of quantum interference, are not operative in quantum systems apart from decoherence. Typically, as was discussed in chapter 2, quantum interference and its undermining of the distributive law[14] has been considered the hallmark feature of quantum mechanics because of the classically counterintuitive implications, captured most infamously by the popular Schrödinger's cat scenario. However, while superpositions of interfering potentia are certainly integral to quantum measurement, the fact that they *always* evolve via decoherence to become non-interfering probability distributions would seem to be even more emblematic of quantum mechanics than superposition itself—i.e., superposition in abstraction from the process of quantum measurement as a whole—a dubious abstraction given that apart from this process, superposition has no meaning. Thus, one could argue that it is not the superposition of potentia but rather the Boolean logical conditioning of potentia, productive of probability valuations such that PNC and PEM are always ultimately satisfied, that is quantum theory's signature feature—i.e., the evolution of potentiality to probability.

The selective evolution of a coherent superposition of potential outcome states, most of which are mutually logically inconsistent in terms of PNC, to a decoherent distribution of logically consistent probable outcome states, is achieved by incorporating the facts constitutive of the measured system's environment[15] into the quantum measurement formalism—that is, making explicit the system-environment relations that have traditionally been treated as implicit and irrelevant in quantum theory. Philosophically, this is justifiable simply because it explicates what is already implicit and presupposed given quantum theory's conventional assessment as a universal theory. But it is also justifiable practically in that it enables one to account for decoherence as a conceptual implication of the standard formalism: Rendering explicit the system-environment relations typically ignored as implicit introduces manifold degrees of freedom into the superposition of potential outcome states. This, in turn, allows for the integration of their respective maximal Boolean subalgebras (in the partial Boolean algebra of properties of the global system) into equivalence classes large enough to accommodate the negative selection process described above, with each class indexed to a different potential outcome state of the locally contextualized measuring apparatus.

As discussed earlier, these alternative potential outcome states are always defined according to the particular orthonormal measurement basis of the local measuring apparatus—that is, according to the Boolean subalgebra by which the local measurement context is defined. In this way, the preferred basis functions as a local Boolean context by which the potentia of the global quantum system are logically integrated—that is, by which the equivalence class of Boolean subalgebras in the partial Boolean algebra defining the global system is revised by a local measurement.

Again, it is the magnitude of environmental degrees of freedom within each equivalence class that produces sufficient logical-mathematical cancellation among potential system-detector-environment outcome states that interfere (i.e., that violate PNC) within each class. A traceover of the environmental degrees of freedom thus eliminates these logically incoherent states represented by the off-diagonal terms in the density matrix. The elimination of these terms effects the reduction of the pure state, with its interfering superposition of potential measurement outcomes, to the mixed state and its logically conditioned, decoherent matrix of mutually exclusive and exhaustive probable outcome states.

As is the case with EPR-type nonlocality experiments, where this nonlocal logical conditioning obviates the need for any proposed superluminal efficient-causal mechanism to account for the quantum correlations among spatially-well-separated systems, recent experiments on precessional decoherence in spin bath dynamics,[16] as discussed in chapter 2.7, likewise entail no energy transfer between the bath and the measured system, unlike previous oscillator bath and other environmental decoherence models.[17] Thus the once conventional description of decoherence-generating system-environment interactions purely as a function of energy exchange—i.e., via an effective Hamiltonian—is arguably incomplete.[18] In light of these experiments, it is clear that the decoherence of global potentia into individuated, locally contextualized probabilities is properly understood not as an efficient casual mechanism, as it is often portrayed, but rather as a logical integration of potential outcome states such that a subset of these can evolve to become probable outcome states. This conditioning is formalized as a revision of the equivalence classes of maximal Boolean subalgebras in the partial Boolean algebra defining the global system, 'logically revised' by the nonmaximal Boolean algebras representing locally contextualized measurements.

3.2 The Problem of Self-Reference in Quantum Systems

In summary of the discussion so far, the evolution of classical to quantum mechanics has carried with it a parallel evolution of the concept of measurement. Classical mechanics depicts measurement as revelatory of objective, context-independent, co-determinate facts such that even the subjective features of measurement can, in principle, be reduced to such facts. By contrast, quantum mechanics depicts measurement as generative of novel objective facts, with this generation conditioned by both [1] the local contextual facts constitutive of the measuring apparatus, and [2] the global facts constitutive of the environment of the measured system.

An important waypoint of this evolution was a refinement, by von Neumann,[19] of the initial conventional interpretation of quantum theory (the so-called 'Copenhagen Interpretation') which rendered the latter a more coherent ontological interpretation. The most pressing impediment to such coherence was Bohr's early insistence that the interaction of measured system with measuring device always be understood in terms of a *classical* detector coupled with a *quantum* system. Instead, von Neumann argued, quantum measurement should always be understood as a quantum mechanically described detector coupled with a quantum mechanically described measured system. In this way, every quantum measurement interaction is properly understood as involving a composite quantum system like that discussed in chapter 2—i.e., as an entanglement of at least two quantum systems. Beyond giving added coherence to an interpretation intended to be universal, the treatment of 'measured system + measuring apparatus' as a composite quantum system would also allow for a more careful exploration of the precise way in which the facts constitutive of the measuring apparatus and its Boolean preferred basis logically condition the potential outcome facts of the measured system, such that the latter *always* correlate with the former both causally and logically.

As noted above, however, prior to this refinement, a quantum mechanical measurement interaction was depicted as a classical measuring apparatus causally interacting with a quantum system—this primarily for practical reasons, since depicting the apparatus as a quantum system entangled with the measured system would make the necessary calculations unwieldy. But this instrumentalist depiction of the measuring apparatus as classical-mechanical, coupled with Bohr's admonition stressing "the impossibility of any sharp separation between the behaviour of atomic objects and the interaction with the measuring instruments which serve to define the conditions under which the phenomena appear"[20] caused many physicists and philosophers of science of the period to lament that quantum theory entailed a gross intrusion of subjectivity into physics—a fatal vitiation of the notion that predication via measurement yields logically sound access to the facts of objective reality.[21] Even today there exists an unfortunate popular shorthand understanding of quantum mechanics whereby 'individual observers create their own realities,' with no reality at all apart from observation.

Against this backward path, in his *Mathematical Foundations of Quantum Mechanics*, von Neumann's conception of predication by measurement as the interaction of quantum measured system and quantum measuring system would provide a way forward for scientific realism by grounding the subjective, contextual features of quantum measurement *in* the objective facts constitutive of the measuring system, and vice versa. His proposal is a crucial step toward the elevation of the Copenhagen Interpretation from a quasi-ontological but mostly instrumentalist interpretation of quantum mechanics to a truly universal and coherent ontological interpretation. But this added coherence generates a new conceptual difficulty that, when acknowledged, poses a serious challenge: A quantum mechanical measurement interaction entails a composite system of measured system + measuring apparatus + unmeasured environment. The potential outcome states of the global composite system always evolve to become probable outcome states relative to the local Boolean measurement context of the measuring apparatus—i.e., its preferred orthonormal measurement basis and associated choice of projection operator. But by incorporating the local apparatus *into* the evolving global composite system, the measurement outcome becomes not only predicative of the measured system, but also predicative of the apparatus performing the measurement.

In other words, if $|\psi\rangle$ = the system state to be measured, $|\phi\rangle$ = the measuring apparatus state, and $|e\rangle$ = the state of the unmeasured environment, then defining the global wave function as:

$$|\Psi\rangle = \alpha_{ijk} \sum_{i,j,k} |\psi\rangle_i |\phi\rangle_j |e\rangle_k$$

is self-referential. This is because the tensor product in this expression is a *totality*, comprising every potential outcome state (i.e., every proposition) subsumed by that totality; yet at the same time, the global totality $|\Psi\rangle$, being a vector of unit length and therefore an actual (but unspecified) state of the world, serves as logical *premise* for the subset of local propositions predicative of $|\phi\rangle$ in the context of its preferred basis. This is well reflected in Bohr's response to the EPR argument,[22] wherein he notes that an *initial actual state* of the composite system must be specified via a preparation measurement prior to quantum mechanical measurement. Recall the earlier discussion of the composite quantum system $S_1 + S_2$. Bohr's point is that a local measurement M_2 on a subsystem S_2 of a composite system $S_1 + S_2$ will only affect the measurement outcomes at S_1 *if* an appropriate preparation measurement of $S_1 + S_2$ is made immediately prior to measurement M_2. It is this 'preparation measurement' that generates the Boolean subalgebra in the partial Boolean algebra by which $S_1 + S_2$ is *initially* defined, thus establishing the system's *initial actual state in terms of this context.*

Of course in the example above there is no way to suitably 'prepare' $|\Psi\rangle$ as an actual initial state in this manner. But by replacing the initial Boolean subalgebra yielded by such a preparation with an equivalence class of Boolean subalgebras[23] (presupposing, per the compatibility condition introduced in chapter 2, that every local context constitutive of $|\Psi\rangle$ is Boolean) one is able to characterize $|\Psi\rangle$ as an actual (but unspecified) state without requiring any particular preparation measurement of the entire system.

If, for example, in the above, $|\psi\rangle$ is a spin ½ system like the one discussed in chapter 2, where $|\Psi\rangle = \alpha\, |\psi\uparrow\rangle\, |\phi\uparrow\rangle\, |e\uparrow\rangle + \beta\, |\psi\downarrow\rangle\, |\phi\downarrow\rangle\, |e\downarrow\rangle$, each eigenstate represents a locally Boolean-contextualized, probability-valuated propositional outcome expressed as a projection of $|\Psi\rangle$ upon each tensor product $|\psi\uparrow\rangle\, |\phi\uparrow\rangle\, |e\uparrow\rangle$, and $|\psi\downarrow\rangle\, |\phi\downarrow\rangle\, |e\downarrow\rangle$. The global wavefunction $|\Psi\rangle$, then, represents not merely a global superposition of vacuous potentia; it represents the global totality of actual facts in a superposition of potential relationships with (projections onto) locally contextualized, potential measurement outcome states. $|\Psi\rangle$ is not a simply a superposition of bare potentia, in other words; it is a superposition of potential *relations* between [a] the uncontextualized *actualities* constitutive of the *global initial state* $|\Psi\rangle$, and [b] the locally contextualized *potential* measurement outcomes $|\psi\uparrow\rangle|\phi\uparrow\rangle|e\uparrow\rangle$ and $|\psi\downarrow\rangle\,|\phi\downarrow\rangle\,|e\downarrow\rangle$ constitutive of the *global outcome state*.

It is an infamous feature of the standard Copenhagen formalism that it cannot account for the actualization of a unique outcome state from among these potential outcome states. It cannot, in other words, account for the existence of facts *consequent* of measurement, but only superpositions of potential facts. But per the discussion above, neither can it account for the existence of the facts *presupposed* by measurement (that is, the facts constitutive of $|\Psi\rangle$, the initial system-detector-environment state) and this is just as crucial. In the expression,

$$|\Psi\rangle = \alpha\,|\psi\uparrow\rangle\,|\phi\uparrow\rangle\,|e\uparrow\rangle + \beta\,|\psi\downarrow\rangle\,|\phi\downarrow\rangle\,|e\downarrow\rangle$$

$|\Psi\rangle$ represents at once the extant world of *actual* facts, *and* a superposition of potential facts ($|\psi\uparrow\rangle\,|\phi\uparrow\rangle\,|e\uparrow\rangle$ and $|\psi\downarrow\rangle\,|\phi\downarrow\rangle\,|e\downarrow\rangle$), locally contextualized as an equivalence class of maximal Boolean subalgebras by which these potential facts are defined (i.e., as projections of $|\Psi\rangle$). The expression represents, in other words, the local potentialization of a *logically antecedent* global system of facts.

Defining the totality $|\Psi\rangle$ by a condition pertaining to $|\phi\rangle$, which implicitly refers to $|\Psi\rangle$ itself, results in a logical vicious circle in any interpretation that fails to depict the generative nature of quantum measurement—i.e., that it is a synthetic process yielding *novel* predicative facts *not contained in the totality undergoing measurement*. For those interpretations that do depict generative predication in this regard, it is understood that every novel local measurement outcome, such as $|\psi\uparrow\rangle\,|\phi\uparrow\rangle\,|e\uparrow\rangle$ or $|\psi\downarrow\rangle\,|\phi\downarrow\rangle\,|e\downarrow\rangle$, necessarily entails two presuppositions:

[1] an *initial* actual state-of-the-world $|\Psi\rangle$ *logically prior in order* to the local measurement outcome.

[2] a *novel* actual state-of-the-world, $|\psi\uparrow\rangle\,|\phi\uparrow\rangle\,|e\uparrow\rangle$ or $|\psi\downarrow\rangle\,|\phi\downarrow\rangle\,|e\downarrow\rangle$, *consequent* of the local measurement outcome.

With every novel fact generated, in other words, a novel state of the universe is likewise generated. $|\Psi\rangle = \alpha|\psi\uparrow\rangle\,|\phi\uparrow\rangle\,|e\uparrow\rangle + \beta|\psi\downarrow\rangle\,|\phi\downarrow\rangle\,|e\downarrow\rangle$ is thus best understood not as an ontologically concrete, classical '*the* state of the world revealed by measurement,' but rather as *a* state of the world in process—a world fundamentally characterized as a serial, state-by-state, continuously punctuated history, with each novel global state $|\Psi\rangle$ indexed to some local concrete quantum fact $|\psi\rangle\,|\phi\rangle\,|e\rangle$.

By such an interpretation of quantum mechanics, the actuality of the world is not some ontologically fixed conjunction of independent brute facts; it is, rather, an evolving complexity of interrelated facts—of quantum actual occasions—more praxiological than ontological in that these facts are generated *by* their logically conditioned interrelation. It is a world always in the process of formation, such that the definition of *a* totality cannot ever entail the definition of *the* totality—i.e., a total order—as it does within the framework of a classical block universe. Rather, the predication of a totality of facts entails the augmentation of that totality with a novel fact—namely, the predicative fact which is the actualized quantum observable. In the words of Whitehead, the only logically coherent physical-mathematical description of our universe must be one in which for every predication, "The many become one and are increased by one."[24]

3.3 Quantum Mechanics and the Theory of Logical Types

The problem of self-reference inherent in the conventional Copenhagen interpretation of quantum mechanics is the chief impediment to its acceptance as a viable ontological interpretation—at least by the standard of logical coherence. For by this metric, quantum mechanics would seem to yield impredicative facts, as will be discussed presently, and therefore be of dubious ontological significance. Of course, there is no reason *prima facie* that self-predication in physics should be viewed as posing an ontological difficulty on the grounds of the epistemic difficulty it caused for set-theoretic mathematics during the decades preceding the advent of quantum theory. But given that theoretical physics is primarily mathematical physics, it is worth exploring the notion that the conceptual innovations devised to relieve the mathematical paradoxes of self-predication might find their use in the relief of related and equally notorious problems in quantum mechanics—among these, the measurement problem, and the incompatibility of local-global relations defined quantum mechanically vs. those defined relativistically.

Bertrand Russell identified the mathematical problem of self-predication as having its genesis in the attempt to predicate a totality, writing: "The notion of 'all propositions' is illegitimate; for otherwise, there must be propositions . . . which are about all propositions, and yet cannot, without contradiction, be included among the propositions they are about."[25] Recall that the quantum mechanical expression,

$$|\Psi\rangle = \alpha_{ijk} \sum_{i,j,k} |\psi\rangle_i |\phi\rangle_j |e\rangle_k$$

reflects precisely this notion of 'all propositions' concerning the universe as a totality.

Russell's Paradox was devised to exemplify the difficulty: "Let *w* be the class of all those classes which are not members of themselves. Then, whatever class x may be, 'x is a w' is equivalent to 'x is not an x.' Hence, giving to x the value w, 'w is a w' is equivalent to 'w is not a w.'"[26] Russell considered his paradox in the context of others, such as the Epimenides paradox discussed earlier, which in the same way can be neither true nor false without internal contradiction, and the Burali-Forti paradox, which shows that the concept of "the set Ω of all ordinal numbers" leads to a contradiction, since Ω itself entails all properties of an ordinal number and would therefore have to be considered an ordinal number itself. Thus, $\Omega < \Omega + 1 \leq \Omega$. "In each contradiction," Russell remarks,

> something is said about *all* cases of some kind, and from what is said a new case seems to be generated, which both is and is not of the same kind as the cases of which *all* were concerned in what was said . . . Whatever we suppose to be the totality of propositions, statements about this totality generate *new* propositions which, on pain of contradiction, *must lie outside the totality*. It is useless to enlarge the totality, for that equally enlarges the scope of statements about the totality. Hence there must be no totality of propositions, and "all propositions" must be a meaningless phrase.[27]

The phrase "must lie outside the totality" is the crux of Russell's proposed solution to the problem of self-predication, and the method by which he proposed to rid mathematics of impredicative statements: the "Theory of Logical Types." Stated simply,

> A function whose argument is an individual and whose value is always a first-order proposition will be called a first-order function. A function involving a first-order function or proposition as apparent variable will be called a second-order function, and so on. A function of one variable which is *of the order next above that of its argument* will be called a *predicative* function.[28]

After several years of refinement, the notion of proper predication via the Theory of Logical Types was given again in Russell and Whitehead's *Principia Mathematica*. They write:

> We are thus led to the conclusion, both from the vicious-circle principle and from direct inspection, that the functions to which a given object can be an argument are incapable of being arguments to each other, and that they have no term in common with the functions to which they can be arguments. We are thus led to construct a hierarchy[29] . . . We will define a function of one variable as *predicative* when it is of the next order above that of its argument, i.e. of the lowest order compatible with its having that argument.[30]

Again, for Russell, "Whatever we suppose to be the totality of propositions, statements about this totality generate new propositions which, on pain of contradiction, must lie outside the totality." It was emphasized in chapter 2 that ontological interpretation of quantum mechanics depicts measurement as generative of facts. In light of the current discussion, it follows that for any ontological interpretation of quantum mechanics that satisfies the desideratum of logical coherence, every predication of a totality, such as

$$|\Psi\rangle = \alpha\,|\psi\uparrow\rangle\,|\phi\uparrow\rangle\,|e\uparrow\rangle + \beta\,|\psi\downarrow\rangle\,|\phi\downarrow\rangle\,|e\downarrow\rangle$$

must be understood as generative of a new fact "outside the totality," and constitutive of a new totality of the "next order above" (or otherwise separated)[31] yet "compatible" with the original argument. As applied to the ontological interpretation of quantum mechanics, the mathematical term 'order,' which in the Theory of Logical Types pertains to functions, can be taken to mean simply 'levels' within the hierarchy of totalities. 'Compatibility' is derived from the asymmetrical internal relatedness of higher to lower levels; that is, each higher level totality contains and is thus conditioned by the facts of the next lowest level, and so on down the series of inclusions. As exemplified by quantum indeterminacy, the conditioning is not determinate, since the predicate is not contained within the order of the argument—i.e., the predication is synthetic. By contrast, the Hilbert space formulation of the conventional Copenhagen interpretation contains no mechanism by which this logical asymmetry can be represented; again, this is the chief impediment to the Copenhagen interpretation's depiction as ontologically coherent.

3.4 Relations as Relata: From Set-Theoretic Classical Material Objects to Category-Theoretic Quantum Relational Events

Beyond exploring its significance to mathematics, Whitehead would go on to rigorously explore the ontological implications of the Theory of Logical Types that he and Russell had developed in the *Principia*. In *Process and Reality*, for example, Whitehead offers a thoroughly systematized cosmology that in many ways can be viewed as a logico-physical exemplification of the thesis of the Theory of Logical Types (though certainly not reducible to the latter). As introduced in chapter 2, in Whiteheadian metaphysics, every predicative fact (alternatively termed 'actual occasion' or 'final real thing') is internally related to, though not determined by, a dative world 'one order below' that of the predicative occasion. Predication is 'dipolar' because each potential quantum actual occasion (i.e., each potential predicative fact / potential measurement outcome) comprises both [1] a *causal-physical pole*, by which the potential novel fact is *spatiotemporally coordinated* with, and efficient-causally influenced by, the actual facts of its *relativistically restricted* dative world;[32] and [2] a *logical-conceptual pole*, by which the potential novel fact and its unique subjective context are *logically coordinated* via asymmetrical internal relation with the *totality* of actual facts constitutive of its dative world—including these facts' *own* subjectively contextualized internal relations to *their* dative worlds.[33] Type-theoretically, these dative worlds are therefore properly understood as lower order totalities relative to the subject occasion. Thus, although Whitehead never formalized in *Process and Reality* the implicit logical structure underlying his philosophical scheme of coordination-by-internal relation, the latter nevertheless clearly entails a type-theoretic, hierarchical structure-preserving mapping, correlated with a scheme of efficient causality; and it is by this correlation that the orders of logical implication and causal relation are coherently integrated in Whiteheadian philosophy.

Indeed, in the context of his earlier work in *Principia Mathematica*, and even going back to Whitehead's first major work, *A Treatise on Universal Algebra*, it is clear that *Process and Reality* was an attempt to define in the language of philosophy what Whitehead could not, in his earlier works, define in the languages of mathematics and logic alone: namely, a realist speculative ontology whose terminal objects are *elemental units of internal relation*, analyzable both in terms of extensive connection to other elemental units (i.e., 'coordinate division' in Whitehead's philosophy), and in terms of structure-preserving relations with other elemental units and *their* relations (i.e., 'genetic division'). But the absence of rigorous logico-mathematical formalism in *Process and Reality*, by which Whitehead's complex philosophical scheme might have been technically represented, ought not to be interpreted as an abandonment of formal logic and mathematics—i.e., in the sense that these are *in principle* counterproductive to the method of speculative philosophy. A more accurate assessment, evinced both in his last published philosophical essay, "Mathematics and the Good," as well as in the previously discussed personal letter to Henry Leonard, is that Whitehead's frustration with the dogmatic positivist restraint of philosophical generalization derived not from some newly acquired pessimism about the enterprise of mathematics itself, but rather from the increasing popularity of extending mathematics beyond its own identifiable limits. But if these limits are explicitly identified and always kept in view, then the task of mathematically formalizing the categorical structure of a speculative philosophical scheme can be undertaken productively (and as argued in the present volume, to great advantage) without succumbing to the temptation of reducing the philosophical scheme *to* the formal structure.

Hints of Whitehead's later philosophical assessment of the limitations of the method of formalism are evident even in his earliest work, *Treatise on Universal Algebra*, as will be discussed presently. Overall, of course, the *Treatise* exhibits an optimism about the method of formalism as a means by which to cross the Platonic chasm separating the order of mathematical necessity and the order of existential contingency—the order of logical implication and the order of causal relation.

> The ideal of mathematics should be to erect a calculus to facilitate reasoning in connection with every province of thought, or of external experience, in which the succession of thoughts, or of events can be definitely ascertained and precisely stated.[34]

Whitehead would, several years later, re-assess the extent to which mathematical formalism, as a conventional enterprise, is able to achieve this goal,[35] shifting his focus next to logic in his work with Russell on *Principia Mathematica*, and then finally to philosophy. Indeed, it is interesting to compare the above quotation from *A Treatise on Universal Algebra* to the following passage from *Process and Reality*:

> Speculative philosophy is the endeavour to frame a coherent, logical, necessary system of general ideas in terms of which every element of our experience can be interpreted. By this notion of "interpretation" I mean that everything of which we are conscious, as enjoyed, perceived, willed, or thought, shall have the character of a particular instance of the general scheme.[36]

But this evolution of focus from algebra, to logic, to philosophy, is all too often characterized as a series of constructions and abandonments rather than a progressive synthesis, wherein the languages of mathematics, logic, and philosophy, each with its own intrinsic limitations, together contribute to a coherent understanding of nature. As mentioned above, Whitehead's early recognition of these limits is evident even in *A Treatise on Universal Algebra*. There, he makes a careful distinction between mathematical truth and philosophical or experiential truth, and explicitly states the former's necessary presupposition of the latter:

> Mathematics in its widest signification is the development of all types of formal, necessary, deductive reasoning. The reasoning is formal in the sense that the meaning of propositions forms no part of the investigation. The sole concern of mathematics is the inference of proposition from proposition. *The justification of the rules of inference in any branch of mathematics is not properly part of mathematics: it is the business of experience or of philosophy. The business of mathematics is simply to follow the rule.*[37]

> Mathematical definitions are always to be construed as *limitations* as well as definitions; namely, the properties of the thing defined are to be considered for the purposes of the argument as being merely those involved in the definitions.[38]

Over four decades later, in his last published philosophical essay, "Mathematics and the Good," Whitehead revisits this notion of the limitation of mathematical definitions, and the necessary presupposition of philosophical truths for the construction and proper synthesis of formalized logical and mathematical truths. In particular, he discusses this limitation and presupposition as they pertain to his work with Russell on the rule of Logical Types (the following passage indeed reads as though it were a natural continuation of the passage above):

> Unfortunately this rule [a function is only predicative when it is of the next order/type above that of its argument] cannot be expressed apart from the *presupposition that the notion of number applies beyond the limitations of the rule*. For the number 'three' in each type, itself belongs to different types. Also each type is itself of a distinct type from other types. Thus, according to the rule, the conception of two different types is nonsense, and the conception of two different meanings of the number three is nonsense. It follows that the only way of understanding the rule is nonsense. It follows that the rule must be *limited to the notion of a rule of safety*, and that the *complete* explanation of number awaits an understanding of the relevance of the notion of the varieties of multiplicity to the infinitude of things. Even in arithmetic you cannot get rid of a sub-conscious reference to the unbounded universe. You are abstracting details from a totality, and are imposing limitations on your

> abstraction. Remember that a refusal to think does not imply the non-existence of entities for thought.[39]

In the context of the current discussion, one can read the last sentence as, "refusal to think *non-mathematically* does not imply the non-existence of non-mathematical entities for thought." But again, the above passage is no disavowal of his earlier work on logic and mathematics, despite its frequent characterization to this effect; it is, rather, an admonition against the folly of extending mathematics and logic beyond their own intrinsic limitations toward the goal of an 'explicatively complete,' 'ultimate' formal structuralist understanding of the world. These limitations had, for Whitehead, become more precisely defined as a result of the progressive synthesis of his work in mathematics, logic, and philosophy throughout his career; but it is clear that even in his earliest work, the idea of such limitations was always central to his thinking. Returning again to *A Treatise on Universal Algebra*, for example, we find the following:

> Mathematical definitions either possess an existential import or are conventional. A mathematical definition with an existential import is the result of an act of pure abstraction. Such definitions are the starting points of applied mathematical sciences ; and in so far as they are given this existential import, they require for verification more than the mere test of self-consistency.

In other words, unlike 'conventional' or 'pure' mathematical definitions, 'existential' mathematical definitions, like the definitions of speculative philosophy, require empirical applicability and adequacy *in addition to* logical consistency and coherence. (As introduced in chapter 1, these are the four desiderata for speculative philosophical definitions given in *Process and Reality*.) By contrast:

> A conventional mathematical definition has no existential import. It sets before the mind by an act of imagination a set of things with fully

> defined self-consistent types of relation . . . *Qua* pure mathematics, mathematical definitions must be conventional.[40]

One of the limitations Whitehead specifies for conventional or pure mathematics, as discussed above, is the necessary presupposition of its underlying logical structure: "The justification of the rules of inference in any branch of mathematics is not properly part of mathematics: it is the business of experience or of philosophy. The business of mathematics is simply to follow the rule."[41] (Cf. the discussion of Gödel's incompleteness theorem in chapters 1 and 2.) A related philosophical limitation of conventional mathematics is that although a conventional mathematical formalism can be understood to represent certain features of nature, the formalization, no matter how elegant its coherence and self-consistency, cannot be taken by these virtues alone as 'scientific' evidence that nature is somehow reducible to these features—let alone reducible to the mathematical formalism by which these features are described.

> In order that a mathematical science of any importance may be founded upon conventional definitions, the entities created by them *must have properties which bear some affinity to the properties of existing things.* Thus the distinction between a mathematical definition with an existential import and a conventional definition is not always very obvious from the form in which they are stated. Though it is possible to make a definition in form unmistakably either conventional or existential, there is often no gain in so doing. In such a case the definitions and resulting propositions can be construed either as referring to a world of ideas created by convention, or as referring exactly or approximately to the world of existing things.[42]

The quantum actual occasions in Whitehead's later philosophy, for example, are existential units of relation rather than classically understood objective entities (i.e., with intrinsic 'primary' qualities). As quanta, these units of relation are indivisible and thus possess an 'objective' character; but they are generated by their internal relations with antecedent dative occasions, and thus also possess a 'subjective' character. This subjective-objective dipolarity of the quantum actual occasion forms the basis of an internal relational structure among all occasions—one that is both inherited contextually (i.e., subjectively) from antecedent dative occasions and ingredient (i.e., objectively) in subsequent occasions via a structure preserving ontological scheme.

Despite the absence in Whitehead's later philosophical work of any attempt at a rigorous mathematical formalization of this structure, his philosophical re-imagining of the classical, material, 'elementary particulate unit' as 'elementary relational unit' clearly derives from his earlier efforts to mathematically formalize the latter. In his paper "On Mathematical Concepts of the Material World," presented to the Royal Society of London in 1905, for example, Whitehead first gives an existential mathematical definition of the classical material point-particle ('Concept I')—and then proceeds to argue that it is not their *particulate* existence that makes them elementary, but rather their *relational* existence ('Concept II'):

> If we abolish the particles (in the "classical" sense) . . . everything will proceed [logically and mathematically] exactly as in the classical concept. The reason for the original introduction of "matter" was, without doubt, to give the senses something to perceive. If a *relation* can be perceived, this Concept II has every advantage over the classical concept.[43]

In this work, Whitehead explicates the properties that essential elementary relations must have if they are to have existential significance. At the time of its writing, for Whitehead this meant properties that would allow them to be schematized via Euclidian geometry. In later years, this would be expanded to metrical geometry more broadly, and by the time of *Process and Reality*, he would argue that even metrical geometry was not the fundamental extensive relational order. But by that time, he had abandoned the effort to mathematically formalize the essential properties of existential relations beyond their mereological structure as internal relations:

> Considered in its full generality, apart from the additional [relativistic] conditions proper only to the cosmic epoch of electrons, protons, molecules, and star-systems, the properties of [the extensive continuum] are very few and do not include the relationships of metrical geometry. An extensive continuum is a complex of entities united by the various allied relationships of whole to part, and of overlapping so as to possess common parts, and of contact, and of other relationships derived from these primary relationships.[44]

It could be argued that because of Whitehead's earlier insistence upon the essential metrical and set-theoretic characteristics of mathematical definitions qualified as 'existential' (and therefore 'meaningful'), he failed to identify in his later philosophical work that these essential characteristics, though more abstract than he had originally supposed, could nevertheless be represented formally. More important, they could be represented formally in a manner *wholly compatible* with his philosophical cosmology, both in terms of its speculative approach and its conceptual structure.

In the present volume, we argue that the category-sheaf theoretic approach to quantum mechanics provides precisely such a formalization, wherein topological relations are shown to be fundamental to higher order metrical relations; and indeed, the essential characteristics of category-sheaf theory's mathematical definitions of elementary relations—i.e., the characteristics by which these definitions can be qualified as 'existential'—are traceable to the very algebraic and Boolean logical structural characteristics of nature that Whitehead had first explored in *Treatise on Universal Algebra*. In a paper appraising the latter's potential reconnection to Whitehead's later philosophical work, Robert Valenza and Granville Henry write:

> It was Whitehead's attitude towards the presupposed radical objectivity of mathematical objects that hindered him from being truly creative algebraically and finally isolated him from the newer relational attitude toward algebra that has found its contemporary expression in category theory . . .

> The surprise, and irony if you like, is that when Whitehead finally gave up mathematics, he discovered a relational viewpoint—which we dare call functorial—that seems to us an ideal philosophical foundation for the emphases of category theory.[45]

First introduced by Samuel Eilenberg and Saunders Mac Lane in the 1940s, category theory was developed by way of algebraic topology in order to explore structure-preserving relations among mathematical objects defined as abstractly as possible—i.e., to the point at which these objects are defined most fundamentally as relations themselves. These are the 'natural transformations' of category theory—the structure-preserving relations whose relata ('functors') are themselves structure-preserving relations among the following:

[1] categories of primitive mathematical 'objects,' A, B, . . .

[2] relations ('morphisms') among these objects, $f: A \to B$, . . .

Compositional structure is preserved throughout this scheme of internal relations via the presupposition of [1] the logical principle of identity (PI), given in the form of the identity morphism $1_A : A \to A$, and [2] binary composition of morphisms, such that if $f : A \to B$ and $g : B \to C$ then the composition $g \circ f : A \to C$ is valid and maintains the associative law—i.e., $h \circ (g \circ f) = (h \circ g) \circ f$.

The fact that composition of morphisms is associative but *not* commutative is central to the applicability of category theory to the interpretation of quantum mechanics and its non-commuting observables—and in particular, to the Whiteheadian relational realist interpretation, given the centrality of asymmetrical internal relations in the latter. However, commutativity in this sense is not to be confused with the commutative diagrams of category theory, which instead show that different compositions might lead to the same result—e.g., if $f : A \to B$ and $g : A \to C$ and $h : B \to D$ and $k : C \to D$, then $h \circ f : A \to D$ and $k \circ g : A \to D$, such that $h \circ f = k \circ g$. This can be shown in the commutative diagram:

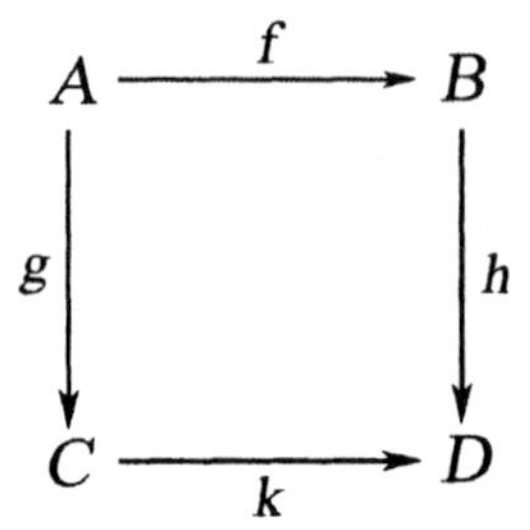

In terms of Whitehead's philosophical schematization of internal relations, the non-commutativity of morphisms has its importance in an actual occasion's asymmetrical relationship to *particular* data—i.e., occasion A's being internally related to dative occasion B means that B is constitutive of A, but A is not constitutive of B. But at the same time, the commutativity represented in category theoretic diagrams has equal importance when considering actual occasions as *totalities*—i.e., as internally related to a totality of dative occasions. With respect to the above diagram, for example, occasion A's internal relation to dative occasion D is mediated by its internal relations to *both B and C*. In this way, A's internal relation to D, mediated by B, *cannot exclude the relationship between* A and C, and by implication, B and C. $A \rightarrow B \rightarrow D$, in other words, must be *compatible* with $A \rightarrow C \rightarrow D$ such that $h \circ f = k \circ g$.

Writ large, the overwhelming complexity of a totality of such equivalences is mitigated in Whiteheadian metaphysics via the formation of equivalence classes ('transmutations') of all potential intermediate internal relations between $A \rightarrow D$. Intermediate internal relations incompatible for integration are those that, unlike $A \rightarrow C \rightarrow D$ and $A \rightarrow B \rightarrow D$, violate commutativity. These are, for Whitehead, potential relations that are incompatible for integration in occasion A, and are thus eliminated via a process of negative selection. As has been discussed previously,[46] and as will be explored further, this method of negative selection via the formation of equivalence classes, and the inductive approximation of *all* potential internal relations relative to a particular locally contextualized quantum measurement outcome, is also central to the relational realist interpretation of quantum mechanics.

Returning again to the asymmetrical (non-commutative) internal relations among morphisms, however, it is important to note that despite their centrality in Whitehead's philosophical scheme, as well as in the Theory of Logical Types in *Principia Mathematica,* this emphasis was not present in his earlier work on algebra. As Valenza and Henry point out,

> The familiar concept of a function [or morphism] is perhaps the most important notion in all of mathematics. That composition of functions is associative but not commutative is almost certainly the reason that associativity has priority over commutativity in the axiomatic development of algebraic systems—a feature characteristically lacking in Whitehead's development of algebraic manifolds.[47]

This is especially notable given that in *A Treatise on Universal Algebra,* Whitehead does explore the relationship between identity and mathematical equality—the fundamental 'truism' and 'paradox' apparent in commutativity (e.g., b = 2+3 and b'= 3+2, such that $b = b'$). With respect to the example of commutativity, he explains:

> The truism is the partial identity of both b and b', their common *B*-ness. The paradox is the distinction between b and b' so that b is one thing and b' is another thing: and these things, as being different, must have in some relation diverse properties.[48]

This 'truism' and 'paradox' also have their reflection in algebraic relations, given that some algebras satisfy the identity of idempotency—that is, their elements, when combined with themselves, do not change (e.g., $a + a = a$); and other algebras entail violations of this identity. In the algebra of real numbers, for example, idempotency is not satisfied ($1 + 1 \neq 1$). But in the case of set theory, for example, idempotency is satisfied (e.g., $A \cup A = A$); and likewise in algebraic topology (e.g., $1_A : A \rightarrow A$, the identity morphism in category theory defined earlier). It is also satisfied in logical conjunction and disjunction ($A \wedge A = A$, and $A \vee A = A$). Indeed, when viewed as an integration of these various contexts, the relevance of the truism and paradox of idempotency to the predication of totalities and Russell's Paradox becomes readily apparent; and in the same way, the relevance of the internal relational type-theoretic solution proposed in *Principia Mathematica* to the internal relational metaphysical scheme proposed in *Process and Reality* becomes quite clear: An actual occasion A is a totality in that it is internally related to the totality of dative occasions constitutive of the world—but this world-as-totality is properly understood topologically as a discrete region bijectively related to occasion A. And each of A's dative occasions is likewise internally related to *its* totality of data through its own bijective relations to the world (topologically a different discrete region), though this totality is one order lower in the type-theoretic sense (such that this region is representable as an open cover of A's region). Thus an actual occasion, as a totality, can be related to itself as a datum without paradox, because it is an asymmetrical internal relation of higher to lower order.[49]

In light of the discussion so far, one can see in Whitehead's later logical and philosophical emphasis of asymmetrical relations an evolution of thought that had important roots in his earlier work on universal algebra. This philosophical emphasis can, therefore, arguably find an appropriate formalized expression via category theory, which was developed as a means of expressing algebraic topological relations. As we have seen, in category theory, one expresses relations among objects, but also relations among relations. One can, for example, define the category **C** as consisting of [1] objects A, B, C . . . and [2] morphisms $f : A \rightarrow B$, and $g : A \rightarrow C$ . . . In the same way that *morphisms* f and g represent relations among *objects* A, B, C . . . , *functors* F and G represent structure-preserving relations among *categories*. If one considers categories **C** and **D**, for example, the functor $F : C \rightarrow D$ entails:

[1] that every object A, B, C . . . in **C** maps to an object $F(A)$, $F(B)$, $F(C)$ . . . in **D**, and

[2] every morphism $f : A \rightarrow B \ldots$ in **C** maps to a morphism $F(f) : F(A) \rightarrow F(B)$ in **D**.[50]

And finally, a natural transformation η is a structure-preserving relation whose relata are functors. In the present example, functors F and G are related (again, such that structural composition of their constitutive morphisms is preserved) via the natural transformation η_A such that:

[1] every object $A, B, C \ldots$ in category **C** is related to a morphism $\eta_A : F(A) \rightarrow G(A)$ in category **D**; and

[2] for every morphism $f : A \rightarrow B \ldots$ in **C**, it is the case that $\eta_B \circ F(f) = G(f) \circ \eta_A$.[51]

As this formalism applies to actual occasions as predicative facts in Whiteheadian metaphysics, recall that predication is 'dipolar' because each potential quantum actual occasion / predicative fact comprises both [1] *a causal-physical pole*, by which the potential novel fact is *spatio-temporally coordinated* with, and efficient-causally influenced by, the actual facts of its *relativistically restricted* dative world ('coordinate division'); and [2] a *logical-conceptual pole*, by which the potential novel fact and its unique subjective context are *logically coordinated* via asymmetrical internal relation with the *totality* of actual facts constitutive of its dative world—including these facts' *own* subjectively contextualized internal relations to *their* dative worlds ('genetic division'). Type-theoretically, as we have seen, these dative worlds are properly understood as lower order totalities relative to the subject occasion. But category-theoretically, it must be recognized that these internal relations, whether analyzed coordinately or genetically, cannot be formalized as 'intrinsic elements' of the actual occasion as some 'terminal' object relatum. Unlike set theory, which *can* provide such terminal internal formalization of an object (e.g., as a Cartesian product, or other 'internal' pair structure), category theory can *only* define objects *via their relations to other objects and other relations*. Formally, this is precisely representative of the philosophical notion that an actual occasion as 'object relatum' is, itself, an integration of internal relations with other dative occasions, which are themselves integrations of internal relations, such that the structure of these dative internal relations, no matter how far down, is preserved throughout.

In other words, the classical and set-theoretically definable conception of an object relatum as possessing some 'objectively and sheerly intrinsic constitution' is as incompatible with the relational realist definition of the dipolar, subjective-objective (i.e., relational) actual occasion as it is with the category-theoretic definition of an object relatum. In the category-theoretic formalism, even the most primitive objects cannot be defined this way; rather, object relata are always fundamentally understood as relations themselves. And as will be introduced in chapter 5 and presented in detail in part II, in quantum mechanics, the category theoretic 'internal' description of a quantum object relatum as a structure of internal relations is formally representable as a Grothendieck topological, sheaf-theoretic structure.

3.5 Summary

Unlike the mutually exclusive dualism of the Cartesian metaphysical scheme, where substance is either a unit of thought or physical extension, in the Whiteheadian, relational realist, dipolar monistic scheme the causal-physical and logical-conceptual poles are mutually implicative within the contextual unity of the actual occasion as a quantum whole; that is, each pole is incapable of definition without reference to the other.

As discussed in chapter 2, for Whitehead, the fundamental unit of substance—the essence of actual existence—is the predicative, dipolar quantum actual occasion, and the essence of 'being' in his ontology is 'becoming'; more specifically, the essence of any fact is its generation ('concrescence') as a predicative fact by its internal relation to the entire dative world comprising a total class one order lower in the logical hierarchy. Thus the scheme of internal relatedness by which a 'becoming' occasion is related to its dative world is asymmetrical, with the higher order always internally related to the lower orders, and the lower orders always externally related to the higher order. This logical asymmetry of internal relations is reflected temporally as the one-way arrow of time in thermodynamics and in the decoherence-based interpretations of quantum mechanics. More generally, it is reflected in the one-way arrow of efficient causality wherein the order of causal relation is always preserved temporally, such that effect can never precede cause—even in the case of relativistic physics. In Whiteheadian metaphysics, then, the past is closed—its actual facts settled—and the future is open, with all predication conditioned, but not determined, by the antecedently established facts of the global history.

Prior to concrescence, the becoming predicative fact / quantum measurement-in-process is only propositional—a potential actuality. But in the Aristotelian sense, and as rehabilitated by Heisenberg[52] in the ontological interpretation of quantum mechanics, both potential facts and actual facts are treated as species of reality. When actualized, the novel quantum fact likewise has its physical and logical causal efficacy in temporally and logically subsequent predication—again, by serving as logical subject whose predication yields a novel global totality of the next higher order.

The becoming of every new quantum actual occasion / predicative fact, then, is most fundamentally a local unification of the global dative totality *relative to that novel, locally contextualized predicative fact-in-process*. For example, $|\Psi\rangle = \alpha\,|\psi\uparrow\rangle\,|\phi\uparrow\rangle\,|e\uparrow\rangle + \beta\,|\psi\downarrow\rangle\,|\phi\downarrow\rangle\,|e\downarrow\rangle$ always evolves relative to an indexical actualization of $|\phi\rangle$, the detector, with its contextual preferred basis; and with the actualization of the novel predicative fact comes the actualization of a novel global totality. With every new fact, then, a new totality—not in the needlessly exotic sense of a 'multiverse' of disconnected alternative universes, but rather in the intuitive sense of a universal history augmented by a novel fact. Writ large, the universe as 'enduring structure' is fundamentally described as a serial historical evolution of totality, discretely punctuated with each new predicative fact, with each fact bijectively related to its discrete level in the logical and mereotopological hierarchy. Revisiting and expanding the earlier quote from Whitehead, "the many become one, and are increased by one. In their natures, entities are disjunctively 'many' in process of passage into conjunctive unity."[53]

Notes

1. This is a necessary presupposition for any ontological interpretation of quantum mechanics, although these might disagree on the definition of 'measurement' productive of eigenvalues (and therefore eigenstates).

2. Simon Kochen and Ernst Specker, "The Problem of Hidden Variables in Quantum Mechanics," *Journal of Mathematics and Mechanics* 17 (1967): 59-87.

3. Unless the system is highly idealized, where observables have only two mutually exclusive possible measurement values (i.e., representable via a Hilbert space of 2 dimensions—cf. the discussion in chapter 2.3) In this case the 0 and 1 eigenvalues can be exhaustively mapped onto these values. The Kochen-Specker theorem pertains to less idealized systems, represented by a Hilbert space of at least 3 dimensions.

4. Werner Heisenberg, *Physics and Philosophy* (New York: Harper Torchbooks, 1958), 53.

5. Heisenberg, *Physics and Philosophy*, 54-55.

6. See, for example, S. Bertaina et al., "Quantum Oscillations in a Molecular Magnet," *Nature* 453 (May, 2008): 203-6.

7. Jeffrey Bub, "Quantum Logic, Conditional Probability, and Interference," *Philosophy of Science* 49 (1982): 402-21. See also: Jeffrey Bub, "The problem of properties in quantum mechanics," *Topoi* 10, no. 1 (1991): 27-34.

8. Bub, "Quantum Logic," 402-21.

9. See, for example: A. Aspect, J. Dalibard, and G. Roger, "Experimental Test of Bell's Inequalities Using Time-Varying Analyzers," *Physical Review Letters* 49, no. 25 (1982): 1804-7.

10. The spontaneous localization approach of Ghirardi, Rimini, and Weber is one example: See G. C. Ghirardi, A. Rimini, and Weber, T. "Unified dynam-

ics for microscopic and macroscopic systems", *Physical Review D* 34 (1986): 470-91.

11. As discussed in chapter 2, exclusive disjunction (XOR) is the appropriate Boolean operator, rather than OR, because PNC is satisfied in addition to PEM.

12. This will be explored further presently, and revisited again in greater detail in chapter 4.2.

13. See, for example: R. T. Cox, "Probability, Frequency and Reasonable Expectation," *American Journal of Physics* 14 (1946): 1-13.

14. G. Birkhoff G. and J. von Neumann, "The logic of quantum mechanics," *Annals of Mathematics* 37 (1936): 823-43.

15. See, for example: Wojciech Zurek, "Decoherence, Einselection, and the Quantum Origins of the Classical," *Reviews of Modern Physics* 75, no. 3 (2003): 715-75.

16. Bertaina, "Quantum Oscillations," 203-6.

17. See, for example: A. Caldeira and A. Leggett, "Quantum Tunneling in a Dissipative System," *Annals of Physics* 149 (1983): 374-456.

18. P. C. E. Stamp, "The Decoherence Puzzle," *Studies in History and Philosophy of Modern Physics* 37 (2006): 485; S. Takahashi, et al., "Decoherence in Crystals of Quantum Molecular Magnets," *Nature* 476 (July, 2011): 76-79.

19. John von Neumann, *Mathematical Foundations of Quantum Mechanics* (Princeton: Princeton University Press, 1996).

20. Niels Bohr, "Discussion with Einstein," in *Albert Einstein: Philosopher-Scientist,* ed. Paul Arthur Schilpp (New York: Harper, 1959), 210.

21. See, for example: Karl Popper, *Quantum Theory and the Schism in Physics: From the Postscript to the Logic of Scientific Discovery* (London: Routledge, 1992).

22. Niels Bohr, "Can quantum-mechanical description of physical reality be considered complete?" *Physical Review* 48 (1935): 696-702.

23. Jeffrey Bub, "Quantum Logic, Conditional Probability, and Interference," *Philosophy of Science* 49 (1982): 402-21. See also: Jeffrey Bub, "The problem of properties in quantum mechanics," *Topoi* 10, no. 1 (1991): 27-34.

24. Alfred North Whitehead, *Process and Reality: An Essay in Cosmology*, Corrected Edition, ed. D. Griffin and D. Sherburne (New York: Free Press, 1978), 21.

25. Bertrand Russell, "Mathematical Logic as based on the Theory of Types," *American Journal of Mathematics* 30 (1908): 224.

26. Russell, "Mathematical Logic," 223.

27. Russell, "Mathematical Logic," 224 (italics added).

28. Russell, "Mathematical Logic," 239.

29. Alfred North Whitehead and Bertrand Russell, *Principia Mathematica to *56* (London: Cambridge at the University Press, 1913), 48.

30. Whitehead and Russell, *Principia Mathematica*, 53.

31. Besides Russell's Logical Theory of Types, there have been a number of set theories constructed to the same purpose. Zermelo-Fraenkel (ZFC) axiomatic set theory is one example, and while it makes no mention of 'logical types,' it contains similar concepts, and some have argued that these constitute an implicitly presupposed type theory similar to Russell's. The 'von Neumann universe' or hierarchy of sets is an example of such a concept in ZFC.

32. This is referred to as 'coordinate division' in Whitehead's philosophical scheme.

33. This is referred to as 'genetic division' in Whitehead's philosophical scheme.

34. Alfred North Whitehead, *A Treatise on Universal Algebra, With Applications* (New York: Cambridge University Press, 1898), vii.

35. Alfred North Whitehead, "The Philosophy of Mathematics." Review of *Mysticism in Modern Mathematics* by Hastings Berkeley. *Science: Progress in the Twentieth Century* 5 (October 1910): 234-39. In his review, Whitehead writes, "I think that the formalist position adopted in [A *Treatise on Universal Algebra*], whilst it has the merit of recognizing an important problem, does not give the true solution."

36. Whitehead, *Process and Reality*, 3.

37. Whitehead, *A Treatise on Universal Algebra*, vi.

38. Whitehead, *A Treatise on Universal Algebra*, vi-vii.

39. Alfred North Whitehead, "Mathematics and the Good" in *The Philosophy of Alfred North Whitehead,* ed. Paul Arthur Schilpp (New York: Tudor, 1951), 671-72.

40. Whitehead, *A Treatise on Universal Algebra*, vii.

41. Whitehead, *A Treatise on Universal Algebra*, vi.

42. Whitehead, *A Treatise on Universal Algebra*, vii.

43. Alfred North Whitehead, "On Mathematical Concepts of the Material World" in *Alfred North Whitehead: An Anthology*, ed. F. S. C. Northrup and M. W. Gross (New York: MacMillan, 1953), 29. Originally published in *Philosophical Transactions of the Royal Society of London, Series A* 205 (1906): 465-525.

44. Whitehead, *Process and Reality*, 66.

45. Granville Henry and Robert Valenza, "Idempotency in Whitehead's Universal Algebra," *Philosophia Mathematica* 1, no. 2 (1993): 171.

46. See, for example: chapters 2.3, 2.4, 2.10, and 3.1.

47. Henry and Valenza, "Idempotency," 168.

48. Whitehead, *A Treatise on Universal Algebra*, 6-7.

49. Topologically, as we will see in chapter 5, this asymmetrical scheme of inclusive, internal relations can be intuitively represented as a nested structure of open covers.

50. Contravariant functors are also possible, where the arrows are reversed.

51. Saunders Mac Lane, *Categories for the Working Mathematician*, Graduate Texts in Mathematics 5 (New York: Springer-Verlag, 1998), 7-16.

52. For Heisenberg, potentia are not merely epistemic, statistical approximations of an underlying veiled reality of predetermined facts; potentia are, rather, ontologically fundamental constituents of nature. They are *things* "standing in the middle between the idea of an event and the actual event, a strange kind of physical reality just in the middle between possibility and reality" (Heisenberg, *Physics and Philosophy*, 41). Elsewhere, Heisenberg writes that the correct interpretation of quantum mechanics requires that one consider the concept of "probability as a new kind of 'objective' physical reality. This probability concept is closely related to the concept of natural philosophy of the ancients such as Aristotle; it is, to a certain extent, a transformation of the old 'potentia' concept from a qualitative to a quantitative idea" (Werner Heisenberg, "The Development of the Interpretation of Quantum Theory," in *Niels Bohr and the Development of Physics*, ed. Wolfgang Pauli (New York: McGraw-Hill, 1955), 12.

53. Whitehead, *Process and Reality*, 21.

CHAPTER 4

Logical Causality in Quantum Mechanics: A Relational Realist Ontology

The process of the quantum mechanical actualization of a potential outcome state (i.e., the predication of an observable) is affected by both [1] the potential actuality's causal-physical relations with the dative actualities 'physically' antecedent to it (prior in time) in terms of metrical, spatiotemporal extensiveness—i.e., within its backward light cone, per the restrictions of relativity theory; and [2] its logical internal relations with those actualities 'logically' antecedent to it (prior in order), per the restrictions of PNC and Boolean material implication, among other logical restrictions. (As discussed in the previous chapter, these are, respectively, the Whiteheadian 'coordinate' and 'genetic' analyses of a quantum actual occasion.[1]) Unlike the causal-physical relations, the logical internal relations are not relativistically restricted, and include both local and nonlocal data. For example, with respect to quantum decoherence, these relations include the unmeasured degrees of freedom environmental to the measured system, which may or may not fall outside the observable's backward light cone. Both kinds of relation, physical and logical, are unified by mutual implication within each predicative fact / quantum actual occasion via its dipolar logico-physical structure, such that all predication is reflective of logical causality.

4.1 Internal Relation and Logical Implication in Quantum Mechanics

The internal relation of local potential predicative fact to global dative actuality, which is the basis of the asymmetrical nature of this scheme, is reflected in the asymmetrical orders of syntactic, Boolean material implication, and semantic, logical implication (i.e., entailment). And with respect to its application to the ontological interpretation of quantum mechanics, this metaphysical and logical asymmetry is crucial. As introduced in chapter 2.4, asymmetrical internal relation, given in the expression $p \Rightarrow q$ (read, '*p only if q*') can be understood as Boolean material implication $p \rightarrow q$ (read 'if *p* then *q*') that is always true semantically—i.e., as a matter of *fact*, not form.

Material Implication		
p	*q*	$\rightarrow$
T	T	T
T	F	F
F	T	T
F	F	T

Internal Relation		
p	*Q*	$\Rightarrow$
T	T	T
F	T	T
F	F	T

As applied to the discussion of Boolean local measurement contextuality, in the truth table above depicting asymmetrical internal relation, *Q* thus refers to the logical context by which *p* can be evaluated. This is denoted by using the capital *Q* in the case of internal relation, versus the lowercase *q* in the case of material implication.

It should be emphasized here that it is because material implication is purely syntactic, i.e., as a truth-functional, that it is representable as a truth table, by which the truth or falsity of the proposition derived simply as a matter of mapping *T* and *F* as values, whether or not these values are *evaluations* of actual facts. Asymmetrical internal relation, by contrast, is syntactic *and* semantic—that is, relating both form *and* fact—and thus its depiction as a truth table here is purely heuristic. It ought not, in other words, be taken to imply that internal relation is reducible to a simple truth-functional.

With respect to its reflection in the relational realist ontology, asymmetrical internal relations, where p necessarily implies Q, presuppose a global 'objectivity' of the implicate Q at the pre-theoretic level that is not presupposed in the material conditional; yet at the same time, it preserves an indeterminacy in the local implicans p (which, as a predicative function of the argument in Q, must be of the next order above Q). This is because the truth value of the potential predicative fact p is not *determined* by the global objective facts constitutive of Q; rather, the latter determine the *conditions* to which all predication must conform. For example, if p = 'Socrates is in Athens' and Q = 'Socrates is in Greece,' Socrates being in Greece does not determine his being in Athens, though it does condition the possibility of the latter. (Conversely, his being in Athens does determine his being in Greece—thus the asymmetry of internal relation via logical implication.) This is analogous to the case in quantum mechanics, such that in the spin ½ system,

$$|\Psi\rangle = \alpha\,|\psi\uparrow\rangle\,|\phi\uparrow\rangle\,|e\uparrow\rangle + \beta\,|\psi\downarrow\rangle\,|\phi\downarrow\rangle\,|e\downarrow\rangle$$

$|\Psi\rangle$ is a vector of unit length, and thus representative of the actual (though indeterminate) state of the composite global system and its facts prior to the measurement outcome. $|\Psi\rangle$, for example, thus represents the *implicate* in the statement $|\psi\uparrow\rangle\,|\phi\uparrow\rangle\,|e\uparrow\rangle \Rightarrow |\Psi\rangle$ (read, $|\psi\uparrow\rangle\,|\phi\uparrow\rangle\,|e\uparrow\rangle$ *only if* $|\Psi\rangle$). These facts subsumed by $|\Psi\rangle$ condition, but do not determine, the novel predicative outcome fact (in this case, either eigenstate $|\psi\uparrow\rangle\,|\phi\uparrow\rangle\,|e\uparrow\rangle$ or $|\psi\downarrow\rangle\,|\phi\downarrow\rangle\,|e\downarrow\rangle$) generated by measurement. This indeterminacy is reflected, for example, in the fact that each eigenstate (i.e., each *implicans* of the internal relation) is always valuated as a probability, via the complex coefficients α and β, respectively—$\alpha\,|\psi\uparrow\rangle\,|\phi\uparrow\rangle\,|e\uparrow\rangle$

and $\beta|\psi\downarrow\rangle\,|\phi\downarrow\rangle\,|e\downarrow\rangle$—with $|\alpha|^2 + |\beta|^2 = 1$. Again, valuation as a probability, which necessarily satisfies PEM, rather than as merely a potentiality, which does not necessarily satisfy PEM, is only possible because the evaluation is relative to a local Boolean context (e.g., that given by the chosen detector, whose pointer reads either exclusively 'up' or 'down'—that is, $\phi\uparrow \veebar \phi\downarrow$—represented in the above expression as $\alpha\,|\phi\uparrow\rangle + \beta\,|\phi\downarrow\rangle$). Further, this local context is itself only definable in quantum mechanics via reference to a global context $|\Psi\rangle$ (represented by Q in the above example).

In other words, though the novel predicative fact p is not determined by the argument Q, it is *contextually dependent* upon Q for its definition. In the same way that Athens cannot be defined without implicit or explicit reference to Greece, no local system can be defined without implicit or explicit reference to a larger system subsuming it. In quantum mechanics, the latter must be a closed system, and interpreted ontologically, the only closed system is the universe itself—i.e., the totality represented by $|\Psi\rangle$. And in $|\Psi\rangle = \alpha|\psi\uparrow\rangle\,|\phi\uparrow\rangle\,|e\uparrow\rangle + \beta|\psi\downarrow\rangle\,|\phi\downarrow\rangle\,|e\downarrow\rangle$, the probability valuated, predicative outcome states are defined as projections of $|\Psi\rangle$ upon each eigenstate. Thus in quantum mechanics, each locally contextualized predicative fact is, by internal relation, dependent upon the global totality for its definition.

Via this conception of logically conditioned, asymmetrical internal relation, then, the ontological interpretation of quantum mechanics described herein posits that every becoming actual occasion/propositional predicative fact is always internally related to a global, objective actual world—an actual system of facts with objective truth values. It is in this way that the objective facts of the actual world serve to condition the local possibilities internally relative to that actual world. Thus every eigenstate p predicative of a local observable with *potential* truth value T or F makes necessary reference to a global *actual* world Q, such that p is internally related to Q, where the implicans p is of the next logical order above that of the implicate Q.

As related to the discussion in chapter 3, Q is always locally Boolean contextualized in quantum mechanics, represented by an equivalence class of maximal Boolean subalgebras in the partial Boolean algebra of observables defining the global system. Recall how a local measurement M_2 on a subsystem S_2 of a global system $S_1 + S_2$ will condition the local measurement outcomes at S_1 via a revision of this equivalence class of Boolean subalgebras; it is only in this way that the facts of Q condition, via *restriction of the local by the global*, the possibilities for p. In the above example of Socrates' location, Q represents the selection of a Boolean context for p among all the potentia referent to the global totality (i.e., $|\Psi\rangle$). But in quantum mechanics, it is not the case that all possible local contexts can be related *to each other* in terms of a global, fully Boolean associative order of inclusion. In the above example, for any two local Boolean contexts $p \Rightarrow Q$ and $r \Rightarrow S$, there are three possible logical internal relations:

[1] $Q \Rightarrow S$ (deductive)

[2] $S \Rightarrow Q$ (deductive)

[3] $(p \Rightarrow Q) \wedge (r \Rightarrow S) \rightarrow \exists(Q \otimes S)$ (inductive)

The third, inductive type is an essential feature of the topological category-sheaf theoretic approach to quantum mechanics, and will be further explored presently, as well as more formally in part II of this volume. For now it is sufficient to note that it depicts a global totality-of-contexts as a *maximally* Boolean overlap among all local contexts, each of which is representable as a local nonmaximal Boolean subalgebra.

As was explored in chapters 2 and 3, the utility of this conceptual framework becomes especially apparent when applied toward the coherent interpretation of quantum phenomena such as EPR-type nonlocality. Against many popular interpretations of the latter, the relational realist interpretation depicts EPR nonlocality as a non-metrical, topologically formalized logical conditioning of potentia; this is in sharp contrast to other interpretations that depict EPR nonlocality as an efficient causal influence requiring a superluminal physical-dynamical mechanism, or as evidence of 'retro-causality' requiring the abandonment of temporal asymmetry and its presupposed correlation with logical asymmetry. In particular, recent experiments in quantum optics have been interpreted as evincing such 'retro-causality,' including the so-called 'delayed choice, double-slit quantum eraser' experiments mentioned in chapter 2.1. In the context of the current discussion of nonlocal logical causality, a brief exploration of quantum eraser experiments is worth exploring here.

As introduced in chapter 2.1, these experiments exploring quantum superposition and entanglement are based on the Young double-slit experiment of 1803, which was originally devised to explore the dual wave-particle nature of light. As we would characterize it today, the experimental arrangement essentially entails photons propagating through either of two parallel slits prior to impinging upon a detector. When either slit is closed, the photons are well localized at the detector, thus exhibiting their classically particulate character; and when both slits are open, the photons form an interference fringe at the detector, thus exhibiting their 'quantum superpositional' wave-like character.

A straightforward example of a double-slit quantum eraser experiment, based on the above, is that of Walborn et al.[2] Many theorists (as well as many science journalists[3]) have interpreted such experiments as 'erasing,' via a kind of 'retro-causality,'[4] individual actualized measurements, and with them the Slit 1 / Slit 2 path information contained in the outcome states recorded by these measurements. This path information is included in the state specification by 'labeling' each path in terms of a combination of linear and circular polarization of the photons particular to each path. Thus, in the same way that electron spin direction served as a Boolean measurement context in the EPR example, linear and circular polarization serve as Boolean contexts in the quantum eraser example. At emission, the photons are initially split into two entangled beams, a 'signal' beam *S* directed at the double-slits, and an 'idler' beam *P* whose linear polarization will be manipulated in order to 'erase' the path information of the signal beam. In front of each slit, a quarter wave plate 'labels' a photon's passage by circularly polarizing it either left or right without otherwise disturbing it as it heads for the detector.

At emission, the linear polarization of both beams is determined, and because they are correlated, direct measurement of the idler beam *P*'s linear polarization indirectly yields, via implicative internal relation, the signal beam *S*'s polarization. Thus, $S_y \Rightarrow P_x$ and $S_x \Rightarrow P_y$. Finally, because of the relationship between linear and circular polarization in this arrangement, it is also a matter of implicative internal relation that depending on the linear polarization of *S*, passage through the quarter wave plates labeling Slit 1 (SL1) and Slit 2 (SL2) will result in either left or right circular polarization of *S* after passage through the slits. Thus, S_{left} (SL1) $\wedge$ S_{right} (SL2) $\Rightarrow S_x$, and S_{right} (SL1) $\wedge$ S_{left} (SL2) $\Rightarrow S_y$. When this arrangement is in place, the signal photon paths are effectively labeled via logical entailment, and detection events are well-localized, with no interference fringe.

The 'eraser' procedure simply involves placing an additional linear polarizer in the idler beam path, oriented so that *both x* and *y* polarized photons will pass through, thus 'erasing' the linear polarization information for the idler beam, which also erases, by logical entailment, the linear polarization information for the signal beam. This, in turn, 'erases' the logically implicative circular polarization labeling by the quarter wave plates at Slit 1 and Slit 2, described above. In this way, erasing the linear polarization information from the idler beam erases the path labeling mechanism for the signal beam, and causes the characteristic interference fringe to appear at the detector.

Finally, it is purported to be significant that the path of the idler beam is longer than the path of the signal beam, such that if the linear polarizer is placed at the end of the idler beam, then for any particular photon pair, the idler polarizer's 'erasure' of path information will occur *after* its counterpart signal photon has already reached its detector.[5] As a result, the characteristic interference fringe reappears as it does with regular erasure, but in this case, *after* the path-labeled signal photon has already been detected. Again, this is often described as 'retro-causal' or 'delayed' erasure, implying that already actualized quantum facts (i.e., facts entailing path information) have somehow been 'erased' from reality.

At once exotic and esoteric, this interpretation has proven understandably tantalizing to both specialists and popular audiences alike. The reason is that the purported disconnection of the asymmetrical order of causal relation and the asymmetrical order of logical implication fatally undermines the foundational principle by which nature is coherently accessible to human reason in general, and to the scientific method in particular: the categorical correlation of these asymmetrical orders—i.e., the presupposition of logical causality. Thus, as was argued in chapter 2, any purported *scientific* invalidation of this presupposition amounts to nothing less than a scientific invalidation of the scientific method itself—which, of course, makes the application of the latter to such an endeavor paradoxical at best.

For this reason alone, one could argue that the only truly coherent scientific interpretations of these double slit, 'quantum eraser' experiments are those that make explicit their reliance on logical causality as a categorical presupposition, on the grounds that the method of science itself is ineluctably rooted in this same presupposition. The relational realist ontological interpretation is one such candidate, and its application to the quantum eraser experiment outlined above is fully consistent with its application to the EPR-type quantum nonlocality experiments previously described, and presented formally in chapter 9.4: The local Boolean measurement contexts of the experimental arrangement are identified, in this case as:

[1] $P_x \veebar P_y$

[2] $S_x \veebar S_y$

[3] $S_{left}(\mathrm{SL1}) \veebar S_{right}(\mathrm{SL1})$

[4] $S_{left}(\mathrm{SL2}) \veebar S_{right}(\mathrm{SL2})$

These measurement contexts, representable as Boolean subalgebras, are mereotopologically internally related such that in the logical order of evaluation (i.e., the order of detection events), every contextualized observable evaluated is internally related to those logically antecedent to it. (This scheme of mereotopological internal relation is formalized sheaf theoretically, as will be introduced in chapter 5, and presented systematically in part II.) Again, this is precisely analogous to the EPR example discussed in chapters 2 and 3, wherein a local measurement M_2 on a subsystem S_2 of a global system $S_1 + S_2$ will condition the local measurement outcomes at S_1 via a revision of the equivalence class of Boolean subalgebras by which *all* operative local measurement contexts are coherently integrated. In the present example, there are four such contexts (listed above) and their coherent integration is evinced by the fact that in all quantum eraser experiments of this kind, when one analyzes the *total* pattern of detected signal photons, it is always a well-localized ensemble, with *no* interference fringe, regardless of any particular manipulations of the idler path. It is only when one retrodictively analyzes the specific subset of signal photons that are correlated with the subset of idler photons whose linear polarization is undefined (i.e., uncontextualized) that an interference fringe is detected *for this subset*. This is simply because for this particular subset, local Boolean contexts [1] and [2] above are undefined, thus altering the equivalence class of Boolean subalgebras representing the totality of local measurement contexts for the signal and idler photons. But this in no way 'erases' the *fact* of each signal photon's individual detection, each of which is always well-localized and free of 'interference.' Rather, these experiments simply demonstrate how the alteration of local Boolean measurement contexts in an ensemble of measurements produces an alteration of the equivalence classes by which these local contexts are logically integrated, and thus an alteration in the probability conditionalization of the correlate measurement outcomes. More important, these experiments demonstrate the presupposition that local measurement contexts are *always* Boolean, and that they are always coherently integrable.

4.2 The Compatibility Condition Revisited

As introduced in chapter 2.5, asymmetrical internal relation among local quantum measurement contexts, combined with the presupposition that individual local measurement contexts are always structurally Boolean, together constitute the *compatibility condition for logical causality* in quantum mechanics. As discussed previously, this is exemplified in physics most generally as the universal, categorical correspondence of [a] the asymmetrical order of material implication and logical consequence (for the purposes of the present discussion, these together can be referred to simply as 'logical implication') with [b] the asymmetrical order of causal relation. Recall that the compatibility condition is built upon two foundational concepts:

[1] *locally*, every measurement context must be Boolean, such that in the mixed state,[6] Boolean material implication holds—e.g., for any measurement context A it will always be the case that for potential outcomes a_1 and a_2, $a_1 \rightarrow \neg a_2$ ('if a_1, then not a_2') and $a_2 \rightarrow \neg a_1$. This, of course, is just PNC for an observable a with only two possible eigenstates (i.e., potential outcome states) a_1 and a_2. This number, however, is potentially infinite in quantum mechanics; again, because the contextual measurement basis is orthonormal (where n mutually exclusive eigenstates—e.g., representing n possible pointer positions on a particular detector—are depicted in the formalism as n mutually orthogonal vectors in a Hilbert space of n dimensions) one can represent the Boolean complement $\neg a_1$ by simply grouping all of the alternative mutually exclusive eigenstates into a subspace $S^{\perp}$ of A, such that $a_1 \rightarrow \neg a_1^{\perp}$. As noted earlier, this presupposition of local Boolean contextuality, yielding mutually exclusive outcome states *regardless of the number of possible outcome states*, is a necessary prerequisite for the probability valuation of these alternative outcome states (the Born rule), which is a categorical presupposition of quantum mechanics.

[2] *globally* (i.e., when local contexts are brought into nonlocal relation), intra-contextual Boolean material implication (that is, *within* individual local measurement contexts) must be relatable inter-contextually

across these local contexts (i.e., 'globally'). In quantum mechanics, as discussed previously, this is expressed as a tensor product relationship of potential outcome states. But the tensor product by itself in the conventional Hilbert space formalism does not explicitly define the underlying Boolean logical structure of the local-global correlations it presupposes. Specifically, it does not define the manner in which a global totality of quantum events, which *cannot* be represented by a global Boolean algebra, can nevertheless be coherently correlated with local Boolean contextualizations of this totality. Recall that globally, the lattice of relations among the totality of quantum events is non-Boolean, evinced by both the non-commutativity of quantum observables, and because PNC and PEM cannot be shown to hold globally in quantum mechanics. That is, one can never, even in principle, evaluate the totality of quantum observables as a comprehensive scheme of mutually exclusive and exhaustive True / False propositions.[7]

However, this *can* be done within local Boolean contextualizations of this global structure; that is, local Boolean sectors of the non-Boolean global lattice can be defined. In this sense, the global quantum event structure, even though it cannot be fully embedded within a global Boolean algebra, can be represented via a partial Boolean algebra—so long as one categorically presupposes that all local measurement contexts are structurally Boolean (i.e., representable as a Boolean subalgebra, or as an equivalence class of such subalgebras).

This presupposition is the first component of the compatibility condition for logical causality in quantum mechanics, from which one can then proceed to define a structure by which Boolean local-global relations can be specified. For even though the totality of facts contained in the global quantum lattice can never be defined completely via deductive analysis, they can be defined approximately via induction from the overlaps of a sufficiently large number of equivalence classes of compatible, or partially compatible, local Boolean subalgebras. It is via this structure that global quantum events can be shown to logically condition local potential measurement outcomes—i.e. *restriction of the local by the global* (see chapter 2.4); likewise, it is via this structure that locally contextualized quantum events can be shown to condition global potentia—i.e., *extension of the local to the global.* As we have already seen, nonlocal probability conditionalization in EPR-type experiments well exemplifies both restriction and extension in this way.

Formally, the structure of these overlaps is not fundamentally metrically extensive, but rather mereotopologically extensive. It is, in other words, a *topological scheme of asymmetrical internal relations*. This characterization is the second component of the compatibility condition for logical causality. The formalization of this mereotopological scheme requires the mathematics of algebraic topology rather than metrical geometry because it entails the representation of a global partially Boolean algebraic structure via a nested covering structure of local Boolean subalgebras. As will be introduced in chapter 5, and fully formalized in part II, this structure is rigorously definable via sheaf theory,[8] by which global algebraic relations can be defined inductively via a topological, internal relational covering scheme of local algebraic relations.

Thus, continuing with the previous example, in a composite quantum system, for local Boolean measurement contexts *A* and *B* (and their associated detectors), if potential outcome a_i is internally related to Boolean context *A*, and potential outcome b_i is internally related to Boolean context *B*, as given in the expression: $(a_i \Rightarrow A) \wedge (b_i \Rightarrow B)$, then the state of the composite quantum system, as we saw in chapter 2.5, is expressed as:

$$\Psi = \frac{1}{\sqrt{2}} (a_1 \otimes b_1 + a_2 \otimes b_2)$$

This conventional tensor product expression presupposes an implicit topological structure by which the local Boolean subalgebras representing *A* and *B* overlap, such that their respective local, intra-contextual Boolean structures are globally relatable inter-contextually. The result is a restricted set of possible correlations:

$$P(a_1 \cap b_1) = 0.5$$
$$P(a_2 \cap b_2) = 0.5$$

$$P(a_1 \cap b_2) = 0$$
$$P(a_2 \cap b_1) = 0$$

In quantum mechanics, however, this topological structure of overlapping local Boolean subalgebras is always asymmetrically internal relational—i.e., it is a mereotopological structure, such that the overlapping regions are always 'nested' in a structure preserving way. This is evinced by the fact that once a measurement outcome has been registered by one of the detectors—say *A*—the integration of potential outcome states at *B* is revised via its internal relation to the outcome at *A*, and this revision is manifest as a probability conditionalization. Thus, it is via the asymmetrical internal relation of *B* to *A* that the asymmetrical structure of material implication within *A* and *B* individually can be correlated inter-contextually via the tensor product relationship.

With respect to the process of decoherence introduced in chapter 3.1, the significance of the compatibility condition becomes particularly acute, and can be more fully described in light of that discussion. For it is via the process of decoherence that *probable* measurement outcomes evolve from *potential* measurement outcomes once equivalence classes of the latter are coarse-grained, via negative selection, into a matrix of mutually exclusive and exhaustive outcome states. In other words, it is via decoherence that classical descriptions of nature in terms of classically meaningful (i.e., mutually exclusive and exhaustive) probable outcome states that exclude violations of PNC, emerge from quantum mechanical descriptions of nature in terms of potential outcome states that initially include violations of PNC.

However, while decoherence is today conventionally acknowledged as a canonical feature of quantum measurement, there as yet exists no conventional understanding of the process of decoherence itself—that is, whether it is primarily a dynamical physical mechanism, or primarily a logical conceptual mechanism. Correspondingly, most of the formal decoherence models proposed over the past decades have been inspired by two basic interpretive approaches: [1] environmental or dynamical decoherence, emphasizing the physical interaction between the observables of a measured system and those of its environment; [2] decoherent histories, emphasizing the logical integration of all potential observable evaluations relative to a given measurement, regardless of their classification as 'system observables,' 'environment observables,' or 'detector observables,' since these divisions are purely arbitrary in quantum mechanics.

The environmental / dynamical decoherence approach,[9] introduced in chapters 2 and 3, is founded upon the principle that every physical system is always in a state of dynamical relation with its environment, such that the observables constitutive of 'measured system,' 'detector,' and 'environment' are in a state of constant physical interaction. The complex structure of the correlate quantum entanglement, while exceeding the possibility of conventional calculation, does allow for an averaging over the immeasurable degrees of freedom subsumed by the entangled system-detector-environment state. This is represented via a trace-over of the off-diagonal terms of the density matrix, eliminating interference terms and yielding the usual mixed state of mutually exclusive and exhaustive probable measurement outcomes. Thus, according to the environmental decoherence approach, it is because of every measured system's constant dynamical relations with its environment that superpositions of counterfactual potential system states are never observed.

One difficulty of this approach, already discussed, is that it typically defines system-environment relations in terms of dissipative energy transfer via an effective Hamiltonian—i.e., it assimilates the logical features of these relations entirely to the physical causal features. The implication is that if systems are sufficiently isolated from their environments, decoherence can be delayed, or perhaps even prevented altogether. Recent experiments[10] have shown, however, that even when system isolation is sufficient to negate any measurable energy transfer between system and environment, decoherence nevertheless persists. Thus, a number of possible environmental sources of *non-dissipative* decoherence have now been proposed—including the notion that these sources are, in principle, unidentifiable because they are 'intrinsic' features of nature.[11]

The other central difficulty, also raised earlier, is that the decomposition of the universe into 'system,' 'detector,' and 'environment,' necessarily presupposed by the environmental decoherence approach, is itself conceptually problematic.[12] On the one hand, quantum theory depicts this decomposition as purely arbitrary and therefore only instrumentally and epistemically significant. However, the process of environmental decoherence is one wherein this subjective, epistemic distinction has objective, ontological consequences; for different decompositions yield very different objective measurement outcomes.

The decoherent histories approach,[13] by contrast, was devised to obviate the global decomposition problem by envisioning alternative potential measurement outcomes as alternative potential histories of the universe itself. Thus, rather than depicting a *physical decomposition* of a global physical-extensive structure into physical-extensive substructures arbitrarily labeled 'system,' 'detector,' and 'environment,' the decoherent histories approach depicts a *logical decomposition* of a global history into alternative potential sub-histories that can be integrated coherently and consistently. In this way, quantum mechanical histories are no different than histories in the conventional sense; that is, despite the existence of fine-grained incompatibilities among potential local sub-histories constitutive of a shared global history, these can always be coarse-grained to the point at which they are mutually consistent and coherently integrable. Robert Griffiths, for example, defines the 'consistency condition' for decoherent histories as the requirement that the probability valuation for each alternative coarse-grained history equal the sum of the probabilities of its constitutive fine-grained histories.[14] Satisfaction of the consistency condition is initially impeded by quantum interference among fine-grained histories, but appropriate coarse-graining effects a logical negative selection process (see chapter 3.1) whereby mutually interfering fine-grained histories are summed over via the trace function described earlier. Note that the presupposition of consistent and coherent integration is justified by the presupposition of a 'global history'—i.e., some 'ultimately' closed system (the universe itself being the only such system), which is a necessary presupposition of quantum mechanics.

This key advantage of the decoherent histories approach to quantum mechanics, however, is unfortunately also its key disadvantage; for as is the case with conventional histories of events, there are many possible coarse-grainings by which maximally fine-grained histories of quantum events can be integrated coherently and consistently. Thus, in a quantum measurement interaction, when mutually exclusive and exhaustive probability outcomes are defined in terms of mutually exclusive and exhaustive alternative coarse-grained histories, there is no criterion by which certain consistent coarse-grainings are preferred over others that might be equally consistent. Even more problematic, one can easily define alternative coarse-grained histories that are consistent (i.e., mutually exclusive and exhaustive, satisfying PNC and PEM) yet nevertheless depict *non-classical sequences of events*. These consistent yet "grotesque" histories, to use Griffiths' term for them,[15] are alien to classical experience. Thus, the probability valuation of potential outcome states guaranteed by the consistency condition is, by itself, a necessary but insufficient criterion for classically meaningful outcome states.

In summary of the discussion thus far, the decoherent histories approach was intended to depict the evolution of classical physical relations from quantum mechanical relations in a manner that avoids the decomposition problem inherent in the environmental decoherence approach—i.e., a purely subjective partitioning of system, detector, and environment that nevertheless has objective, measurable consequences. This problem is avoided by denying altogether the quantum mechanical significance of any physical distinction between system and environment, instead reducing their relation to a purely mathematical integration in the form of consistent global histories. But this sheer mathematization of relations, conditioned by the consistency condition, though it does account for the classicality of probabilities in quantum mechanics (satisfaction of the sum rules, distributive law, PNC, PEM, etc.), does not account for the classicality of the actual measurement outcomes generated by quantum mechanics. Likewise, the environmental decoherence approach attempts a similar reduction but in the opposite direction—a reduction of the logical features of system-environment relations to purely physical-dynamical features.

As will be shown presently, the central deficiencies of both the decoherent histories and environmental decoherence approaches stem from their shared portrayal of the physical-causal and mathematical-logical aspects of decoherence as mutually exclusive, with their proper mode of relation being the sheer assimilation of one aspect to the other. By contrast, the relational realist approach, in emphasizing the mutually implicative relationship between the logical and physical aspects of quantum decoherence, is reflective of both the environmental decoherence and decoherent histories approaches. For example, in emphasizing the ontological significance of local measurement contextuality, the relational realist approach does not dismiss the distinction of local system and global environment as purely epistemic. But at the same time, by emphasizing that locally contextualized measurement interactions are always internally related to a global totality of dative facts, the relational realist approach likewise does not sheerly assimilate the global logical aspects of these relations to local efficient causal aspects—i.e., such that it is only via energy exchange that these relations are understood to be physically relevant.

Both of these emphases are together embodied by the compatibility condition for logical causality proposed herein (and formalized mathematically in chapters 9 and 10), which can be viewed as a supplement to the consistency condition. As discussed in chapter 3.1, in order to properly formalize system-environment relations such that the inadequacy of the consistency conditions is remedied, one must first address the fact that there is a structural difference between [a] the Boolean subalgebras by which quantum observables are *locally contextualized* and [b] the partial Boolean algebras by which the latter are *globally related.*

In other words, there is a difference between the algebraic structure of quantum observables and the algebraic structure of macroscopic observables at coarse-grained scales, and this difference is ignored in the consistent histories approach. But by augmenting the Griffiths consistency conditions with the compatibility condition introduced herein, one is able to sustain the concept of classical local-global relations as *emergent from* quantum local-global relations. Again, as will be further elaborated in part II, this is achieved via a sheaf-theoretic mereotopological description of the local-global relation of Boolean algebras. In this regard, the compatibility condition essentially depicts a transition morphism from one local Boolean context to another, generating asymmetrical logical and mereotopological revisions[16] of equivalence classes of local contextual Boolean algebras.

Topologically, these revisions yield partial compatibility on their overlapping regions (e.g., partial Boolean compatibility of coarse-grained position and momentum observables, up to the limit of Heisenberg uncertainty relations). This 'inductive limit,'[17] by which local compatible families of Boolean algebras can be extended globally, is constructed via the formation of a set of equivalence classes of partially compatible Boolean subalgebras, representing partially compatible Boolean contextualized observables, on *all* possible overlaps. The compatibility condition thus requires as a categorical presupposition that all local measurement contexts are structurally Boolean, and that the asymmetrical Boolean structure of any local context is preserved when extended globally, via internal relation, to other contexts.

The implementation of the compatibility condition in quantum measurement via this sheaf-theoretic topological method, the rudiments of which will be introduced in chapter 5 as a propaedeutic to part II, allows one to formally depict, mereotopologically, this structure-preserving extension of local Boolean contextuality globally. It is via this mereotopological depiction of decoherence that one can characterize classical observables as emergent from quantum observables in a manner that is ontologically consistent and coherent, and which requires no ad hoc additions to (or dispensations from) the standard quantum theory; it merely makes explicit use of the same categorical logical presuppositions already present implicitly in the standard formalism, employing them in a novel, sheaf theoretic description of quantum measurement.

4.3 The Evolution of Potentiality to Probability

In the relational realist interpretation of quantum mechanics described thus far, predicative facts / measurement outcome events are fundamentally quantum units of logico-physical internal relation. In the process of actualization—e.g., quantum mechanical state evolution—the novel predicative fact / measurement outcome in process (or in Whiteheadian terms, the 'actual occasion' undergoing 'concrescence') is internally related to its dative world of actualized facts. With respect to the earlier discussion of the expression:

$$|\Psi\rangle = \alpha\,|\psi\uparrow\rangle\,|\phi\uparrow\rangle\,|e\uparrow\rangle + \beta\,|\psi\downarrow\rangle\,|\phi\downarrow\rangle\,|e\downarrow\rangle$$

by 'internal relation,' it is meant that each datum constitutive of the global totality $|\Psi\rangle$ is also *internally constitutive* of the quantum process by which the novel predicative fact $|\psi\uparrow\rangle\,|\phi\uparrow\rangle\,|e\uparrow\rangle$ or $|\psi\downarrow\rangle\,|\phi\downarrow\rangle\,|e\downarrow\rangle$ is actualized. This process is fundamentally a logically conditioned integration of potential relations between [a] the local *propositional* measurement outcome / predicative fact, i.e., $\alpha\,|\psi\uparrow\rangle\,|\phi\uparrow\rangle\,|e\uparrow\rangle$ or $\beta\,|\psi\downarrow\rangle\,|\phi\downarrow\rangle\,|e\downarrow\rangle$, internally related to [b] the global data constitutive of $|\Psi\rangle$. This integration takes the form of an evolution from potential relations to probable relations in that the latter satisfy both PNC and PEM. This evolution is reflected in the reduction of the pure state to the mixed state in quantum mechanics.

But as discussed in the previous section, an evolution of potential relations to probable relations is only possible if one presupposes a local Boolean context for each of the relata—i.e., both [a] the actual occasion-in-process (the propositional predicative fact/measurement outcome)—e.g., $\alpha\,|\psi\uparrow\rangle\,|\phi\uparrow\rangle\,|e\uparrow\rangle$ or $\beta\,|\psi\downarrow\rangle\,|\phi\downarrow\rangle\,|e\downarrow\rangle$, where the selection of 'local context' is determined by the particular measuring apparatus $|\phi\rangle$ and defined by its preferred orthonormal basis—in this case, $|\phi\uparrow\rangle + |\phi\downarrow\rangle$; and [b] the individual data constitutive of $|\Psi\rangle$. And while a local Boolean context is presupposed for each datum in $|\Psi\rangle$, quantum mechanics does not permit the construction of a 'global' Boolean context for $|\Psi\rangle$ as a concatenation of all local contexts.[18] This is well evinced in the quantum indeterminacy relations, according to which a complete state specification of $|\Psi\rangle$—i.e., a total order of its data—is not possible. Quantum mechanics can, however, yield an inductively limited *maximal* specification by defining the global state $|\Psi\rangle$ as an overlap of equivalence classes of all compatible local Boolean contexts / subalgebras. This is the inductive limit described at the end of the previous section and which will be described in more formal detail in part II. Most important, however, is that in quantum mechanics, the global data contextualized via this overlap can only be induced by reference to the local Boolean context (preferred orthonormal basis) of a particular measurement interaction. This local measurement context, though it is contained in the overlap, is 'indexical' in that it restricts the potential relations between [a] and [b] (above) to its preferred basis. Apart from this restriction, potential outcomes cannot evolve to become classically meaningful probable outcomes.

Decoherence, as discussed earlier, refers to the outcome of a process whereby a coherent pure state of alternative potential outcomes is distilled into the 'decoherent' set of mutually exclusive and exhaustive probable outcomes characteristic of the mixed state. Quantum mechanics always describes the evolution of pure-to-mixed state relative to a single, locally contextualized propositional measurement outcome / predicative fact—an 'indexical eventuality' locally Boolean contextualized by the orthonormal basis of the detector. This indexical eventuality is, in the Whiteheadian / relational realist scheme, the actual occasion in process. The correlation of the measured system $|\psi\rangle$ and the detector $|\phi\rangle$ with the unmeasured environment $|e\rangle$ and its manifold degrees of freedom presupposes a global logical structure by which the *potential* relations constitutive of the expression

$$|\Psi\rangle = \alpha_{ijk} \sum_{i,j,k} |\psi\rangle_i |\phi\rangle_j |e\rangle_k$$

can be integrated such that they evolve to become *probable* relations constitutive of the expression $|\Psi\rangle = \alpha|\psi_\uparrow\rangle\,|\phi_\uparrow\rangle\,|e_\uparrow\rangle + \beta|\psi_\downarrow\rangle\,|\phi_\downarrow\rangle\,|e_\downarrow\rangle$.

Again, this global logical structure is defined in quantum mechanics as a maximal overlap or equivalence class of all local Boolean contexts indexed to the local Boolean context (preferred orthonormal basis) of some particular indexical event / actual occasion-in-process belonging to $|\phi\rangle$. But beyond the integration of local contexts, decoherence pertains to the integration of the dative content of these contexts—or more precisely, the integration of the *relations among these data*, to which the indexical eventuality / actual occasion-in-process is internally related. As discussed earlier, the evolution of a coherent superposition of potential outcome states, most of which are mutually logically inconsistent in terms of PNC, to a decoherent probability distribution of logically consistent probable outcome states satisfying PNC and PEM is achieved by incorporating the unobserved facts of the measured system's environment—or alternatively, the manifold fine-grained histories integrating these environmental facts with those of the measured system—into the quantum measurement formalism. Making explicit these system-environment relations introduces manifold degrees of freedom into the superposition of potential outcome states, allowing for their integration into coarse grained equivalence classes, with each class indexed to a different potential outcome state (Whiteheadian 'subjective form') of the indexical event / actual occasion-in-process.

With the imposition of the Boolean compatibility condition discussed earlier, the magnitude of environmental degrees of freedom within each equivalence class produces a great deal of mathematical cancellation among potential system-detector-environment outcome states that interfere (i.e., that violate PNC) within each class. A traceover of the environmental degrees of freedom thus represents the elimination of these logically incoherent states represented by the off-diagonal terms in the density matrix. The elimination of these terms effects the reduction of the pure state, with its interfering superposition of *potential* measurement outcomes, to the mixed state and its logically conditioned, decoherent matrix of *probable* outcome states.

This crucial function of the environment in quantum mechanical measurement interactions has, for the most part, been ignored over the years, though its significance was alluded to even in the early days of the quantum theory. Recall the earlier quote from Heisenberg, who wrote that the superposition or interference of potentia, "which is the most characteristic phenomenon of quantum theory, is destroyed by the partly indefinable and irreversible interactions of the system with the measuring apparatus and the rest of the world."[19]

Since quantum mechanics can be applied only to closed systems, any ontological interpretation of quantum mechanics must recognize the universe as the only truly closed system, such that as noted earlier, the decomposition of $|\Psi\rangle$ into 'system' 'detector' and 'environment' subsystems is purely arbitrary. Thus, the evolution of the state of any particular measured system is, in fact, the evolution of the state of the universe itself; and its evolution relative to a given indexical event, Boolean-contextualized by a particular measuring apparatus, is merely its evolution relative to a particular fact belonging to itself. The paradoxes of self-reference discussed in chapter 3 are avoided, as we saw earlier, by acknowledging that every quantum mechanical predication is generative of a novel fact internally related to the global dative totality during the process of actualization, and *augmenting* that totality once actualized. The correlation among facts belonging to 'system,' 'detector,' and 'environment,' then, is easily comprehended by virtue of the fact that as the closed system of the universe evolves, so must all the relational structures subsumed by it, however they might be grouped and whatever they might be named.

Quantum theorist Wojciech Zurek has described decoherence as a consequence of the universe's role as the only truly closed system, which guarantees the ineluctable 'openness' of every subsystem within it, and likewise the relational structure integrating the totality of these open subsystems. "This consequence of openness is critical in the interpretation of quantum theory," Zurek continues, "but seems to have gone unnoticed for a long time."[20] By correlating system $|\psi\rangle$ and detector $|\phi\rangle$ with the environment $|e\rangle$ and its manifold degrees of freedom, one thus gives an explicit representation of *all* potential internal relations inherent in any particular measurement interaction.

For example, in the evolution of,

$$|\Psi\rangle = \alpha_{ijk} \sum_{i,j,k} |\psi\rangle_i |\phi\rangle_j |e\rangle_k$$

to

$$|\Psi\rangle = \alpha |\psi\uparrow\rangle |\phi\uparrow\rangle |e\uparrow\rangle + \beta |\psi\downarrow\rangle |\phi\downarrow\rangle |e\downarrow\rangle$$

those potential relations capable of logical integration, conditioned by the Boolean compatibility requirement discussed earlier, evolve to become decoherent, mutually exclusive and exhaustive, probable relations (Whiteheadian 'valuated subjective forms'):

$$\alpha |\psi\uparrow\rangle |\phi\uparrow\rangle |e\uparrow\rangle \quad \text{and} \quad \beta |\psi\downarrow\rangle |\phi\downarrow\rangle |e\downarrow\rangle.$$

These are represented in the density matrix as the following terms, respectively:

$$|\alpha|^2 |\psi\uparrow\rangle\langle\psi\uparrow||\phi\uparrow\rangle\langle\phi\uparrow|| e\uparrow\rangle\langle e\uparrow|$$

$$|\beta|^2 |\psi\downarrow\rangle\langle\psi\downarrow||\phi\downarrow\rangle\langle\phi\downarrow|| e\downarrow\rangle\langle e\downarrow|$$

Other potential relations are incapable of such integration without violating PNC. These are represented by the off-diagonal terms:

$$\alpha\beta^* |\psi\uparrow\rangle\langle\psi\downarrow||\phi\uparrow\rangle\langle\phi\downarrow|| e\uparrow\rangle\langle e\downarrow|$$

$$\alpha^*\beta |\psi\downarrow\rangle\langle\psi\uparrow||\phi\downarrow\rangle\langle\phi\uparrow|| e\downarrow\rangle\langle e\uparrow|$$

The potential relations represented by these terms are eliminated from the integration via a negative selection process, mathematically expressed as a trace-over of the unmeasured degrees of freedom supplied by the environment[21]:

$$\begin{aligned}\mathrm{Tr}_E\,|\Psi\rangle\langle\Psi| &= \Sigma_i\,\langle e_i|\Psi\rangle\langle\Psi|e_i\rangle \\ &= |\alpha|^2\,|\psi\uparrow\rangle\langle\psi\uparrow||\phi\uparrow\rangle\langle\phi\uparrow| + |\beta|^2\,|\psi\downarrow\rangle\langle\psi\downarrow||\phi\downarrow\rangle\langle\phi\downarrow|\end{aligned}$$

Again, by the categorical imposition of the compatibility condition, this process of negative selection rests upon a fundamental presupposition of a global logical structure by which nonlocal potential relations among system, detector, and environment can be integrated into probable relations. But this global logical structure can only be defined in terms of local Boolean contexts and their associated Boolean subalgebras, such that a properly formed internal relational structure of overlapping equivalence classes of these contexts/subalgebras can be employed to represent an inductive approximation of the global.

During the process of the actualization of a measurement outcome, the dative content of this approximation—that is, the potential relations constitutive of the expression:

$$|\Psi\rangle = \alpha_{ijk}\sum_{i,j,k}|\psi\rangle_i|\phi\rangle_j|e\rangle_k$$

is thus essentially coarse-grained by this Boolean localization scheme. Whitehead describes a very similar negative-selection mechanism in the process of concrescence—a mechanism he termed 'transmutation,' wherein potential relations incompatible for integration into a logically coherent and consistent history ('nexus') are eliminated via

> a massive average objectification of a nexus, while eliminating the detailed diversities of the various members of the nexus in question. This method, in fact, employs the device of blocking out unwelcome detail. It depends on the fundamental truth that objectification is abstraction. It utilizes this abstraction inherent in objectification so as to dismiss the thwarting elements of a nexus into negative prehensions.
>
> This mode of solution requires the intervention of mentality [i.e., activity of the logico-conceptual pole of concrescence—not conscious or anthropic mentality] operating in accordance with the Category of Trans-

> mutation. It ignores diversity of detail by overwhelming the nexus by means of some congenial uniformity which pervades it. The environment may then change indefinitely so far as concerns the ignored details—so long as they can be ignored.[22]

This 'massive average objectification' is possible logico-mathematically for the same reason it is possible conceptually in Whitehead's scheme; it is prefaced by an explicit integration of the *totality* of potential relations between [a] the propositional measurement outcome / predicative fact internally related to [b] the data constitutive of $|\Psi\rangle$. And this evolution of potential relations to probable relations is only possible if one presupposes a global logical order by which a Boolean localization scheme can be constructed. Whitehead writes:

> The irrelevant multiplicity of detail is eliminated, and emphasis is laid on the elements of systematic order in the actual world . . .[23] In this process, the negative prehensions which effect the elimination are not merely negligible . . . The actual cannot be reduced to mere matter of fact in divorce from the potential.[24]

> Unless *some systematic scheme of relatedness* characterizes the environment, there will be nothing left whereby to constitute vivid prehension of the world.[25] Transmutation is the way in which the actual world is felt as a community, and is so felt in virtue of its prevalent order.[26]

4.4 Quantum Events as Dipolar Units of Relation: Their Subjective and Objective Features Are Mutually Implicative

In the same way that the preferred orthonormal measurement basis / Boolean context of a detector is understood as constitutive of the detector, the local context of the Whiteheadian actual occasion is understood as constitutive of the actual occasion. In this way the indexical context / 'subjective standpoint' of the actual occasion-in-process is 'self-determined.' This seems an odd notion when applied to a classical conception of 'detector' since the latter is obviously chosen by the experimenter. But if the detector is conceived ontologically as a quantum system itself, as suggested by von Neumann, it can only be understood as a serial route or history of actualizations—a so-called 'Von Neumann chain' terminating, he proposed, in the mind of the experimenter. But if the mind is likewise describable quantum mechanically (at least to some degree, as it must be in any coherent ontological interpretation of quantum mechanics) then the detector-mind system for *any* given experiment is likewise properly understood as constitutive of a global system,

$$|\Psi\rangle = \alpha_{ijk} \sum_{i,j,k} |\psi\rangle_i |\phi\rangle_j |e\rangle_k$$

As such, the 'choice' of any particular measurement context for any particular actual occasion in the global system cannot simply be derived in the classical sense of an experimenter choosing a particular device. The most one can say is that every quantum actual occasion has its own particular local context, which is constitutive of itself. Indeed, to the extent that quantum mechanical predication is generative of novel facts, and thus novel totalities (i.e., with every novel fact, a novel $|\Psi\rangle$), it must also be seen as generative of novel local contexts. The problem of self-reference with respect to predication of a totality will likewise rear its head if one attempts to simply derive a particular local measurement context from some other local context.

This 'self-determination' of an actual occasion's subjective standpoint / local context is a signature feature of Whiteheadian metaphysics. But it is also the case in both Whiteheadian metaphysics and quantum mechanics that the local measurement context of a novel actual occasion-in-process, though not *externally determined*, is nevertheless *internally conditioned* by virtue of its internal relatedness to its dative world. In quantum mechanics, this is evinced by the fact that the presupposed global logical structure is defined by a Boolean localization scheme, where compatible local Boolean contexts overlap, and the indexical local measurement context must be part of this overlap.[27] In other words, potential relations among the data of $|\psi\rangle$, $|\phi\rangle$, and $|e\rangle$, via their individual local contexts, presuppose a partial compatibility among these local contexts in the expression $|\Psi\rangle = \alpha\,|\psi\uparrow\rangle\,|\phi\uparrow\rangle\,|e\uparrow\rangle + \beta\,|\psi\downarrow\rangle\,|\phi\downarrow\rangle\,|e\downarrow\rangle$. This internal relational compatibility can be expressed mereotopologically as a covering system of local Boolean algebras, heuristically representable in classical topology as a localization system of open covers, for example, where each open cover represents a local Boolean context.[28] Though the local indexical measurement context (i.e., the particular orthonormal measurement basis of the detector) cannot simply be derived from the Boolean overlap, it must be compatible with this overlap—i.e., compatible for integration with the scheme of local contexts constitutive of the induced global context.

Once actualized, the quantum actual occasion thus becomes a novel datum constitutive of a novel global totality, with its local context integrated into the internal-relational, mereotopological Boolean covering scheme. Whitehead writes:

> The atomic actualities individually express the genetic unity of the universe. The world expands through recurrent unifications of itself, each, by the addition of itself, automatically recreating the multiplicity anew.[29]

In this way, every quantum actual occasion-in-process is both [a] subjective in that its local context, *internally self-determined*, is also *internally conditioned* by the global facts to which it is internally related; and [b] objective in its actualization as a datum that is logically relatable to subsequent quantum occasions-in-process internally related to it (i.e., subsequent measurement interactions). As noted in chapter 2.8, and as will be discussed further in chapter 5, 'subsequent' here does not necessarily mean 'temporally subsequent,' but rather logically subsequent in the mereotopological hierarchy of internal relations. It will be argued that any metrical, spatiotemporal expression of subsequence is a specialized case of, and dependent upon, a more fundamental mereotopological expression of logical subsequence. For any ontological interpretation of quantum mechanics, then, [a] and [b] above are evinced by the fact that any measurement outcome must itself be measured—i.e., 'related' as a datum to a subsequent actual occasion-in-process.

Of the subjective aspect of quantum actual occasions, Whitehead writes:

> An entity is actual when it has significance for itself. By this it is meant that an actual entity functions in respect to its own determination . . . It is self-creative; and in its process of creation transforms its diversity of roles into one coherent role . . . This self-functioning is the real internal constitution of an actual entity. An actual entity is called the 'subject' of its own immediacy.[30]

> An actual entity, by functioning in respect to itself, plays diverse roles in self-formation without losing its self-identity. It is self-creative; and in its process of creation transforms its diversity of roles into one coherent role. Thus 'becoming' is the transformation of incoherence into coherence, and in each particular instance ceases with this attainment.[31]

Of the objective features of the actual occasion, and its function in relation to the systematic totality of occasions constitutive of its dative world (i.e., $|\Psi\rangle$), Whitehead writes:

> to 'function' means to contribute determination to the actual entities in the nexus of some actual world. Thus the determinateness and self-identity of one entity cannot be abstracted from the community of diverse functionings of all entities.[32]

> *The real potentialities relative to all standpoints are coordinated as diverse determinations of one extensive continuum.* This extensive continuum is one relational complex in which all potential objectifications find their niche . . . All actual entities are related according to the determinations of this continuum; and all possible actual entities in the future must exemplify these determinations in their relations within an already actual world.[33]

In quantum mechanics, as in Whiteheadian metaphysics, quantum actual occasions are irrelevant to the world apart from their relational function as data (i.e., the 'facts of the world') in a subsequent actualization / concrescence. For Whitehead, the actualization of an actual occasion-in-process (its 'satisfaction') entails "the notion of the 'entity as concrete' abstracted from the 'process of concrescence'; it is the outcome separated from the process, thereby losing the [dipolar] actuality of the atomic entity, which is both process *and* outcome."[34] Thus, the probability valuations of quantum mechanics describe probabilities that a given potential outcome state will be actual *upon evaluation*—implying a subsequent evolution and an interminable evolution of such evolutions. Every fact or system of facts in quantum mechanics, then, subsumes and implies both an *initial state of potential relations* with its dative world, and a *final state of actual relations* with its dative world via its incorporation, as novel datum / predicative fact, into that world. In other words, there can be no state specification of a system *S* without reference, implicit or explicit, to S_{initial} and S_{final}. This is reflected in Whitehead's scheme by referring to the quantum actual occasion-in-process as the 'subject-superject,' where both subjective and objective features are mutually implicative aspects of every quantum occasion.

> The 'satisfaction' is the 'superject' rather than the 'substance' or the 'subject.' It closes up the entity; and yet is the superject adding its character to the creativity whereby there is a becoming of entities superseding the one in question.[35]
>
> An actual entity is to be conceived as both a subject presiding over its own immediacy of becoming, and a superject which is the atomic creature exercising its function of objective immortality[36] . . . It is a subject-superject, and neither half of this description can for a moment be lost sight of[37] . . . [The superject is that which] adds a determinate condition to the settlement for the future beyond itself.[38]

Thus, the process of the actualization of a quantum measurement outcome—the evolution of potential relations to probable relations via decoherence—is never terminated by actualization, but only punctuated by it. The many facts constitutive of $|\Psi\rangle$ and their associated potential relations to the actual occasion-in-process, become one novel predicative fact—a predication of the global totality—and are thus increased, historically, by one. "The oneness of the universe, and the oneness of each element in the universe, repeat themselves to the crack of doom in the creative advance from creature to creature, each creature including in itself the whole of history and exemplifying the self-identity of things and their mutual diversities."[39]

4.5 Quantum Mechanics Presupposes Logically Related Actual Occasions

In both the decoherence-based interpretations of quantum mechanics and the relational realist philosophical cosmology, with its foundations in Whiteheadian metaphysics, the world is characterized not as a fixed totality, but rather as a global mereotopological process generative of internally related and logically structured facts. Both attempt to describe the relational forms of these presupposed facts; they do not attempt to explain the existence of these facts as in other interpretations of quantum mechanics—e.g., the various quantum cosmologies discussed in chapter 2, as well as those interpretations of quantum mechanics that posit a physical dynamical mechanism (stochastic field fluctuations, etc.) as 'causative' of wave packet collapse to a unique actual state.[40] In other words, neither the decoherence-based interpretations nor the relational realist cosmology proposes a mechanism that 'accounts for' the existence of quantum events, the 'objective facts' of reality. Unlike other ontological interpretations of quantum mechanics that consider this a deficiency or a problem to be solved, the decoherence-based interpretations, whether implicitly or explicitly, acknowledge that facts are necessarily presupposed by quantum mechanics and indeed—as argued earlier—by the scientific method in general. Focus is instead placed upon the *forms of relations* among quantum events—the logically conditioned evolution of potential relations to probable relations—and the conception of novel quantum events as predicative facts generated by these relations.

Beyond the necessary presupposition of facts in quantum mechanics, then, the decoherence-based interpretations of quantum mechanics and the relational realist philosophy rest upon an equally fundamental presupposition of relation among facts. In the same way that the concept of relation necessarily presupposes object relata (even when these objects are, themselves, relations), it is also the case that in quantum mechanics, the fundamental objects—quantum events as *res vera* or 'final real things'—presuppose relation as a generative process.

Furthermore, as will be explored in the next chapter, although the quantum actual occasions themselves are discrete, their *potential relations* are continuous and therefore metrically describable in terms of their extensive features. Thus, in the same Aristotelian sense that potentia always presuppose actualities, the continuous metrical descriptions of extension deriving from these potential relations presuppose a discrete, mereotopological extensive order of actual facts. But at the same time, in the relational realist philosophical scheme it is also true that actualities always presuppose potentia, since potential relations are always generative of novel actualities—viz., the quantum mechanical evolution of potential fact to probable fact to novel actual fact. In this regard, the various attempts by the physical sciences to depict physical reality as either fundamentally discrete or fundamentally continuous (reflected, for example, by the tension between quantum mechanics and general relativity) neglect the fact that quantum actual occasions cannot be abstracted from, or even defined without reference to, their actual *and* potential relations, both logical / mereotopological and extensive / metrical.

Potentiality and actuality are thus mutually implicative in the relational realist philosophy, whose fundamental objects are quantum mechanical units of logico-physical relation rather than simply units of physical relata. Objects are therefore always understood as relata, and likewise relations are always understood objectively.

4.6 Summary

Both the decoherence-based ontological interpretations of quantum mechanics and the relational realist philosophical cosmology, the latter with its roots in Whiteheadian philosophy, describe a relational process productive of predicative quantum actual occasions / measurement outcomes in three stages:

[1] An *initial phase* consisting of the potentialization of the global totality of facts $|\Psi\rangle$ relative to a particular, indexical potential quantum measurement outcome / quantum actualization-in-process. In terms of the standard quantum formalism, this phase is represented by the expression:

$$|\Psi\rangle = \alpha_{ijk} \sum_{i,j,k} |\psi\rangle_i |\phi\rangle_j |e\rangle_k$$

where $|\phi\rangle$ represents the state of the detector according to its particular local Boolean contextualization, defined by its particular orthonormal measurement basis, and $|\psi\rangle$ and $|e\rangle$ represent the state of the measured system and environment, respectively, relative to the particular contextualization given in $|\phi\rangle$.

Ontologically, this expression represents the internal relation of the indexical actualization-in-process to the dative global totality—i.e., the global lattice of quantum events—conditioned by presupposed logical first principles: the Principle of Identity (PI), the Principle of Non-Contradiction (PNC), the Principle of the Excluded Middle (PEM), and the Boolean relational forms of conjunction, disjunction, implication, complement, and exclusive disjunction by which PI, PNC, and PEM are defined.

These requirements are satisfied via the presupposition of a Boolean compatibility condition and localization scheme, by which the underlying, *implicit* Boolean structure of the global pure state $|\Psi\rangle$—i.e., implicit in virtue of the presupposition that each and every one of its constituent local contexts is Boolean and thus representable as a Boolean subalgebra—is rendered *explicit* via its definition as an overlap of compatible local Boolean contexts, expressible as an equivalence class of Boolean subalgebras. Potential internal relations of the quantum actual occasion / measurement event-in-process to the dative content of these overlapping local contexts are always indexed to the local context (Boolean subalgebra) of the former—i.e., such that it is understood as an indexical event, which in quantum mechanics is always associated with the detector.

Thus, a quantum actual occasion-in-process entails the evaluation of a general Boolean algebra to a two-valued (*T*,*F*) Boolean algebra that is defined by the locally contextualized indexical quantum event, classically represented by the detector. This is exemplified in the conventional Hilbert space formalism by its requirement of an orthonormal measurement basis by which the local indexical Boolean context is defined, and the fact that quantum mechanical measurement outcomes are *always* actualized as contextualized by this basis.[41]

[2] A *supplementary phase*, whereby the coherent superposition of potential internal relations among the data contextualized by this Boolean localization scheme are integrated into equivalence classes, each class indexed to a potential outcome state defined by the local Boolean context of the actual occasion / measurement event-in-process. In the spin ½ system discussed earlier, this is reflected in the expression:

$$|\Psi\rangle = \alpha|\psi\uparrow\rangle\,|\phi\uparrow\rangle\,|e\uparrow\rangle + \beta|\psi\downarrow\rangle\,|\phi\downarrow\rangle\,|e\downarrow\rangle$$

Thus, in the supplementary phase, global potential internal relations in the initial phase are restricted by the local indexical context; but at the same time, the local indexical context is necessarily contained within the overlap of all compatible Boolean contexts constitutive of the global totality. Potential internal relations incapable of integration within this Boolean logical structure (those that violate PNC) are eliminated via negative selection. This is represented in the decoherence-based interpretations of quantum mechanics as a traceover of environmental degrees of freedom and elimination of off-diagonal terms in the density matrix. The result is a decoherent reduced density matrix of mutually exclusive and exhaustive, logical integrations of potential internal relations—i.e., a reduced matrix ρ^r of probable predicative facts / measurement outcomes.[42]

$$\begin{aligned} \mathrm{Tr}_E\,|\Psi\rangle\langle\Psi| &= \Sigma_i\,\langle e_i|\Psi\rangle\langle\Psi|e_i\rangle \\ &= |\alpha|^2\,|\psi\uparrow\rangle\langle\psi\uparrow||\phi\uparrow\rangle\langle\phi\uparrow| + |\beta|^2\,|\psi\downarrow\rangle\langle\psi\downarrow||\phi\downarrow\rangle\langle\phi\downarrow| \end{aligned}$$

[3] The presupposed and anticipated *actualization* of one of these predicative facts / measurement outcomes $|\alpha|^2\,|\psi\uparrow\rangle\langle\psi\uparrow||\phi\uparrow\rangle\langle\phi\uparrow|$ or $|\beta|^2\,|\psi\downarrow\rangle\langle\psi\downarrow||\phi\downarrow\rangle\langle\phi\downarrow|$ according to the probability valuations α and β qualifying each. This anticipated actualization reflects a presupposition of PNC and PEM, justified by the fact that as probability valuations, $|\alpha|^2 + |\beta|^2 = 1$. By PEM, *at least one* outcome must be actualized. And by PNC, *at most one* outcome must be actualized.

In the most general terms, then, these three stages together depict fundamental substance as quantum 'becomings' (Whiteheadian 'actual occasions') whose actualization ('concrescence') is predicative via internal relations with ('prehensions' of) a totality of antecedently settled becomings—i.e., 'concrete facts'—the global data of the settled world. Internal relations to these data, as a totality, are perpetually re-potentialized with each novel quantum actual occasion-in-process and its internal relation to that totality.

These relations are both causal-physical and logical-conceptual in that their causal-physical features are logically integrated or 'objectified' in the evolution from potentiality to probability; thus each novel quantum predicative fact when considered as a substantial 'thing-in-itself' is fundamentally a discrete unit of relational evolution—from the facts of $|\Psi\rangle$ to potential internal relations to $|\Psi\rangle$, to probable internal relations to $|\Psi\rangle$, to concrete actual internal relation to $|\Psi\rangle$, and thereby inclusion within it as a novel fact constitutive of a novel totality. The internal relations integrated in this evolution are thus both physically and logically significant, such that every quantum actual occasion / predicative fact is properly understood as dipolar; that is, its physical and logical features, as well as its objective (i.e., dative) and subjective (i.e., contextual) features, are understood as mutually implicative aspects of every quantum actual occasion / predicative fact.

The relational realist scheme of internal relation among dipolar quantum actual occasions thus reflects a correlation of the order of logical implication and the order of causal relation—what Whitehead terms, respectively, the 'genetic division' and the 'coordinate division' of the predicative fact / actual occasion. The history of modern physics has revealed with abundant clarity that the proper relation of the casual and logical orders as mutually implicative is crucial to the coherent integration of fundamental physical theories and their associated models—e.g., quantum theory, relativity theory, the standard cosmological model, etc. For example, it is by the physico-logical dipolarity of a quantum actual occasion's internal relation to its dative world that the inability to causally order spacelike separated occasions per the restrictions of relativity theory—i.e., in terms of a total order—does not in any way imply that spacelike separated occasions likewise lack a logical order. In relativity theory, for example, this notion of a logical order is clearly evinced in the *partial* ordering among events constitutive of individual spacetime intervals. The challenge of reconciling quantum theory and general relativity is deeply rooted in this issue, and after nearly a century of investigation, a satisfactory integration has yet to be realized. To this end, the relational realist ontological interpretation of quantum mechanics suggests an alternative pathway forward that is grounded in the mutual implication of logic and causality, providing both a conceptual philosophical and formal mathematical framework that elucidates how individual, local partial orders, defined on a continuum, are connectible logically and physically within a unified hierarchy of discrete, internally related, predicative facts, each bijectively related to its global totality in the hierarchy.[43]

Notes

1. Alfred North Whitehead, *Process and Reality: An Essay in Cosmology*, Corrected Edition, ed. D. Griffin and D. Sherburne (New York: Free Press, 1978), 283-93.

2. S. Walborn et al., "Double-slit Quantum Eraser," *Physical Review A* 65, no. 3 (2002): 033818.

3. See, for example: Patrick Barry, "What's done is done . . ." *New Scientist* 191, no. 2571 (2006): 36.

4. See, for example: Daniel Sheehan, ed., *Frontiers of Time: Retrocausation – Experiment and Theory* (Proceedings of the Pacific Division of the American Association for the Advancement of Science, American Institute of Physics, San Diego, California, June 20-22, 2006).

5. Correlated *S* and *P* detection events are registered via a coincidence counter linking both *S* and *P* detectors—the same method of correlating detection events at the separate detectors in a typical EPR experiment.

6. Introduced in chapter 2.1.

7. Simon Kochen and Ernst Specker, "The Problem of Hidden Variables in Quantum Mechanics," *Journal of Mathematics and Mechanics* 17 (1967): 59-87.

8. For an introduction to algebraic topology and sheaf theory, see: David Eisenbud, *Commutative Algebra: With a View Toward Algebraic Geometry*. Vol. 150 (New York: Springer, 1995); Sergei Gelfand and Yuri Manin, *Methods of Homological Algebra*, 2nd edition (New York: Springer, 2003); Saunders MacLane and Ieke Moerdijk, *Sheaves in Geometry and Logic: A First Introduction to Topos Theory* (New York: Springer, 1992); Saunders MacLane, *Categories for the Working Mathematician* (Graduate Texts in Mathematics), 2nd ed. (New York: Springer, 1998).

9. See, for example: Wojciech Zurek, "Decoherence, Einselection, and the Quantum Origins of the Classical," *Reviews of Modern Physics* 75, no. 3 (2003): 715-75. See also: Maximilian Schlosshauer, "Decoherence, the Measurement Problem, and Interpretations of Quantum Mechanics," *Reviews of Modern Physics* 76, no. 4 (2005): 1267; and Maximilian Schlosshauer, *Decoherence and the Quantum-to-Classical Transition* (New York: Springer, 2007).

10. S. Bertaina et al., "Quantum oscillations in a molecular magnet," *Nature* 453 (May 2008): 203-6; S. Takahashi et al., "Decoherence in Crystals of Quantum Molecular Magnets," *Nature* 476 (July, 2011): 76-79.

11. Pritiraj Mohanty, E. M. Q. Jariwala, and R. A. Webb, "Intrinsic Decoherence in Mesoscopic Systems," *Physical Review Letters* 78, no. 17 (1997): 3366; P.C.E. Stamp, "The Decoherence Puzzle," *Studies in History and Philosophy of Modern Physics* 37 (2006): 490.

12. See, for example: Olimpia Lombardi, S. Fortin, and M. Castagnino, "The Problem of Identifying the System and the Environment in the Phenomenon of Decoherence," *The European Philosophy of Science Association Proceedings: Amsterdam 2009*, ed. H. W. de Regt, S. Hartmann, and S. Okasha (Berlin: Springer, 2012), 161-74; Wojciech H. Zurek, "Decoherence, Einselection and the Existential Interpretation (the Rough Guide)," *Philosophical Transactions of the Royal Society A: Mathematical, Physical and Engineering Sciences* 356, no. 1743 (1998): 1820.

13. See, for example: Robert Griffiths, "Consistent Histories and the Interpretation of Quantum Mechanics," *Journal of Statistical Physics* 36, no. 1 (1984): 219-72; Roland Omnès, "Logical Reformulation of Quantum Mechanics I.-IV," *Journal of Statistical Physics* 53, nos. 3-4 (1988): 893-957; Murray Gell-Mann and James Hartle, "Quantum Mechanics in the Light of Quantum Cosmology," *Proceedings of the 3rd International Symposium on the Foundations of Quantum Mechanics in the Light of New Technology*, Tokyo, Japan (August 1989): 321-43.

14. Robert Griffiths, "Consistent Histories and the Interpretation of Quantum Mechanics," *Journal of Statistical Physics* 36, no. 1 (1984): 219-72.

15. Griffiths, "Consistent Histories and the Interpretation of Quantum Mechanics," 233.

16. See chapter 3.

17. Introduced in chapters 2 and 3 and formalized mathematically in part II.

18. As discussed in chapter 2, per the Kochen-Specker Theorem, the partial Boolean algebra of projection operators characteristic of quantum mechanics cannot, in general, be embedded into a Boolean algebra.

19. Werner Heisenberg, *Physics and Philosophy* (New York: Harper Torchbooks, 1958), 143.

20. Wojciech Zurek, "Letters," *Physics Today* 46, no. 4 (1993): 84.

21. Zurek, "Decoherence and the Transition from the Quantum to the Classical," 36-44.

22. Whitehead, *Process and Reality*, 101.

23. Whitehead, *Process and Reality*, 254.

24. Whitehead, *Process and Reality*, 226-27.

25. Whitehead, *Process and Reality*, 254.

26. Whitehead, *Process and Reality*, 251.

27. Elias Zafiris, "Boolean Coverings of Quantum Observable Structure: A Setting for an Abstract Differential Geometric Mechanism," *Journal of Geometry and Physics* 50, no. 99 (2004).

28. This classical topological covering scheme is heuristic representation only. A proper mathematical representation in terms of Grothendieck topology will be discussed in detail in part II of this volume.

29. Whitehead, *Process and Reality*, 286.

30. Whitehead, *Process and Reality*, 25.

31. Whitehead, *Process and Reality*, 25.

32. Whitehead, *Process and Reality*, 25.

33. Whitehead, *Process and Reality*, 66 (italics added).

34 . Whitehead, *Process and Reality*, 84.

35. Whitehead, *Process and Reality*, 84.

36. Whitehead, *Process and Reality*, 45.

37. Whitehead, *Process and Reality*, 29.

38. Whitehead, *Process and Reality*, 150.

39. Whitehead, *Process and Reality*, 228.

40. For example, the Spontaneous Localization theory of Ghirardi, Rimini, and Weber. See, for example: G. C. Ghirardi, A. Rimini, and T. Weber, "Unified dynamics for microscopic and macroscopic systems," *Physical Review D* 34 (1986): 470-91, and G. C. Ghirardi, *Sneaking a Look at God's Cards: Unraveling the Mysteries of Quantum Mechanics* (Princeton: Princeton University Press, 2004), 406.

41. As discussed earlier, the conventional Copenhagen interpretation with its Hilbert space formalism gives no reason why this should be so.

42. Zurek, "Decoherence and the Transition from the Quantum to the Classical," 39.

43. See chapter 11.6; see also: A. Mallios and E. Zafiris, "The Homological Kähler-De Rham Differential Mechanism—Part I: Application in the General Theory of Relativity," *Advances in Mathematical Physics*, 2011 (2011), doi:10.1155/2011/191083, and A. Mallios and E. Zafiris, "The Homological Kähler-de Rham Differential Mechanism—Part II: Sheaf-Theoretic Localization

of Quantum Dynamics," *Advances in Mathematical Physics*, 2011 (2011), doi:10.1155/2011/189801.

CHAPTER 5

Integrating Logical Relation and Extensive Relation: Mereotopology and Quantum Mechanics

The internal relational framework by which the discrete, logical, 'genetic' relations among quantum events are correlated with their continuous physical 'coordinate' relations is presented by Whitehead in part IV of *Process and Reality:* It is his mereotopological scheme of 'extensive connection,' whose fundamental features are the relations among locally defined actual occasions and their internally related global regions. In earlier works, this was a set-theoretic mereological scheme, whose fundamental units were 'extensive wholes' and 'extensive parts' rather than the internal relation morphisms connecting actual occasions to their regions, and their regions to other regions via internal relation to the global totality. "This defect of starting-point," wrote Whitehead, "revenged itself in the fact that the 'method of extensive abstraction' developed in those works was unable to define a 'point' without the intervention of the theory of 'duration.'"[1] His earlier works, in other words, defined 'point' as the discretization of an 'actual' continuum. But defining a point as an extensive element of the continuous real line implies that the point, too, must be continuously divisible extensively (i.e., in terms of its coordinate analysis); thus the actual occasion, when represented coordinately as a point, loses its quantum character.

By contrast, Whitehead's later argument in *Process and Reality* is more firmly anchored to his commitment to the Aristotelian notion of "*infinitum actu non datur*": There is no 'actual' super-denumerable infinite—only infinite *potentia* for relations among actual occasions. In the relational realist scheme, for example, the apparent infinite divisibility of a finite length is properly understood as an infinite number of potential relations among the denumerable actual occasions constitutive of that length; in this way, 'length' is properly understood as a higher order, metrical abstraction derived from these underlying discrete relations. The potential relations are infinite ontologically, not merely epistemically, because predication is synthetic,[2] such that the actualization of any potential relation yields a novel actual relatum—and thus novel potential relations. This is easily depicted via the real number line, a finite representation of the infinite, wherein the totality of relations among numbers implies augmentation by either infinite extrapolation or interpolation.

Likewise, recall the mathematical analogy suggested in chapter 2: Consider the relationship between $\mathbb{R}$ (the set of real numbers—a continuum and thus uncountable), $\mathbb{Q}$ (the set of rational numbers—discrete and thus countable), and $\mathbb{Z}$ (the set of integers—also discrete and countable). According to classical intuition, we usually take the 'foundational' set to be the continuum $\mathbb{R}$; and in extending this mathematical intuition to nature more broadly, we likewise consider an 'extensive continuum' as a fundamental spatiotemporal background against which all physical relations, including quantum events, might be related metrically via coordinate division. The mathematical analog representative of such relations (e.g., as ratios) would be $\mathbb{Q}$, defined as a discretized subset of $\mathbb{R}$ (i.e., we quantize the continuum $\mathbb{R}$). Likewise, the mathematical analog of the quantum objects related might be $\mathbb{Z}$, defined as a subset of $\mathbb{Q}$.

But one could just as intuitively begin with $\mathbb{Z}$ as the fundamental background of discrete quantum units and proceed in the other direction, deriving $\mathbb{Q}$ as a set of discrete relations (i.e., ratios) among these discrete units of $\mathbb{Z}$, and likewise deriving $\mathbb{R}$ as a set of relations among the units of $\mathbb{Q}$ (viz., the Cauchy sequence), with *relations of relations*, etc. By proceeding in this direction, one could extend the mathematical representation even further: Since each quantum actual occasion is both unique and internally relatable to the totality of occasions constitutive of its dative world, one could mathematically represent actual occasions via the prime numbers—each unique and indivisible (i.e., quantum), but *relatable* via their shared property of belonging to $\mathbb{Z}$.

Given that the conceptual foundation of Whitehead's ontological scheme is the quantum actual occasion, the latter order, from $\mathbb{Z}$ as fundamental to $\mathbb{R}$ as a higher-order derivation, would seem to be the more appropriate in defining extensive connection among quantum *res verae*. Whitehead writes, for example:

> The concrescence presupposes its basic region, and not the region its concrescence. Thus the subjective unity of the concrescence is irrelevant to the divisibility of the region. In dividing the region we are ignoring the subjective unity which is inconsistent with such division. But the region is, after all, divisible, although in genetic growth it is undivided.[3]

The key impediment confronting Whitehead in his attempt to formalize the conceptual relationships described in the above quote—quantum actual occasions internally related genetically (via a logically asymmetrical structure) but extensively related coordinately (via a continuous, symmetrical structure)—is his presupposition that extensive, regional divisibility is at once fundamentally set-theoretic and continuous. However, since quantum actual occasions are, by definition, discrete units, it is clear that the *regional contexts* by which they are defined *must also be discrete* if internal relation and extensive connection are to be coherently integrated. The idea of internal relations among fundamentally discrete regions cannot be adequately formalized via the language of set theory, the conventional language available to Whitehead in his time, because it is grounded in a continuum (i.e., the real line); however, as introduced in chapter 3, structure-preserving internal relations among discrete regions *can* be adequately formalized topologically in the language of category theory, as will be explored presently.

By making the structure-preserving internal relations among occasions *and* their regions the fundamental objects of his scheme of extensive connection, rather than just the related quantum objects themselves, Whitehead attempted to construct a rudimentary, quasi-set-theoretic framework of extensive connection in part IV of *Process and Reality*; but because of the inherent limitations of set theory, this framework—an abstractive shift from a sheerly set-theoretic structure, but still firmly grounded in the latter—lacked the formal rigor necessary for serious application to the fundamental physics of his time.[4]

Less than two decades later, Samuel Eilenberg and Saunders Mac Lane would develop their category theoretic approach to algebraic topology in a similar abstractive shift of focus from the purely set-theoretic framework, whose fundamental units are the elements of sets, to the category-theoretic framework, whose fundamental units are the structure-preserving relations (morphisms) among discrete objects that are *themselves* relational structures. Indeed, one finds in category theory, and in particular as pertaining to sheaf theory (to be discussed presently), a rigorous formalism that would seem to satisfy all the demands Whitehead made of his 'theory of extensive connection' in part IV of *Process and Reality*, but was not able to fully satisfy via the set-theoretic conceptual framework he had earlier developed with Russell.

For example, Russell and Whitehead's original method of bypassing the paradoxes of set-theoretic self-reference, as discussed in chapter 3, was the Theory of Logical Types; but as part IV of *Process and Reality* well demonstrates, the set-theoretic framework of Logical Types was difficult to apply to the extensive features of physical relations. By contrast, the category-sheaf theoretic formalism, as a mereotopological schematization of quantum mechanics,[5] provides a uniquely appropriate and formally rigorous method of translating the asymmetrical logical and mereological features of quantum mechanical relations ('genetic division') to the extensive features of these relations ('coordinate division') without difficulty. Self-reference is avoided in this scheme analogous to the way that it is avoided in the Theory of Logical Types—that is, via a hierarchical structure of logical contexts internally related asymmetrically. But again, unlike the quasi-set-theoretic framework of extensive connection given in part IV, category-sheaf theoretic regional contexts are fundamentally topological, and therefore discrete, rather than continuous.

Several key aspects of the sheaf-theoretic structure were introduced in chapters 3 and 4—namely, those aspects by which the logical integration of potential relations to probable relations could be defined *apart* from the extensive features of these relations, such that even when relata are spacelike separated (i.e., nonlocally related, as in the EPR experimental arrangement), the logical integration of potential relations to probable relations is unimpeded. Again, this can only be accounted for via the categorical presupposition that all local contexts of related data are Boolean, such that each context is representable in quantum mechanics as a Boolean subalgebra. Thus, the totality of local contexts, to the extent that they are compatible for integration into an induced, approximated global context, can be formalized as a mereotopological framework of nested (i.e., internally related) contextual inclusions. It is through this framework that the internal relation of the local novel actual occasion / predicative quantum measurement outcome to the global dative world is logically coordinated such that the latter can be formalized topologically. Thus the sheaf-theoretic Boolean localization scheme has both logical and ontological significance in the relational realist philosophy, via the concepts of logical implication and internal relation, respectively.

In exploring the extensive aspects of the logical concept of 'local Boolean context' discussed above, it is clear that in any ontological interpretation of quantum mechanics, these aspects are fundamentally topological rather than metrical; for the concept of 'local context' in quantum mechanics has no ontological meaning within the metrical framework of the relativistic spacetime continuum, where 'local' refers to points connected by continuously divisible spatiotemporal intervals, not discrete regions connected by discrete logical internal relations. In a mereotopological continuum, however, one can represent a discrete 'local context,' in precisely this latter way, as an open cover. A scheme of internally related local contexts can then be represented topologically via a scheme of open covers of an open set. But as applied to quantum mechanics, there is an additional level of abstraction required, in that the open set must be understood as representing a Boolean algebra. This further abstraction is also required in order to accommodate the fact that potential relations in quantum mechanics, defined over a framework of inclusively related local contexts, are always indexed to the local Boolean context / Boolean subalgebra of the measuring device—i.e., the orthonormal basis of the detector. Apart from this restriction, potential outcomes cannot evolve to become probable outcomes.

Classical topological structures like open sets and covering systems of open sets are heuristically useful in exploring the extensive features of quantum mechanical relations in a manner consistent with its logical relations. In this way, an open set is an intuitive representation of a local context, and a scheme of inclusively related open covers of the open set is an intuitive representation of a global scheme of internally related local contexts. However, to the extent that [a] local contexts in quantum mechanics are represented by Boolean subalgebras, and [b] equivalence classes of these subalgebras require reference to an indexical local Boolean context in order for potential relations to evolve to become probable relations, the more abstractive sheaf-theoretic scheme of mereotopological extension is required to fully capture these logico-extensive aspects of quantum mechanical relation.

In part II of this volume, this scheme will be presented in detail, both in terms of its technical formalism and its application to quantum mechanics from the standpoint of philosophical foundations of relational realism presented in part I. In preface to that presentation, the remainder of this chapter will be devoted to exploring some key formal connections between the category-sheaf theoretic framework of mereotopologically extensive relations and the relational realist interpretation of quantum mechanics. In order to mediate between these two frameworks, the mathematical concepts used will be restricted to those of classical topology as a heuristic representation of the more technical Grothendieck topological concepts introduced in part II.

5.1 Mereotopological Notions of Internal Relation and Logical Implication

The first point of intersection between category-sheaf theory and Whitehead's theory of extensive connection is the concept of internal relations between objects.[6] For example, consider an actual occasion α, asymmetrically internally related to its dative world Ψ_α (where Ψ_α is one logical type lower than α) such that α is predicative of certain data constitutive of Ψ_α.[7] Next, consider actual occasion β, internally related to its dative world Ψ_β, which includes occasion α and *its* dative world Ψ_α. For Whitehead, the internal relation of β to α does not merely include α as datum, but also the *entire structure of* α's *internal relation to its dative world* Ψ_α. In other words, the internal relation of β to α must be a structure-preserving relation, one which transmits information from dative fact to novel predicative fact.

Whitehead writes:

> These extensive relations do not make determinate *what* is transmitted; but they do determine conditions to which all transmission must conform. They represent the systematic scheme which is involved in the real potentiality from which every actual occasion *arises*. This scheme is also involved in the attained fact which every actual occasion *is*. The 'extensive' scheme is nothing else than the generic morphology of the internal relations which bind the actual occasions into a nexus, and which bind the prehensions of any one actual occasion into a unity, coordinately divisible.
>
> For Descartes the primary *attribute* of physical bodies is *extension;* for the philosophy of organism the primary *relationship* of physical occasions is *extensive connection*. This ultimate relationship is *sui generis*, and cannot be defined or explained. But its formal properties can be stated. Also, in view of these formal properties, there are definable derivative notions which are of importance in expressing the morphological structure.[8]

With respect to sheaf theory over topological spaces, there are a number of formal concepts central to the 'morphological structure' of Whitehead's theory of extensive connection, particularly as the latter pertains to the relational realist program. The term 'section' in sheaf theory refers to a potential internal relation between an actual occasion-in-process (i.e., a potential quantum measurement outcome) and the dative occasions constitutive of its locally contextualized world (i.e., the global state $|\Psi\rangle$ locally Boolean contextualized by some particular orthonormal basis.) This local context is represented topologically as an open set, such that a locally contextualized potential internal relation is represented as a section over an open set. (Again, as applied to quantum mechanics, representing a local context as an open set is only an informal representation of local context as Boolean subalgebra. The formal Grothendieck topological representation of this and related concepts is given in part II.)

A 'sheaf' represents a bundle of sections over an open set that are 'glued' to other logically compatible sections over other open sets. Gluing of sections across open sets is only possible when the open sets are inclusively related to each other as overlapping open covers—i.e., representing a mereotopological scheme of internally relatable local contexts. A sheaf of glued sections thus represents an integration of logically compatible potential internal relations among data across multiple, internally related local contexts.

The sheaf is an excellent tool for the formalization of the internal relation of the local to the global[9] because its structure represents the conditioning of local potential measurement outcomes (i.e., the 'subjective forms' of an actual occasion-in-process) by the global data to which the actual occasion-in-process is internally related. This conditioning via internal relation, in other words, is a 'restriction' of these potentia to those that are logically compatible with the global data—i.e., those compatible for gluing. At the same time, the global represented sheaf theoretically is an inductive approximation—a structure of inclusively related local contexts represented as open covers of an open set. In quantum mechanics, the open set represents the indexical local context—i.e., the

Boolean context of the measuring device, defined by its orthonormal basis. The formation of equivalence classes indexed to this basis in the decoherence-based interpretations of quantum mechanics described earlier, or equivalence classes of maximal Boolean subalgebras[10] similarly indexed, exemplifies this logically conditioned, inductive integration.

The quantum actual occasion in-process is globally conditioned by (versus classically 'determined' by) a historical hierarchy / covering structure of internally related local contexts, referred to as a 'covering sieve'—essentially a 'global context' defined as a totality of internally related local contexts. Likewise, the totality of internally related facts contextualized via the covering sieve is referred to as a 'Boolean germ'—an equivalence class of compatible sections over the covering sieve of compatible underlying Boolean contexts. The Boolean germ, in other words, can be thought of as a kind of 'global history' correlate with the global context defined by the sieve.

It is only via these two structures together that local sections (potential relations among data) can be integrated with (glued to) those of all other local contexts, such that the ontological significance of locally contextualized facts is always *extendable globally*, and at the same time, the ontological significance of global facts is always *restrictive locally*. Once actualized as a measurement outcome, the actual occasion augments the Boolean germ, serving as a datum for all future actual occasions; likewise, its local context becomes included within, and thereby augments, the global covering sieve. In this way, the Boolean germ and covering sieve are historical-progressive, mereotopological, internal relational structures of fact and form, respectively, depicting the ontological formation of the quantum continuum as unending process. They are the two structures by which the quantum actual occasion as discrete, locally contextualized datum becomes constitutive of the global totality as extensive quantum continuum.

The application of these topological terms and related concepts from category-sheaf theory to Whitehead's theory of extensive connection can be easily explored beginning with the following passage from *Process and Reality*. Whitehead writes:

> The real potentialities relative to all standpoints are coordinated as diverse determinations of one extensive continuum. This extensive continuum is one relational complex in which all potential objectifications find their niche. It underlies the whole world, past, present, and future. Considered in its full generality, apart from the additional [relativistic] conditions proper only to the cosmic epoch of electrons, protons, molecules, and star-systems, the properties of this continuum are very few and *do not include the relationships of metrical geometry.*[11]

> An extensive continuum is a complex of entities united by the various allied relationships of whole to part, and of overlapping so as to possess common parts, and of contact, and of other relationships derived from these primary relationships. This extensive continuum *expresses the solidarity of all possible standpoints throughout the whole process of the world.* It is not a fact prior to the world; it is the first determination of order—that is, of real potentiality arising out of the general character of the world . . . All actual entities are related according to the determinations of this continuum; and all possible actual entities in the future must exemplify these determinations in their relations within an already actual world.[12]

The concepts given in this passage can be formalized by applying the language of category-sheaf theory to the relational realist / Whiteheadian interpretation of quantum mechanics explored thus far. One begins by considering the category theoretic structures introduced in chapter 3.4 as topological structures. For example, starting with a topological space *X*, the open sets of this space are heuristically representative of local Boolean contexts by which potential relations among data can be integrated. The structure by which the extensive continuum is defined as a manifold of local contexts—each internally related to a global totality and to each other in a coherent manner exemplifying the full meaning of the above passage—is given in the relationship between sheaf, germ, and covering sieve: An open set *U* on the topological space *X* represents the local context of an actual occasion-in-process, i.e., a quantum mechanical measurement interaction prior to measurement outcome. In the Category of Sets **C**, each open set *U* of the topological space *X* is thus represented by a set *F(U)*, which is a set of sections over *U*, representing the actual occasion's internal relations with its dative world.

The assignment *F(U)* for all open sets *U* of the topological space *X* is considered to be a 'presheaf' if it satisfies the following three properties: First, for each inclusion of open sets (local *dative* contexts) to which *U* is internally related, such as $V \subseteq U$, there is a 'restriction morphism' in **C** that takes the following form: $res_{V,U} : F(U) \rightarrow F(V)$, read 'the restriction of *U* by (internal relation to) *V*.' Second, the uniqueness of each becoming actual occasion is guaranteed by the restriction morphism $res_{U,U}$: $F(U) \rightarrow F(U)$, which is the 'identity morphism' on *F(U)*. Further, open sets other than *U*, and not included in *U* as an open cover, are considered externally related to *U*. Whitehead terms actual occasions with such local contexts as occasions 'in unison of becoming' and thus mutually externally related. It should be noted also that metrical spacelike separation among occasions is *not* sufficient to deem them mutually externally related mereotopologically, though it is sufficient to deem them contemporaries if coordinated temporally. As will be discussed presently, spatiotemporal extensiveness is a higher order type of continuous extensiveness which presupposes the more fundamental mereotopological scheme of discrete extensive connection discussed here.

The third condition that *F(U)* must satisfy to be a presheaf—and of central importance to the relational realist interpretation of quantum mechanics, and Whiteheadian metaphysics in general—is that restriction morphisms can take the form of composite inclusions. For example, a serial order of open inclusions $W \subseteq V \subseteq U$ yields the following restriction: $res_{W,V} \circ res_{V,U} = res_{W,U}$. Thus a quantum actual occasion-in-process, prior to measurement outcome, internally related to (restricted by) its dative world, is also restricted by those data's restrictions by *their* dative worlds, and so on, such that the serial order of open inclusions $W \subseteq V \subseteq U$ can be thought of as the structure of internally related global data constitutive of the Boolean germ, and their correlate internally related local Boolean contexts constitutive of the covering sieve. In this way, composite restriction morphisms yield internal relations that are *entirely structure-preserving*.

For Whitehead, too, internally related data each include their own serially-inclusively related global regions. Each dative global region, as described in part IV of *Process and Reality*, is thus bijectively correlated with its respective quantum predicative fact / actual occasion. However, consistent with the restrictions of relativity theory, these data are not serially ordered 'in time'—i.e., in terms of their metrical relations. They are, rather, serially ordered in terms of their mereotopological relations of inclusion, which is an asymmetrical logical order that is *reflected* (incompletely) in the temporal order yielded by the derivative metrical description of relativistic spatiotemporal extension.[13] Thus the concrescence of an actual occasion (i.e., the actualization of a potential quantum predicative fact) is not properly understood as occurring 'in time'—that is, within some more fundamental temporal continuum. As noted in chapter 2.1, time is not itself an 'observable' in quantum theory—i.e., it is not a quality intrinsic to fundamental quantum objects. It is, rather, a relational inference, quantified via an external parameter by which an initial actual state of a measured system is metrically related to a subsequent actual state. Thus, the continuity of the temporal order is, as described above, a derivative and emergent metrical-extensive relational feature of the fundamentally discrete quantum process by which actual occasions are generated. But at the same time, this generation is driven via the becoming occasion's internal relation with its dative world, and this internal relation is mereotopologically extensive. "There is a becoming of continuity," writes Whitehead, "but no continuity of becoming. The actual occasions are the creatures which become, and they constitute a continuously extensive world."[14]

5.2 Mereotopological Extensive Relations

In the Whiteheadian philosophical framework, spatiotemporal extension is a higher-order form of extensiveness, and presupposes the first-order logical / mereotopological extensiveness discussed here. He writes:

> In this general description of the states of extension, nothing has been said about physical time or physical space . . . These are notions which presuppose the more general relationship of extension. They express additional facts about the actual occasions. The extensiveness of space is really the spatialization of extension; and the extensiveness of time is really the temporalization of extension.[15]

> Extension, apart from its spatialization and temporalization, is that general scheme of relationships providing the capacity that many objects can be welded into the real unity of one experience. Thus, an act of experience has an objective scheme of extensive order by reason of the double fact that its own perspective standpoint has extensive content, and that the other actual entities are objectified with the retention of their extensive relationships. These extensive relationships are more fundamental than their more special spatial and temporal relationships. Extension is the most general scheme of real potentiality, providing the background for all other organic relations.[16]

Likewise, in the relational realist philosophical scheme, this "more general relationship of extension" is mereotopological, mathematically formulable via category-sheaf theory. Thus for both the Whiteheadian and relational realist frameworks, the super-denumerability of relativistic, continuous spatiotemporal extensive relations based on the real line has nothing to do with reality as concrete—i.e., as a history of actual occasions. It pertains, rather, to the *potential* internal relations of every locally contextualized quantum actual occasion-in-process to its global data. Again, the global conditioning of locally contextualized internal relations is represented in sheaf theory by the restriction morphisms discussed above. Their analog in Whitehead's theory of extension is given in his Definition 11 from part IV of *Process and Reality*:

> An abstractive set α is said to 'cover' an abstractive set β, when every member of the set α includes some members of the set β. It is to be noticed that *each abstractive set is to be conceived with its members in serial order, determined by the relation of inclusion.* The series starts with a region of any size, and converges indefinitely towards smaller and smaller regions, without any limiting region. When the set α covers

> the set β, each member of α includes all the members of the convergent tail of β, provided that we start far enough down in the serial arrangement of the set β.[17]

By the restriction given in Definition 10, below, this inclusion is both infinite and non-tangential:

> A set of regions is called an 'abstractive set,' when (i) any two members of the set are such that one of them includes the other *non-tangentially*, and (ii) there is no region included in every member of the set . . . Since every region includes other regions, and since the relation of inclusion is transitive, it is evident that every abstractive set must be composed of an infinite number of members.[18]

The principal advantage of category-sheaf theory, in connection with the above definitions from *Process and Reality*, is that it provides a rigorous mathematical formalism that accommodates not only the restrictive features of internal relation—i.e., how global data condition the generation of individual local potential predicative facts; it also represents how locally contextualized actual occasions / quantum measurement outcomes are predicative of the global totality, thus generating a new totality via augmentation:

> The atomic actual entities individually express the genetic unity of the universe. The world expands through recurrent unifications of itself, each by the addition of itself, automatically recreating the multiplicity anew.[19]

Most important, the category-sheaf theoretic approach depicts this global internal relational structure in a manner that is inductively limited, such that it does not necessitate the specification of a total ordering among all facts—i.e., some unique global seriality that defines the evolution of the universe, generated-fact by generated-fact. Analogously, in the decoherence interpretations of quantum mechanics, there is no fine-grained definition of the global totality; rather, this totality is always coarse grained into equivalence classes, each class indexed to a particular locally contextualized potential predicative fact (e.g., a particular eigenstate of the Boolean orthonormal basis of a particular detector) such that the latter is always internally related to the global totality. But whereas the decoherence formalism offers little by way of accounting for the extensive features of its processes—particularly the nonlocal extensive features—category-sheaf theory provides a more rigorous formalism to that end, and one that is fully compatible with the formalism by which decoherence is described.

Specifically, this is given in the *transition from presheaf to 'sheaf.'* This occurs via the satisfaction of additional compatibility conditions by which *locally contextualized sections can be joined together across manifold local contexts.* As discussed earlier, the joining process is referred to as the 'gluing' of sections, and is described thus: The presheaf F defined above comprises sections of F over each open cover U. With respect to the relational realist framework, a section over U represents a particular potential relation between [a] a particular potential measurement outcome—i.e., a particular subjective form of an actual occasion-in-process—whose local context is represented by the open set U; and [b] the data of the world. But each datum, as discussed before, has its own local context, structured by its own internal relations to each datum of *its* world; thus the dative local contexts to which U is internally related are represented as a series of open covers of U—i.e., the covering sieve—similar to the inclusion relations depicted by Whitehead in the above passages. The sections of the presheaf F are restricted via the structure-preserving restriction morphisms discussed earlier. In this sense, then, each section is a *potential* (though not yet *probable*) history relating a particular *potential* measurement outcome, locally contextualized by U, to its dative world. The evolution from *potential* histories to *probable* histories is formalized in the evolution of a *presheaf* to a *sheaf.*

The presheaf F becomes a sheaf when two additional axioms are satisfied: First, the axiom of 'local identity' which is related to the 'identity morphism' discussed earlier, only here applied to sections. It is defined thus: V_i is an open cover of an open set U. If sections s and t are such that $s,t \in F(U)$ and $s|_{Vi}$ (read s restricted by Vi) $= t|_{Vi}$ for each set V_i of the cover, then $s = t$. Second, the 'gluing axiom,' defined thus: If V_i is an open cover of an open set U, and if for each i there is a section s_i of F over V_i such that for each pair V_i, V_j of the covering sets the restrictions of s_i and s_j agree on the overlaps: $s_i|_{V_i \cap V_j} = s_j|_{V_i \cap V_j}$, then there is a section $s \in F(U)$ such that $s|_{V_i} = s_i$ for each i.[20]

The gluing axiom simply guarantees that among the manifold compatible sections of $F(U)$—that is, those sections satisfying the axiom of local identity—there will be a unique section s which is a 'gluing' or concatenation of sections s_i. Each concatenation of compatible presheaf sections—that is, each 'glued' sheaf section—is thus comparable to an internally consistent, *probable* history by which each probable measurement outcome / Whiteheadian subjective form (whose local context is represented by the open set U) is internally related to the global totality of data. Again, this totality is represented *inductively* via a covering structure of internally related Boolean contexts—the covering sieve; and the correlate totality of internally related facts contextualized via the covering sieve is represented inductively via the Boolean germ—an equivalence class of compatible sections over the covering sieve of compatible underlying Boolean contexts. A sheaf, then, is a bundle of glued sections representing mutually exclusive and exhaustive, logically non-interfering probable measurement outcomes spanning all internally related open covers of the open set / local measurement context. Each probable outcome thus entails its own consistent history by which it is internally related to its dative world.

5.3 Summary

The relational realist interpretation of quantum mechanics provides a mereotopological formulation of extensive connection in quantum theory. Among its many conceptual advantages, this formulation accommodates the phenomenon of nonlocal spatiotemporal extensive connection by representing continuous metrical-extensiveness via a more fundamental quantum mereotopological extensiveness. This is crucial to the coherent interpretation of quantum mechanics because the mathematical rigidity of the conventional Hilbert space formalism (as a topological vector space) does not allow a relativization analogous with the one of classical relativity theory on smooth manifolds. By contrast, and as will be discussed in part II, a mereotopological, category-sheaf theoretic formulation of quantum theory that depicts generative quantum predication in terms of internally related Boolean localization systems, achieves precisely such an objective by generalizing the smooth metrical manifold construction in generic algebraic / categorical terms.

This makes it possible to formulate a quantum mechanical framework of local / global or part / whole relations free from the paradoxes of self-reference discussed in chapter 3, and without the intervention of a classically conceived spatiotemporal continuum. On the contrary, the usual spacetime manifold and its metrical relations (imposing extra conditions like the light-cone causality relations) appear as emergent at a higher level than the more fundamental level of logical and algebraic part-whole internal relations.

Further, in the relational realist interpretation of quantum mechanics, the distinction between [a] physical / metrical / efficient causal relations, and [b] mereotopological / logical internal relations, is reflected in the dipolar nature of each quantum predicative fact / actual occasion. Likewise in the latter, the mutually implicative relationship between [a] the extensive pole of efficient causal relation and [b] the conceptual pole of logical internal relation is evinced empirically by the ineluctable correlation of the orders of causal relation and logical implication in physics—i.e., logical causality.

Among the key issues to be further explored in part II is the problem of localization and the problem of passing from the local to the global in an extensive continuum. Per the relational realist framework, these problems are treated mereotopologically rather than metrically. For example, the emergent metrical relations of relativity theory presuppose a pointwise localization on a smooth manifold and the metric is a local expression of the simultaneity conventions imposed by the constancy of the speed of light. But at the quantum theoretic level, the assumption of point-wise localization of events is not feasible unless a local Boolean measurement context has been specified and also a specific type of measurement has been performed.

In Whitehead's language this is reflected in the distinction between genetic division (presupposing a generic / algebraic type of localization) and coordinate division (like the one of the relativistic spacetime manifold endowed with a semi-Riemannian metric and dictating the light-cone causality relations). By demonstrating the latter's dependency upon the former, the tension between the fundamental mereotopological order of the extensive continuum (modeled in terms of the sheaf-theoretic topos framework) and the more special spatiotemporal causal extensive order pertinent to a point-wise localized and metricized continuum, can be relieved. This has fundamental consequences for the interpretation of the mechanism of differential extensive analysis, which is crucial for physics since it captures dynamical relations in an extensive continuum of events.[21]

In further explication of these consequences, as we turn to a more technical discussion of the mathematical formalism in part II, it must again be emphasized that despite the philosophy's capacity for formal rendering, relational realism is not intended to be interpreted as an exercise in ontic structural realism, which both conflates form and fact (e.g., relation and relatum) and at the same time attempts to abstract form *from* fact, such that form alone is elevated to ontological primacy. Fact and form are mutually implicative in the relational realist philosophy, as are the several dualities discussed earlier, including 'local and global,' 'physical and conceptual,' 'actuality and potentiality,' and 'subject and object.' Embodying these mutually implicative features of nature, each quantum predicative fact / actual occasion is fundamentally a dipolar, elementary *unit of logico-physical internal relation* by which these features are coherently schematized, and by which the orders of efficient causal relation and logical implication are demonstrably cooperative in every generative quantum predication.

Notes

1. Alfred North Whitehead, *Process and Reality: An Essay in Cosmology*, Corrected Edition, ed. D. Griffin and D. Sherburne (New York: Free Press, 1978), 287.

2. Discussed in chapters 3 and 4.

3. Whitehead, *Process and Reality*, 283-84.

4. That is, special and general relativity and the 'early' quantum theory of Planck and Einstein. Whether he was versed in the latter's continued development by Bohr, Heisenberg, Dirac, etc., contemporaneous with his own philosophical work, is a matter of debate.

5. See, for example: Elias Zafiris, "A sheaf-theoretic topos model of the physical 'Continuum' and its cohomological observable dynamics," *International Journal of General Systems* 38, no. 1 (2009): 1-27; Elias Zafiris, "Quantum Event Structures from the Perspective of Grothendieck Topoi," *Foundations of Physics* 34, no. 7 (2004); Elias Zafiris, "Boolean Coverings of Quantum Observable Structure: A Setting for an Abstract Differential Geometric Mechanism," *Journal of Geometry and Physics* 50, no. 99 (2004); Elias Zafiris, "Generalized Topological Covering Systems on Quantum Events' Structures," *Journal of Physics A: Mathematical and General* 39 (2006); Elias Zafiris, "Sheaf-Theoretic Representation of Quantum Measure Algebras," *Journal of Mathematical Physics* 47 (2006).

6. See chapter 2.4.

7. Again, by the statement "α is asymmetrically internally related to Ψ_α," it is meant that Ψ_α is constitutive of α, but α is *not* constitutive of Ψ_α—that is, Ψ_α is *externally* related to α. Analogously, in the conventional understanding of time, the present is internally related to the past, while the past is externally related to the present.

8. Whitehead, *Process and Reality*, 288.

9. See chapters 2.4 and 4.2.

10. Jeffrey Bub, "Quantum Logic, Conditional Probability, and Interference," *Philosophy of Science* 49 (1982): 402-21. See also: Jeffrey Bub, "The problem of properties in quantum mechanics." *Topoi* 10, no. 1 (1991): 27-34.

11. Indeed, they are, as we argue, fundamentally topological rather than metrical. See chapters 2.9 and 3.4; for a formal mathematical discussion, see Part II.

12. Whitehead, *Process and Reality*, 66 (italics added).

13. The logical asymmetry of this process of generation by internal relation is reflected temporally as the one-way arrow of time in thermodynamics and in the decoherence-based interpretations of quantum mechanics. Most generally, it is reflected in the one-way arrow of efficient causality wherein the order of causal relation is always preserved temporally, such that effect can never temporally precede cause—even in the case of relativistic physics.

14. Whitehead, *Process and Reality*, 35.

15. Whitehead, *Process and Reality*, 289.

16. Whitehead, *Process and Reality*, 67.

17. Whitehead, *Process and Reality*, 298.

18. Whitehead, *Process and Reality*, 297-98.

19. Whitehead, *Process and Reality*, 286.

20. Cf. the tensor product relationship discussed in chapters 2 and 3.

21. See chapter 11.6; see also: A. Mallios and E. Zafiris, "The Homological Kähler-De Rham Differential Mechanism – Part I: Application in the General Theory of Relativity," *Advances in Mathematical Physics*, 2011 (2011), doi:10.1155/2011/191083, and A. Mallios and E. Zafiris, "The Homological Kähler-de Rham Differential Mechanism – Part II: Sheaf-Theoretic Localization of Quantum Dynamics," *Advances in Mathematical Physics*, 2011 (2011). doi:10.1155/2011/189801.

INTERLUDE

As applied to the ontological interpretation of quantum mechanics, the convergence of category-sheaf theory and the relational realist philosophy, with its grounding in the theoretical perspectives offered by Whitehead's process theory, is based on the conception of a mereotopological physical continuum of events in a process of continuous ontological formation via extensive connection. This is achieved by explicitly considering potentiality as ontologically significant—that is, in terms of potential predicative facts (i.e., observables) whose actualization is an ontological process rather than merely an epistemic analysis. This process entails a logically conditioned integration of potentia via a selection mechanism by which certain potential predicative facts evolve to become probable predicative facts.

In the category-sheaf theoretic formalism of relational realism, this mechanism is described by the functionality of a uniform fibration of the global totality of quantum events and their relations in terms of local Boolean reference frames. The crucial requirement of sheaf theory in this conception of the quantum event continuum is that the selection mechanism by which certain potential relations evolve to become probable relations is ultimately local and describable in a precise topological sense. With respect to quantum theory this means that the selection of an observable to be measured, by means of a corresponding preparation procedure, locally instantiates a physical context—a local Boolean reference frame for the individuation of measurement outcome events. This is conceived as analogous to the subjective standpoint of a Whiteheadian actual occasion undergoing concrescence, and its associated mereotopological, internal relational selection mechanism by which potential relations are logically integrated to become probable relations—that is, mutually exclusive and exhaustive, valuated subjective forms that satisfy the principle of non-contradiction and the principle of the excluded middle.

While it is true that the notion of a local Boolean reference frame has no practical utility in the usual metrical point-spacetime continuum of classical and relativistic physics, it is absolutely essential when depicting physical systems in a topological continuum where the distinction between local and global is made explicit. In the usual metrical continuum, for example, 'local' refers merely to a point and not to a region. In a topological continuum, however, regions are the focus, and points are derived from them.

The definition of 'local topological region' in quantum mechanics is far from trivial. In classical topological theory, for example, a region is easily defined as an open set, and while this is heuristically useful for an introductory understanding of quantum mechanical topology, it is not formally adequate. This is because the local evolution of potentiality to probability in quantum mechanics (the reduction of the pure state to the mixed state) is enacted via a Boolean logic constraint—i.e., a constraint that subsumes the logical function of the preparation procedure for the measurement of some observable. That is to say, the local selection of a Boolean reference frame is instantiated via the usual preparation measurement, always required in quantum mechanics, by which the 'selected-to-be-measured' observable is initially defined. It is this Boolean contextualization via the standard preparation procedure that enables the probabilistic inference of the actualization of an outcome event contextualized *as* this measured observable. Thus, the topological notion of a local region in this regard—i.e., the measured object's local Boolean contextualization—cannot be adequately represented simply via an open set in a classical topological continuum; rather, it must be represented topologically as a Boolean algebra. By using Grothendieck topology to this end, which generalizes the notion of a topological space, it is possible to use a sheaf-theoretic topological conceptual framework to progress from [a] local or partial information over a local Boolean algebra to [b] a global non-Boolean (i.e., partially Boolean) algebra of events, via the definition of local contexts as Boolean algebras.

Proceeding in this direction, having defined the concept of a local topological region as a local Boolean measurement context in quantum mechanics, one must likewise define the concept of a 'quantum event': In the framework of quantum theory, events are identified as measurement outcomes referent to corresponding observables, and the theory provides the means of correlating these events. In this respect, the conceptual complexity of any ontological interpretation of quantum theory stems from two factors: First, the existence of an event referring to a quantum system can be inferred only probabilistically with respect to the local Boolean measurement context of a selected observable, and moreover, can be affirmed only after a measured result has been registered by a corresponding measuring device. Second, if we agree that a quantum event corresponds to an evaluation of an observable with respect to a measuring device as above, then the totality of events related to the behavior of a quantum system cannot be actualized within the same local Boolean measurement context due to the property of global non-commutativity of quantum observables. Each quantum event, then, is a novel actualized fact conditioned by the selection constraints of the local context by which it is individuated.

Quantum-theoretically this means that locally, each event is a Boolean event. This is important because locally in the quantum continuum, all the rules of Boolean logic for logical inferences are operative. As discussed extensively in part I, this is due to the categorical specification of every local context as a local Boolean algebra which, in turn, allows for the local reduction of potentiality to probability in quantum theory. More significantly, this local Boolean reduction obeys the rules of restriction in the definition of a presheaf for nested sequences of Boolean algebras. This is exactly what is meant by 'uniformity' in the continuum, and how this is modeled in terms of a uniform fibration. In the same vein of ideas, the notion of a Boolean algebra thought of as a 'local context' refers to a variable-elastic topological concept; it is, in other words, not rigid like a point.

In order to correlate the above with the relational realist philosophical notion of an extensive event continuum discussed in chapter 5, it is instructive to stress the following: In the Whiteheadian, relational realist scheme, the notion of 'continuum' captures the concept of all potential extensive relations. In the sheaf-theoretic scheme, this is captured by means of a uniform fibration—that is, by means of a presheaf of observables defined over a category of local contexts. The local contexts thus provide the base category by which potential extensive relations can be defined as 'relations among local sections.' These potential extensive relations, in other words, are induced by the relations among local contexts. Thus, one must distinguish between: [a] a horizontal dimension, which contains the base objects (local contexts identified with local Boolean algebras) and [b] a vertical dimension, which contains the potential extensive relations (in terms of local predicative facts or local sections, or local observables). In a pictorial sense, then, the vertical dimension is the display space of potential relations induced by the structure of a uniform fibration over the base category of local contexts. Thus global observables are locally graded with respect to local contexts—i.e., structured in terms of distinctive levels of potential relations forming a category.

In this way, local observables can be related only by means of the relations induced from their underlying local contexts. The notion of a continuum, then, is captured by the totality of potential relations defined over the underlying category of local contexts according to the requirements of a uniform fibration. Thus, a quantum event—i.e., a quantum measurement outcome or 'actual occasion' in Whitehead's terminology—is a local, Boolean context-dependent, true-false evaluation of a section. In other words, a section, which represents a local quantum observable contextualized according to its underlying local Boolean algebra, is defined via a morphism of that Boolean algebra to the bivalent logical (true / false) evaluation. Thus, the evaluation of a local section is a basic formalization of the concrescence of an actual occasion. As discussed in part I, according to the Kochen-Specker theorem, one cannot evaluate simultaneously all observables on a single two-valued Boolean algebra.

This is possible only locally—that is, only with respect to some local Boolean algebra corresponding to some selected observable defined by the standard preparation procedure in quantum measurement.

A natural question, then, is, "How do we manage to reconcile all these varied forms of 'local' and avoid inconsistencies?" First, we proceed by obviating the metrical spacetime point continuum as the conventionally assumed fundamental order, recognizing it instead as a higher order abstraction which we can recover when appropriate at the limit where all local elastic variability is contracted to rigid points. There, we can subsequently impose a Euclidean or Lorentzian or Riemannian metric inducing the well-known causality conditions. Second, we introduce in its place a category-theoretic, mereotopological continuum in a process of ontological formation (rather than predefined existence as in the case of the metrical continuum) apprehended locally via the Boolean contexts covering it. Third, we invoke a process of gluing topologically from the local level to the global level.

The proper understanding of the topological term 'gluing' is crucial, here, because this is the key step to progress mereotopologically from a local level to a higher local level (keeping in mind that the notion of local is variable or elastic) and, ultimately, to an inductive approximation of the global level. Gluing of sections, as introduced in chapter 5, takes place over partially compatible local Boolean contexts, such that gluing is essentially a compatibility relation between local sections defined over overlapping local Boolean contexts. Thus gluing is the means of extending a local section over a local context to a higher level local context, and so on, towards an inductive approximation of the global level.

With respect to classical physics, wherein a single local Boolean context is also considered global, the evaluation of a section (observable) would simply represent a measurement outcome—i.e., an actual occasion—such that the latter could be reduced analytically to that formal structure. Thus, classically we might think of a becoming actual occasion as the process of evaluating a section, and likewise think of the result of the evaluation—the measurement outcome—as the actualized occasion. But quantum theoretically, this cannot be achieved via analytical reduction the way it can be classically, because the global cannot be defined as a simple composite of local contexts (again, per the Kochen-Specker theorem). Quantum mechanically, we are limited to local relations. So how is it possible to extend this procedure from the local to the global in the quantum case? The answer, provided by the sheaf-theoretic scheme, is by forming Boolean germs—that is, equivalence classes of compatible sections over compatible underlying Boolean contexts. The notion of a germ of sections, in other words, is defined by means of inducing the same contextual information if sections are restricted to an overlap of local Boolean contexts. The germ thus cannot be defined in abstraction from its local contextuality—i.e., wherein local sections agree; likewise, at the same time, the local context associated with a germ is never considered apart from the local information content contextualized.

For example, in an EPR experiment for spin ½ systems (see chapter 2.8 and chapter 9.4) we have two local Boolean algebras correlated and glued together into a higher level, which is manifested by their tensor product. The latter thus essentially represents equivalence classes of compatible local observable information contained in the separated Boolean algebras. The tensor product, in other words, incorporates the compatible correlated information of two local Boolean contexts overlapping partially due to the common origin of the two separated systems before the emission. It is this Boolean correlative context with respect to the global level that defines the nonlocal correlations that are so problematic from the perspective of a metrical spatiotemporal point continuum. But in a topological continuum it is still a legitimate context, formalized via the tensor product of the respective Boolean contexts. Again, this is not possible in a metrical point continuum. Thus, what we gain in the topologically defined continuum is the fact that the tensor product, when understood *as a local context itself*, can be glued further to an even higher level if we entangle it with the Boolean algebra of a third system correlated initially with the two former ones, and so on. What has been described in pictorial terms, then, is the mechanism of extensive connection in a topological continuum through the transition from a local context to a higher local context and so on towards the global, such that inversely, the reduction of the global into different levels of local contexts is consistent, compatible, uniform and non-paradoxical.

But how is this relational framework consistent with the notion of a quantum event? This is the crux of the distinction between the local and the global, and the necessity of extensive connection. As discussed in part I, and as will be explored further in part II, we argue that each quantum measurement event is a *novel* actualized fact, but one whose novelty is nevertheless conditioned by the selection constraints of the local Boolean context by which it is individuated. Thus, quantum measurement events as elementary units of relation are always individuated and actualized with respect to local Boolean contexts. But in that regard, how can we think of a locally contextualized quantum event as a globally relevant, ontologically significant constituent of nature, such that it can have the capacity to change the continuum globally? Again, the application of sheaf theory to quantum theory provides the answer: Globally, a quantum event is not specified solely by the evaluation of an observable, as is the case with a classical event, but rather is specified by the evaluation of a corresponding Boolean observable germ. That is to say, a quantum event is specified in a global sense by the equivalence class of all compatible Boolean contexts, conceived in all different local levels, 'nested' within each other by internal relation, with respect to which observable information can be glued. Thus a Boolean germ is an internal relational progression in the ontological formation of the quantum continuum; it is the relational structure of extensive connection from the local to the global.

It is instructive now to correlate concisely the above sheaf-theoretic viewpoint with the corresponding Whiteheadian, relational realist viewpoint explicated in part I. This correlation rests upon two central concepts asserting the ontological primacy of quantum events / facts / actual occasions: [1] The subjective aspect of each quantum actual occasion, whereby the latter determines its *own* subjective standpoint or local context, and thus its own logical integration of potential internal relations contextualized by this standpoint; [2] The objective ('superjective') aspect of each quantum actual occasion, whereby the latter, once actualized, serves as a datum to which subsequent actual occasions-in-process will be internally related. With respect to both the subjective and objective aspects of the actual occasion, as discussed in chapter 5.1, it is useful to think of a Boolean germ as representing the structure of contextualized internal relations between an actual occasion-in-process and its dative world—i.e., representing the internal constitution of an actual occasion in its process of concrescence. As discussed in chapter 4, this structure pertains to the supplementary phase of Whiteheadian concrescence, and presupposes a primary phase whereby the subject occasion's local context / subjective standpoint is initially defined via internal relation to a dative totality by way of a uniform fibration of this totality.

Indeed, the particular structure of connectivity definitive of quantum actual occasion / Whiteheadian 'concrescence' here is reflected in the Latin root of the word, which translates as 'growing together.' From a sheaf-theoretic viewpoint, this is exactly what a Boolean germ represents with respect to the actualization of a quantum measurement outcome: It integrates compatible information at higher and higher local Boolean levels via equivalence classes, and in this way it progresses from the local to the global. Most important, the integration is possible only through the uniform restriction of potential internal relations among quantum facts via their associated internally related local Boolean contexts. Thus, a global internal relational covering structure—viz., a Boolean localization scheme of the quantum continuum—is needed for a coherent and consistent understanding of how a locally ('subjectively') contextualized quantum event is objectively constitutive of a global totality of such events.

We can thus conceive of the generative progression of a Boolean germ as a process formative of a coherent and consistent global history of internally related quantum events, each event carrying with it its own local contextualization. Thus the locally contextualized actual occasion-in-process is internally related to its dative actual occasions *via their own local contextualizations*, such that it is not only the dative occasions that are internally related, but also their local contexts. It is an internal relation of the local to the global, in other words, wherein the global is always incorporated as a nested series of local, partially compatible Boolean contexts extending to higher and higher levels, thus inductively constitutive of the global-in-formation. This inductive characterization of the global via internal relation of locally contextualized data is defined category-theoretically as the inductive limit (introduced in part I and mathematically formalized in part II). Note that since the Boolean germ is a generative structure—i.e., it is serially increased with each quantum actual occasion—it cannot be conceived in the set-theoretic sense of an already extant, predefined totality; rather, it is a synthetic totality, continuously in extensive formation via a process of discrete, internally related, predicative actualizations.

On the basis of this conceptualization of the functional role of Boolean germs in the formation of the quantum continuum, the formation of an equivalence class constitutive of a Boolean germ always takes place relative to the indexical subjective standpoint / local context of a particular actual occasion-in-process. Equivalently put, the standpoint of an actual occasion-in-process instantiates the local context over which some germ of compatible sections (an equivalence class of sections) may be defined. In this sense, an actual occasion in the process of concrescence (i.e., a quantum actualization-in-process) can be understood sheaf-theoretically as the synthetic evaluation of a germ—that is, the actualization of some potential augmentation of the germ. In this way, the subjective and objective features of the quantum actual occasion are always cooperative. Its subjective aspect is reflected in its own local contextualization; and once actualized, its objective aspect is reflected in its role as a datum to which other actual occasions-in-process will be internally related. Thus, one can trace an actual occasion's genetic internal relational structure by defining germs of compatible sections over partially overlapping dative contexts. The compatibility of these sections is always relative to the indexical local context of the actual occasion-in-process; in this way, the latter is always considered the starting point of its own concrescence, constituted internally via the formation of a corresponding germ.

It is important to emphasize again that the formation of a germ involves [a] the localization of sections and [b] the formation of equivalence classes of compatible sections by which a local context of agreement among these sections (i.e., inter-contextual compatibility) is always indexed to the local context of the actual occasion-in-process. In this way, the latter with its own local contextualization always defines a germ of local potential predicative facts / local observables / local sections, one of which will become actualized (e.g., as a quantum measurement outcome.) Thus a germ describes the internal constitution of an actual occasion-in-process in terms of its *potential* internal relations with the dative world. Likewise, once actualized, the occasion is specified in terms of its *actual* internal relations with the dative world—i.e., a specification whereby the germ is augmented by the fact of the occasion's actualization.

In this way, the sheaf-theoretic explication of Whitehead's dictum "the many become one and are increased by one" is as follows: The many become one by compatible interconnection or gluing of their information content relative to the local 'indexical' context of a particular actual occasion-in-process. This interconnection entails the internal relation of the actual occasion-in-process to the global totality of already actualized, dative occasions in their own respective local, partially compatible contexts, thus forming a germ. Once the process of concrescence is complete and the occasion is actualized, the dative totality is thereby 'increased by one'; that is, the novel occasion, constitutive of a novel totality, will serve as a datum to which a subsequent actual occasion-in-process will be internally related according to the latter's own local context, and so on. The conceptual framework by which this process was elucidated in part I can be rigorously formalized in terms of the mathematical framework of sheaf theory, as will be demonstrated in part II.

In preface to this more technical discussion, it will be useful to emphasize two key aspects of the relational realist philosophical cosmology as it will be formalized sheaf theoretically: The first of these aspects refers to the issue of self-reference, discussed in chapter 3, wherein the same theoretical entity can assume more than one functional role within the same theory. An example of this is provided by the dipolar 'subject-superject' role characteristic of the Whiteheadian actual occasion. In the present account, self-reference is bypassed via the transition from a set-theoretic universe of discourse to a topological (or better, mereotopological) category-theoretic universe, which is structured into distinct, internally related, extensive local levels. In this way, it is possible to move from level to level keeping the adjective 'local' during the transition, because what is local is variable-elastic; that is, its structure is preserved under extensive connection via germs (equivalence classes of potential internal relations) and is thus coherently constitutive of the global—not as a predefined totality, but rather as a totality in process—i.e., the global in-formation. A reader trained in formal logic might simply say that self-reference is bypassed via making identity locally variable.

The second essential aspect refers to the novel idea of the mutually implicative internal relation of local and global. On the one hand, every local observable (potential fact) is internally related, via the congruence and compatibility relations induced by a Boolean localization scheme, to a global quantum event structure of actualities. Thus, as noted earlier, a global quantum event 'continuum' is not formalized as a completed, already extant set-theoretic structure; rather, it is constituted in a continuous mereotopological process of ontological formation via the actualization of potential facts conditioned by local Boolean contexts. Thus it is also the case that the global is internally related to the local. As will be elaborated throughout the course of part II, with respect to the functional role of Boolean germs as internally constitutive of becoming actual occasions, this process of formation via the mutually implicative internal relation of local and global is fully logically coherent and consistent in the sense of respecting the rules of transition from the local and partial to the global and total, and conversely.

PART II

Mathematical Foundations of Quantum Relational Realism

A sheaf-theoretic description of internally related quantum event structures in terms of Boolean covering systems

ELIAS ZAFIRIS
M.Sc., Ph.D. (Imperial College, University of London)

CHAPTER 6

Notion of Localization Processes

6.1 Localization in Physical Theories

The operational foundations of physics as a natural science are based on observation and measurement. The abstraction of the measurement process gives rise to the notion of *coordinatization* or *arithmetization*. In this sense the formation of algebraic structures of 'number-like' quantities of any particular operational form can be thought of as solutions to corresponding physical measurement problems. In the interpretation of physical theories, our empirical access to the world is coordinatized through the concept of observables. *Observables* denote physical quantities that, in principle, can be measured in the context of appropriate experimental arrangements. The crucial further assumption is that quantities admissible as measured results must be *real numbers*, since it is accepted that the resort to real numbers has the advantage of making our empirical access to the world secure. More concretely,the model adopted to represent the measured values of observables is the real line $\mathbb{R}$ and its powers, specified as a set-theoretic structure of points that are independent and possess the property of infinite distinguishability with absolute precision. In this way, the semantics of the *physical continuum* in the standard interpretation of physical systems theories is associated with the codomain of valuation of observables, that is, the set-theoretic real line model $\mathbb{R}$ and its powers. Due to this assumption, in any experiment performed by an observer, the propositions that can be made concerning an observable are of the type which asserts that the value of an observable lies in some *measurable set* of the real numbers. A proposition of this kind corresponds to an *event* as it is apprehended by an observer using his measuring device.

The stipulation of the set-theoretic real line model $\mathbb{R}$ as the semantic information carrier of the notion of physical 'continuum' has posed enormous interpretational problems for quantum theory, and in particular quantum measurement. The root of this problem is situated on the different notions of *localization* characterizing *classical* and *quantum* observables. Overlooking this subtle difference, referring to localization properties of observables in different physical regimes, is a source of paradoxes and misinterpretations.

First, it is important to clarify the meaning of the notion of localization. Physical observation and measurement presupposes, at the fundamental level, the existence of a *localization process* via which we may extract information related to the local behavior of a natural system. Only on the basis of a localization process does it become possible to discern events and assign an individuality to them. Generally, a localization process is co-implied by the preparation of suitable local reference contexts for measurement. These contexts identify concretely the kind of reference frames used for observation of events. The methodology of observation is effectuated by the functioning of event-registering measurement devices, which operate locally within the context of prepared reference frames. In this general setting, it is important to notice that registering an event, which has been observed in the context of a reference frame, is not always equivalent to conferring a numerical identity to it by means of a real value corresponding to a physical attribute. On the contrary, the latter is only a *limited case* of the localization process, when, in particular, it is assumed that all reference contexts can be contracted to points. This is exactly the crucial assumption underlying the employment of the set-theoretic structure of the real line as a model of 'physical continuum.' The consequences of this common assumption have posed enormous problems for the interpretation of quantum theories, which are characterized by different localization schemes for the individuation of events. More concretely, set-theoretic axiomatizations of quantum event structures tend to hide the intrinsic significance of *Boolean* localizing contexts in the formation of these structures.

The operational procedures followed in quantum measurement are based explicitly on the employment of appropriate Boolean environments. The construction of these contexts of observation is related to our specific measurement arrangements and can be metaphorically considered as Boolean pattern recognition mechanisms. In this way, we may argue that the real significance of a quantum event structure proves to be, not at the level of events, but at the level of the specific form of interlocking or gluing together overlapping Boolean localizing contexts of observation into a globally coherent whole. From a logical perspective, the *gluing mechanism* accompanying quantum observable localization schemes is effectively a novel form of *logical internal relation*, which is absent classically, but plays a fundamental role in conceptualizing quantum-theoretic predication in topological terms. This is crucial for modern physics because topology is uniquiely appropriate to the depiction of the transition from the *local* to the *global*, and vice versa, without using any metrical notions. For example, the significance of localization and gluing is of paramount importance for making sense of some unusual *topological quantum effects*, like quantal phase factors (the Berry phase) and the Aharonov-Bohm effect, as well as for understanding *decoherence* in the framework of the consistent histories approach to quantum theory, via coarse-graining localization processes.[1]

A natural question arising in this setting, inspired by the various interpretational problems of quantum theory, is the following: Is there an alternative conception of the 'physical continuum' that does not rely on an assumed preexisting set-theoretic structure of points on the real line? This question properly challenges the hypothesis that our form of observation is tautological with set-theoretic real number representability. In order to realize the necessity of an evolution in the semantics of the 'physical continuum' we emphasize that we have to consider carefully the different notions of localization associated with physical observables in classical and quantum measurement processes, respectively. The elucidation of this difference induces a natural evolution in the semantics of the physical continuum from a set-theoretic to a sheaf-theoretic one, as will be explained presently.

6.2 Localization Schemes

Before discussing the functioning of localization processes or schemes related with physical observables, it is necessary to include a brief intuitive description of ideas related with the notion of a physical continuum. For this purpose, the following quotations from Henri Poincare[2] and Alfred North Whitehead,[3] respectively, are particularly helpful:

> We cannot say that our element is without extension, since we cannot distinguish it from neighboring elements and it is thus surrounded by a sort of haze. If the astronomical comparison may be allowed, our "elements" would be like nebulae, whereas the mathematical points would be like stars.

> The ultimate facts of nature are events . . . The constitutive character of nature is expressed by the "contingency of appearance" and the "uniform significance of events" . . . By extension we mean that quality by virtue of which one event may be part of another, or two events may have a common part. Nature is a continuum of events so that any two events are both parts of some larger event.

The primary conception of a 'physical continuum' constitutes an inexhaustible complex of overlapping and non-overlapping events. Such an *event ontology* lies at the basis of our physical theories and necessitates a careful analysis of its premises before the acceptance of a particular mathematical model as a representational device for its functioning. The consideration of the notion of 'event' as a primary concept in the comprehension of the 'physical continuum' poses immediately the following question: How are events to be *related* to each other? If the continuity of nature is to be ascribed to the relations among events, then the fundamental relation is *extension*. The relata in the relation of extension are the events, such that each event is part of a larger whole and each event encompasses smaller events. Extension is also inextricably tied with the assumption of *divisibility* of events signifying a *part-whole* or *local-global* type of relation. Thus, the crucial role in the conceptualization of event ontology is played by the topological notion of a localization scheme, inevitably endowed with an internal relational gluing structure, such that there exists a consistency in the *inverse processes* of *extending* or *inducing* observables from the local or partial to the global, and conversely, *restricting* or reducing observables from the global to the local. The internal relational gluing structure of a localization scheme expresses the fundamental idea that "the primary relationship of physical occasions is extensive connection."[4]

It is important to emphasize, again, that only via a localization scheme does it become possible to discern events and assign an individuality to them. The formulation of the relations of extension and restriction via the physical and mathematical notion of a localization scheme elucidates the following excerpts from Whitehead's *Process and Reality*, seen earlier in chapter 5:

> The real potentialities relative to all standpoints are coordinated as diverse determinations of one extensive continuum. This extensive continuum is one relational complex in which all potential objectifications find their niche. It underlies the whole world, past, present, and future. Considered in its full generality, apart from the additional conditions proper only to the cosmic epoch of electrons, protons, molecules, and star-systems, the properties of this continuum are very few and do not include the relationships of metrical geometry. An extensive continuum is a complex of entities united by the various allied relationships of whole to part, and of overlapping so as to possess common parts, and of contact, and of other relationships derived from

> these primary relationships. This extensive continuum expresses the solidarity of all possible standpoints throughout the whole process of the world. It is not a fact prior to the world; it is the first determination of order—that is, of real potentiality—arising out of the general character of the world . . . All actual entities are related according to the determinations of this continuum; and all possible actual entities in the future must exemplify these determinations in their relations within an already actual world.[5]

> Extension, apart from its spatialization and temporalization, is that general scheme of relationships providing the capacity that many objects can be welded into the real unity of one experience. Thus, an act of experience has an objective scheme of extensive order by reason of the double fact that its own perspective standpoint has extensive content, and that the other actual entities are objectified with the retention of their extensive relationships. These extensive relationships are more fundamental than their more special spatial and temporal relationships. Extension is the most general scheme of real potentiality, providing the background for all other organic relations.[6]

The general purpose of a localization scheme amounts to filtering the information contained in a global structure of partially ordered physical events, through an appropriate *algebraic structure of localizing frames*, which can be determined by a *homologous* operational physical procedure of measurements. In this way, the substantial constituents of a localization scheme should specify the kind of loci of variation (or equivalently, reference contexts) to be used for observation of events. These contexts should instantiate generalized *reference frames* such that reference to concrete events of the specified kind can be made possible with respect to them. The requirements accompanying their instantiation are the following: First, they should constitute expressions of event ontology respecting the inverse relations of extension and restriction. Second, they should *not* be based on an underlying structure of set-theoretic points on the real line. This equivalently means that localization schemes used for the individuation of events should not depend on the existence of points. Third, due to their determination by a homologous operational physical procedure of measurements the *internal relations* of these varying reference contexts should play a crucial role in the coherent action of a localization scheme, and thus should not be neglected in its initial specification. According to Whitehead:

> The extensive scheme is nothing else than the generic morphology of the internal relations which bind the actual occasions into a nexus, and which bind the prehensions of any one actual occasion into a unity, coordinately divisible.[7]

Interpreted mathematically, this essentially means that the reference contexts of a localization scheme should not only form a class, but they should form a *category*. More concretely, *the reference contexts together with their structural transformations should form a mathematical category*[8] of localizing frames. For example, a localization scheme, which is based on the process of coarse-graining, should be implemented by a partial ordering relation among the localizing frames, which is obviously a category, where the only possible arrow between any two localizing frames is an inclusion. It is necessary to emphasize that the kind of reference contexts or loci of variation, together with their structural internal relations, determines precisely the concrete categorical environment employed operationally by a homologous operational physical procedure of measurements. An instance of this is a category of open sets, ordered by inclusion, in a topological measurement space. Another instance, of particular significance in quantum theory, is a category of Boolean localizing algebraic contexts of quantum measurements, which may be considered as local Boolean subalgebras of a global quantum observable algebra.[9]

6.3 The Conceptual Framework of Category Theory

6.3.1 The Necessity of a Categorical Framework

Category theory provides a general theoretical framework for dealing with systems formalized through appropriate mathematical structures emphasizing their *mutual relations* and *transformations*. The central focus of the categorical way of rethinking basic notions can be described as a shift in the emphasis of what is considered to be fundamental for the formation of structures. In the set-theoretic mode of thinking, structures of any conceivable form are defined as sets of elements endowed with appropriate external relations. In the category-theoretic mode, the emphasis is placed on the transformations among the objects of a category devised to represent its internal relational structure by means of appropriate constraints on the collection of these transformations. In this sense, the notion of structure does not refer exclusively to a fixed universe of sets of predetermined elements, but acquires a *variable reference*.

The major foundational significance of category theory regarding our present enquiry on the problem of modeling the physical continuum, according to the requirements posed previously, is related to the following realization: The effect of casting a mathematical construct for modeling the physical continuum in category-theoretic language is equivalent to conferring a degree of ambiguity of reference on this construct. This is an appealing aspect for our purposes since it is exactly the doctrine of employing these constructs to depict a fixed, absolute universe of sets that we wish to overthrow. According to category theory the mathematical constructs used for our modeling purposes should be regarded as possessing meaning only in relation to local frameworks, thus replacing the notion of unique reference to the universe of sets with other legitimate and varying frameworks of interpretation.[10] This means, in turn, that a model of the physical continuum can be determined only in relation to a category and, most important, this category may vary.

We claim that it is exactly the referential ambiguity of the construct representing the physical continuum, interpreted in varying categorical environments, that is needed in order to accomplish the critical evolution away from the conventionally assumed preexisting structure of points representing a true 'physical continuum' of events—an evolution that is of paramount importance to modern fundamental physics. Of course, the ambiguity of reference after casting the construct in categorical language should be fixed by its applicability in concrete categorical environments. Thus, the instantiation of these reference contexts should, from a physical viewpoint, depend on the *operational procedures* devised for the individuation of events in the continuum. In this sense, the notion of a localization scheme becomes a fundamental ingredient in modeling the physical continuum because it depicts the appropriate categorical environment suited to express its meaning, in agreement with the operational procedures followed by distinct physical theories.

Apart from determining the appropriate categorical environments for the operational modeling of the physical continuum, localization schemes referring directly to events likewise play the equally important role of representing a *base structure of varying loci* of the categorical kind depicted—a base structure over which the *partial order* of the extended continuum of events should *fiber*. Notice that in the set-theoretic real line model the role of such a locus is played only by a point, since a point conferring numerical identity to an event is the unique idealized localization measure. Consequently, from a topological perspective, adopting this assumption of an *atomic topology* on the continuum set is responsible for the conception of observable events as sharply distinguishable in classical theories. Indeed, from such a perspective, the fibering of an event continuum over a base structure of varying loci, representing concrete localization schemes in a topological context, can provide for such distinguishability and yet account for possible variations in the topology as well.

It is instructive to make clear that the process of casting a concept in categorical language is achieved by the characterization of the concept in terms of an *arrow-theoretic* formulation, and not by a specification in terms of a set equipped with certain operations and satisfying axioms expressible as relations among its elements. In this sense, the adoption of the categorical syntax involves at least two important conceptual shifts away from typical thinking about mathematical constructs considered for the modeling of the physical continuum. The first shift concerns the primary significance of *structure-preserving mappings*, referred to as arrows in category theoretic language. This shift reflects an opposition to the usual prevailing significance of sets with structure, referred to as objects, correspondingly. The second shift concerns the meaning of a *universal construction*, which is made precise in the categorical formalism, and moreover constitutes the main objective of the latter in the syntactical terms of this theory.

6.3.2 Categorical Principles and Terminology

The basic categorical principles that we adopt for this application are summarized as follows:

[i] To each kind of mathematical structure used to represent systems, there corresponds a *category* whose objects have that structure, and whose morphisms preserve it.

[ii] To any natural construction on structures of one kind, yielding structures of another kind, there corresponds a *functor* from the category of the first specified kind to the category of the second. The implementation of this principle is associated with the fact that a construction is not merely a function from objects of one kind to objects of another kind, but must preserve the essential relationships among objects.

[iii] To each natural translation between two functors having identical domains and codomains there corresponds a *natural transformation.*

[iv] To any canonical bidirectional functorial correlation between two kinds of mathematical structures there corresponds an *adjunction* (pair of *adjoint functors*) between the corresponding categories.

Categories:

A category $\mathcal{C}$ is a class of objects and morphisms of objects such that the following properties are satisfied:

[1] For any objects X,Y all morphisms $f : X \to Y$ form a set denoted $Hom_{\mathcal{C}}(X,Y)$;

[2] For any object X an element $id_X \in Hom_{\mathcal{C}}(X,X)$ is distinguished; it is called the identity morphism;

[3] For arbitrary objects X, Y, Z, the set mapping is defined $Hom_{\mathcal{C}}(X,Y) \times Hom_{\mathcal{C}}(Y,Z) \to Hom_{\mathcal{C}}(X,Z)$;

For morphisms $g \in Hom_{\mathcal{C}}(X,Y)$, $h \in Hom_{\mathcal{C}}(Y,Z)$ the image of the pair (g,h) is called the composition; it is denoted $h \circ g$. The composition operation is associative;

[4] For any $f \in Hom_{\mathcal{C}}(X,Y)$ we have $id_Y \circ f = f \circ id_X = f$.

For an arbitrary category $\mathcal{C}$ the opposite category $\mathcal{C}^{op}$ is defined in the following way: the objects are the same, but $Hom_{\mathcal{C}^{op}}(X,Y) = Hom_{\mathcal{C}}(Y,X)$, that is, all arrows are inverted. A category $\mathcal{C}$ is called small if the classes of its objects and morphisms form genuine sets respectively.

Functors:

Let $\mathcal{C}$, $\mathcal{D}$ be categories. A covariant functor $\mathbf{F} : \mathcal{C} \to \mathcal{D}$ is a class mapping that transforms objects to objects and morphisms to morphisms preserving compositions and identity morphisms:

$$\mathbf{F}(id_X) = id_{\mathbf{F}(X)}; \mathbf{F}(g \circ f) = \mathbf{F}(g) \circ \mathbf{F}(f)$$

A contravariant functor $\hat{\mathbf{F}} : \mathcal{C} \to \mathcal{D}$ is, by definition, a covariant functor $\mathbf{F} : \mathcal{C}^{op} \to \mathcal{D}$.

Natural Transformations:

Let $\mathcal{C}$, $\mathcal{D}$ be categories, and let further $\mathbf{F}$, $\mathbf{G}$ be functors from the category $\mathcal{C}$ to the category $\mathcal{D}$. A natural transformation τ from $\mathbf{F}$ to $\mathbf{G}$ is a mapping that assigns to each object A in $\mathcal{C}$ a morphism τ_A from $\mathbf{F}(A)$ to $\mathbf{G}(A)$ in $\mathcal{D}$, such that for every arrow $f : A \to B$ in $\mathcal{C}$ the following diagram in $\mathcal{D}$ commutes:

$$\begin{array}{ccc} \mathbf{F}(A) & \xrightarrow{\tau_A} & \mathbf{G}(A) \\ {\scriptstyle \mathbf{F}(f)}\downarrow & & \downarrow{\scriptstyle \mathbf{G}(f)} \\ \mathbf{F}(B) & \xrightarrow{\tau_B} & \mathbf{G}(B) \end{array}$$

That is, for every arrow $f : A \to B$ in $\mathcal{C}$ we have:

$$\mathbf{G}(f) \circ \tau_A = \tau_B \circ \mathbf{F}(f)$$

Natural Isomorphisms:

A natural transformation $\tau : \mathbf{F} \to \mathbf{G}$ is called a natural isomorphism if every component τ_A is invertible.

Adjoint Functors:

Let $\mathbf{F} : \mathcal{C} \to \mathcal{D}$ and $\mathbf{G} : \mathcal{D} \to \mathcal{C}$ be functors. We say that $\mathbf{F}$ is left adjoint to $\mathbf{G}$ (and correspondingly that $\mathbf{G}$ is right adjoint to $\mathbf{F}$), if there exists a bijective correspondence between the arrows $\mathbf{F}(C) \to D$ in $\mathcal{D}$ and $C \to \mathbf{G}(D)$ in $\mathcal{C}$, which is natural in both C and D.

$$\mathbf{F}:\mathcal{C}\rightleftarrows\mathcal{D}:\mathbf{G}$$

Pictorially we have:

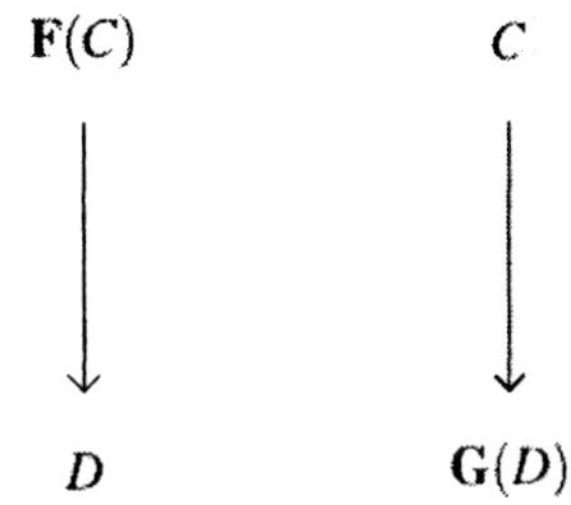

where the left part is in $\mathcal{D}$ and the right in $\mathcal{C}$. Then, we say that the above pair of adjoint functors constitute a categorical adjunction.

Diagrams:

A diagram $\mathbf{X}=(\{X_i\}_{i\in I},\{F_{ij}\}_{i,j\in I})$ in a category $\mathcal{C}$ is defined as an indexed family of objects $\{X_i\}_{i\in I}$ and a family of morphism sets $\{F_{ij}\}_{i,j\in I}\subseteq Hom_{\mathcal{C}}(X_i,X_j)$.

Cocones:

A *cocone* of the diagram $\mathbf{X}=(\{X_i\}_{i\in I},\{F_{ij}\}_{i,j\in I})$ in a category $\mathcal{C}$, consists of an object X in $\mathcal{C}$, and for every $i\in I$, a morphism $f_i:X_i\rightarrow X$, such that $f_i=f_j\circ f_{ij}$ for all $j\in I$, that is, such that for every i, $j\in I$, and for every $f_{ij}\in F_{ij}$ the diagram below commutes:

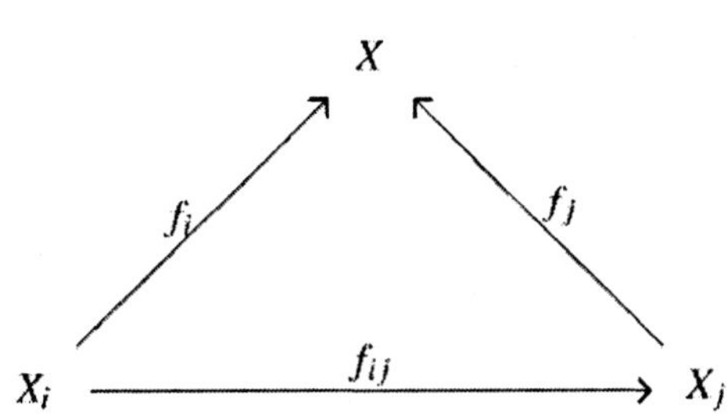

Colimits:

A colimit (*inductive limit*) of the diagram $\mathbf{X} = (\{X_i\}_{i\in I}, \{F_{ij}\}_{i,j\in I})$ is a cocone with the property that for every other cocone given by morphisms $f_{i'} : X_i \to X'$, there exists exactly one morphism $f : X \to X'$, such that $f_{i'} = f \circ f_i$, for all $i \in I$ (universality property).

Reversing the arrows in the above definitions of cocone and colimit of a diagram $\mathbf{X} = (\{X_i\}_{i\in I}, \{F_{ij}\}_{i,j\in I})$ in a category $\mathcal{C}$, results in the dual notions called *cone* and *limit* (*projective limit*) of $\mathbf{X}$ respectively. Moreover, starting with a diagram $\mathbf{X} = (\{X_i\}_{i\in I}, \{F_{ij}\}_{i,j\in I})$ in a category $\mathcal{C}$ that consists only of the objects X_i, $i \in I$, as nodes but without morphisms, that is all $F_{ij} = \varnothing$, we obtain the notion of the categorical *coproduct*, $\coprod_{i\in I} X_i$ (as a special colimit) and *product*, $\prod_{i\in I} X_i$ (as a special limit) respectively. The morphisms f_i in the corresponding definitions are called *canonical injections* of the coproduct and canonical projections of the product, respectively. We emphasize that we can derive special notions of limits and colimits, corresponding to the *shape* of the base diagram $\mathbf{X}$. In this sense we obtain the following; an *initial object* is the colimit of the diagram consisting of the empty set. A *coequalizer* is the colimit of a diagram consisting of two parallel arrows $A \rightrightarrows B$. A *pushout* is the colimit of a diagram of the form:

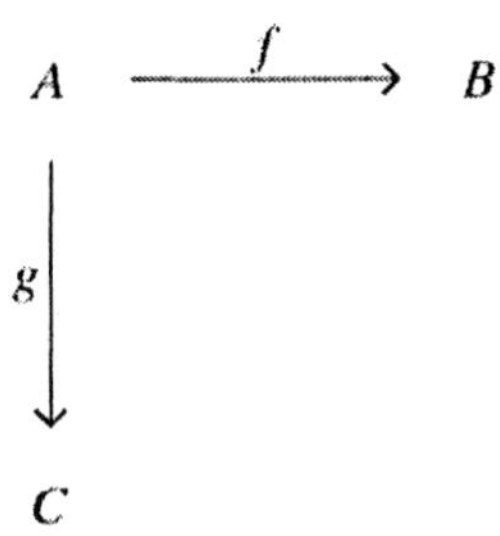

The dual notions are the following: a *terminal object* is the limit of the diagram consisting of the empty set. An *equalizer* is the limit of a diagram consisting of two parallel arrows $A \rightrightarrows B$. A *pullback* is the limit of a diagram of the form:

$$\begin{array}{ccc} & & B \\ & & \downarrow f \\ C & \xrightarrow{g} & A \end{array}$$

6.3.3 Universality and Equivalence

The definition of a category of algebraic objects of some kind with arrows being *homomorphisms* (structure-preserving morphisms) between them may be thought as an abstraction of the behavior of functions closed under the associative operation of composition. More precisely, the notion of a function is generalized to the notion of a homomorphism, whereas the associative operation of composition becomes an operation on sets of homomorphisms between algebraic objects of the same kind satisfying the same properties that functions and compositions satisfy. Notice that the composition of two functions *f* and g, denoted as $f \circ g$, is defined only in the case that the codomain of g is the domain of *f*. Moreover, the composition of a function *f* with the identity of either its domain or codomain gives *f* again.

In a nutshell, the notion of a category of algebraic objects of some kind is a conception based on the behavior of *functions* closed under the associative operation of composition, abstracted in terms of homomorphisms, which in turn, have been idealized as algebraic 'generalized elements' (or their duals) determining completely the algebraic objects themselves. Because there exists an obvious *duality* between *incoming* and *outgoing* homomorphisms with respect to an algebraic object, this is built into the definition of a category so that the operation of *arrow reversal* leaves *invariant* the concept of a category, meaning that this operation gives again a category being in dual or opposite relation in comparison to the given one. A consequence of this fact is that all categorical constructions come in *dual pairs* corresponding to the dual viewpoints of considering incoming or outgoing arrows with respect to a constituted algebraic object.[11]

The foundational significance of category theory for our purposes stems from the fact that it provides a precise mathematical framework to express the basic Whiteheadian structure of the relational realist philosophical scheme. In particular, the two most fundamental concepts epitomizing the philosophical function of the categorical framework in relation to our objectives are the concepts of *universality* and *equivalence*. More concretely, category theory provides a framework of identification of universals in mathematical terms—that is, the instances of a property which exemplify the property in such a paradigmatic way that all other instances have this property by virtue of factorizing uniquely through the corresponding universal. In this sense, objects are specified categorically via some universality property and only up to equivalence (canonical isomorphism).

The universal for a property represents the essential characteristic of a property in the abstractive interpretational sense of the Aristotelian *aphairein*. For example, the categorical notion of the inductive limit is a universal, viz., it is characterized by a universal mapping property, which depicts it *uniquely up to equivalence*. It is instructive to think about the categorical notion of inductive limit as a universal process which exemplifies algebraically Whitehead's method of *extensive abstraction*. We note that the essence of the latter was its formulation of a mathematical procedure by which the *approximation to an ideal limit* is consistently achieved depending on the properties of the relation of extension. The precise term coined by Whitehead in order to describe universals of this categorical nature is "eternal object" or "pure potential for the specific determination of fact." An eternal object is always a *potentiality* for actual entities, and thus has a metaphysical status which is distinct from actuality, although it cannot exist separately from it. In particular, each eternal object / pure potential is characterized by its mode of *ingression* into some concrete actual occasion. Thus its functioning as a potentiality for actuality reveals its *relational essence*. Following Whitehead's description:

> A member of this species can only function relationally: by a necessity of its nature it is introducing one actuality, or nexus, into the real internal constitution of another actual entity. Its sole avocation is to be an agent in objectification. It can never be an element in the definiteness of a subjective form. The solidarity of the world rests upon the incurable objectivity of this species of eternal objects.[12]

The crucial semantic convergence between the category-theoretic notion of a *universal* with Whitehead's notion of an *eternal object* comes from the realization that both of them act as agents in the objectification of a multiplicity of actual occasions as a unified relational structure. For Whitehead, this is a categorical feature of relations among actual occasions, referred to as 'transmutation.'[13]

In category-theoretic terms the operation of the category of transmutation is subsumed by the action of pairs of adjoint functors forming adjunctions. This is the case because all *categorical universals* may be thought of as defined up to canonical isomorphism by the existence of pairs of *adjoint functors* between appropriate categories. In a well-defined sense adjoint functors should be understood as generalized *conceptual inverses* to each other. For example, the conceptual inverse of an 'extension functor' from one algebraic category to another is called a 'restriction functor' pointing in the inverse direction. Likewise, an adjunction between two categories gives rise to a family of *universal morphisms*, one for each object in the first category and one for each object in the second. In this way, each object in the first category induces a certain property in the second category and the universal morphism carries the object to the universal for that property; that is, it transmutes it to an agent of objectification in the second category exemplifying this property in a paradigmatic way. Every adjunction extends to an *adjoint equivalence* of certain subcategories of the initially correlated categories.

In general, the equivalence of two categories $\mathfrak{D}$ and $\mathfrak{A}$ is defined as follows: An *equivalence* of two categories $\mathfrak{D}$ and $\mathfrak{A}$ is defined by means of a pair of opposite directing functors; $\mathbf{F}:\mathfrak{D}\to\mathfrak{A}$, $\mathbf{G}:\mathfrak{A}\to\mathfrak{D}$ and a pair of *natural isomorphisms*; $\tau:\mathbf{1}_{\mathfrak{D}}\to\mathbf{G}\circ\mathbf{F}$, $\rho:\mathbf{1}_{\mathfrak{A}}\to\mathbf{F}\circ\mathbf{G}$ where τ is a *natural isomorphism of the identity functor* on $\mathfrak{D}$, and ρ is a *natural isomorphism of the identity functor* on $\mathfrak{A}$. In this case, the functorial process $\mathbf{F}:\mathfrak{D}\to\mathfrak{A}$ is called *strictly inverse* to the functorial process $\mathbf{G}:\mathfrak{A}\to\mathfrak{D}$. If the pair of opposite directing functors is adjoint, then the equivalence is called an adjoint equivalence. The notion of equivalence of categories thus transcribes the general concept of *congruence* in a categorical/functorial context.

Now we make the observation that given any category $\mathfrak{A}$, the morphisms in $\mathfrak{A}$, or equivalently the morphisms in $\mathfrak{A}^{op}$, can be defined as the elements in the values of the $\mathbf{Hom}_{\mathfrak{A}}$ - *bifunctor*:

$$\mathbf{Hom}_{\mathfrak{A}}(-,-)=\mathbf{y}^{(\mathcal{A})}_{(\mathcal{B})}=\mathfrak{A}^{op}\times\mathfrak{A}\to\mathbf{Sets}$$

from the product category $\mathfrak{A}^{op}\times\mathfrak{A}$ to the category of sets. This is because $\mathfrak{A}$ and $\mathfrak{A}^{op}$ are considered to have the same objects and reversed morphisms. The $\mathbf{Hom}_{\mathfrak{A}}$ -bifunctor is such that:

i. By fixing an object $\mathcal{B}$ in $\mathfrak{A}^{op}$, $\mathbf{y}_{\mathcal{B}}:\mathfrak{A}\to\mathbf{Sets}$ is the *covariant representable functor*, represented by $\mathcal{B}$, and defined as follows:

[1] For all objects $\mathcal{X}$ in $\mathfrak{A}$, $\mathbf{y}_{\mathcal{B}}(\mathcal{X}):=Hom_{\mathfrak{A}}(\mathcal{B},\mathcal{X})$.
[2] For all homomorphisms $f:\mathcal{X}\to\mathcal{Y}$ in $\mathfrak{A}$,

$$\mathbf{y}_{\mathcal{B}}(f):Hom_{\mathfrak{A}}(\mathcal{B},\mathcal{X})\to Hom_{\mathfrak{A}}(\mathcal{B},\mathcal{Y})$$

is defined as post-composition with *f*, viz., $\mathbf{y}_{\mathcal{B}}(f)(g):=f\circ g$.

ii. Correspondingly by fixing an object A in A, $\mathbf{y}^{A} : \mathsf{A}^{op} \to \mathbf{Sets}$ is the *contravariant representable functor*, represented by A, and defined as follows:

[1] For all objects X in A^{op}, $\mathbf{y}^{A}(X) := Hom_{\mathsf{A}^{op}}(X, A)$.
[2] For all homomorphisms $f : Y \to X$ in A^{op},

$$\mathbf{y}^{A}(f) : Hom_{\mathsf{A}^{op}}(X, A) \to Hom_{\mathsf{A}^{op}}(Y, A)$$

is defined as pre-composition with *f*, viz., $\mathbf{y}^{A}(f)(g) := g \circ f$.

The fundamental fact is that an object A in A is *entirely retrievable up to equivalence* (canonical isomorhism) from knowledge of the representable functor $\mathbf{y}^{A} : \mathsf{A}^{op} \to \mathbf{Sets}$ (or equivalently from $\mathbf{y}_{A} : \mathsf{A} \to \mathbf{Sets}$). This fact, a consequence of a result known as the *Yoneda Lemma*,[14] can be expressed this way:

Let A, B be objects in a category A. Suppose we are given an isomorphism of their associated functors: $\mathbf{y}^{A} \cong \mathbf{y}^{B}$. Then there is *a unique isomorphism of the objects themselves*, that is $A \cong B$, which gives rise to this isomorphism of functors. The conceptual significance of this result is that it has the following relational process-theoretic interpretation: An object A of a category A is *determined uniquely up to equivalence by the network of all internal relations* that the object A has with all the other objects in A. Guerino Mazzola puts it in the following way:

> The revolutionary idea of "understanding by changing perspective" was introduced . . . In its common sense rendering, it states that in order to understand an object, you just have to walk round the object. This seemingly trivial insight was put into a rigorous mathematical shape by Nobuo Yoneda [Yoneda 1954]. His famous Yoneda lemma states that a mathematical object can be classified up to isomorphisms by its so-called functor; the latter is precisely the system of all "views" or "perspectives" of the given object from all other objects of the same

structure type (category). This statement even allows to construct objects via their functors.[15]

6.4 The Methodology of Uniform Fibrations: Variable Set Presheaves

From the preceding it is clear that a localization scheme can be precisely conceived as a *functorial* generalization of the notion of *functional dependence.* In the trivial case, the only locus is a point serving as a unique idealized measure of localization, and moreover the only kind of reference frame is the one attached to a point. Thus, in mathematical terms, the localization process should be understood as an *action* of some category of local or partial reference contexts on a set-theoretic global structure of physical events. The latter is thus *partitioned into sorts* parameterized by the objects of the category of contexts. In this sense, the functioning of a localization scheme can be represented by means of a *fibred construct*, understood geometrically as a *variable set over the base category of local reference contexts.* Analogous to the case of the action of a group on a set of points, the *fibers* of this construct may be thought of as the '*generalized orbits*' of the action of the category of local contexts. The notion of functional dependence incorporated in this action forces the partially ordered structure of physical events to fiber over the base category of local reference contexts.

From a physical perspective, the meaningful representation of a global ordered structure of events as a fibred construct, expressed in terms of an action of some localization scheme as depicted above, should also incorporate the requirement of *uniformity.* By uniformity we mean that for any two events observed over the same domain of measurement, the structure of all reference frames that relate to the first *cannot be distinguished* in any possible way from the structure of frames relating to the second. According to this principle, all the localized events within any particular reference context of a localization scheme should be *uniformly equivalent* to each other. The compatibility of the localization process with the requirement of uniformity demands that the relation of (partial) order in a global set-theoretic universe of events is induced by appropriately *lifting* a structured family of arrows from the base category of local reference contexts to the fibers. In technical terms, the representation of a global ordered structure of events as a fibred construct via the action of some localization scheme incorporating the requirement of uniformity is modeled by the concept of a *presheaf*, and in particular by the category of its elements[16] as it will be precisely formalized later in Part II.

The meaning of uniformity with respect to the above is depicted nicely by Nicholas Rescher in the following passage:

> With concreta and existence in a world, this feature of uniformity and homogeneity must come to terms with differentiation and distinctiveness. Here every constituent substance must indeed look out upon exactly the same overall scene—exactly the same world—but it now does so from its own distinctively characteristic point of view. Like a line or plane, there is a complete uniformity and homogeneity, but it is not the uniformity of undifferentiated sameness but rather that of synoptic harmonization. And it is the feature of uniformity and homogeneity—of coordinating overall all of those different points of view—that is critical. For just this cosmic feature of reciprocal adjustment, of [Leibniz's] harmonie universelle is one of the definitive factors for this world's perfection.[17]

6.5 The Localization Role of Topology: Covering Sieves and Sites

Having established the concept of a local reference context, defined according to the operational requirements of a corresponding homologous and uniform physical procedure of measurement, we can now proceed to describe the mechanism by which observable information, defined with respect to overlapping local reference contexts, can be glued together by appropriate means. The gluing procedure is the topological manifestation of Whitehead's notion of extensive connection:

> The "extensive" scheme is nothing else than the generic morphology of the internal relations which bind the actual occasions into a nexus, and which bind the prehensions of any one actual occasion into a unity, coordinately divisible . . . The primary relationship of physical occasions is extensive connection.[18]

The topological character of the gluing procedure representing the process of extensive connection justifies its priority over the metrical notions of space and time. Likewise, for Whitehead:

> In this general description of the states of extension, nothing has been said about physical time or physical space, or of the more general notion of creative advance. These are notions which presuppose the more general relationship of extension. They express additional facts about the actual occasions. The extensiveness of space is really the spatialization of extension; and the extensiveness of time is really the temporalization of extension.[19]

Mathematically, the requirement of a consistent gluing procedure is implemented by the methodology of *completion* of a presheaf, or equivalently, its *sheafification*.[20] In this way, the structure of a sheaf arises by imposing upon the corresponding presheaf the following two requirements: [i] *Compatibility of observable information under restriction from the global to the local level, and* [ii] *Compatibility of observable information under extension from the local to the global level*. If we take together the requirements enforcing a representation of a global partially ordered structure of physical events as a uniform fibred construct over a base category of local reference contexts, localization schemes of the former are precisely modeled in the syntactical terms of *sheaves*. In this perspective, the evolution in the semantics of physical events from a set-theoretic description to a sheaf-theoretic one is completely justified, as will become clear throughout the remainder of part II. Moreover, the technical categorical notion of a topological sheaf captures mathematically Whitehead's conception of an extensive continuum as given in the passage quoted earlier:

> The extensive continuum is a complex of entities united by the various allied relationships of whole to part, and of overlapping so as to possess common parts, and of contact, and of other relationships derived from these primary relationships . . . It is the first determination of order. . . The properties of this continuum are very few and do not include the relationships of metrical geometry . . . These extensive relationships are more fundamental than their more special spatial and temporal relationships . . . Extension, apart from its spatialization and temporalization, is that general scheme of relationships providing the capacity that many objects can be welded into the real unity of one experience.[21]

Up to now, we have qualified reference contexts in the base category as 'local' in a rather informal way in order to convey the intuitve nature of this qualification. In order to further justify the latter and clarify the precise functionality of a localization scheme, it is necessary to define a proper notion of *topology on the base category* of reference contexts. Then, the meaning of local, as well as the meaning of the transition from the local level to the global level acquires a precise connotation.

The notion of a *categorical or Grothendieck topology*[22] on a base category of reference contexts B is a categorical generalization of a system of set-theoretic *covers* on a topology *T*, where a cover for $U \in T$ is a set $\{U_i : U_i \in T, i \in I\}$ such that the union $\bigcup U_i = U$. The generalization is achieved by noting that the topology ordered by inclusion is a poset (partially ordered set) category and that any cover corresponds to a collection of inclusion arrows $U_i \circ \quad U$. The notion of a Grothendieck topology on a base category B, consisting of reference contexts *B*, can thus be presented in terms of appropriate covering devices admitting a functorial interpretation.

The concept of a categorical topology requires, first of all, the abstraction of the constitutive properties of localization in appropriate categorical terms, and then, the effectuation of these properties for the definition of localization schemes. Thus, the concept of a categorical topology epitomizes the meaning of a localization scheme. Regarding these objectives, the requisite abstraction is being implemented by means of covering devices on the base category of reference contexts, called in categorical terminology *covering sieves*. The constitutive properties of localization being abstracted categorically in terms of sieves, qualified as *covering* sieves are the following: [i] The covering sieves should be *covariant* under pullback operations, viz., they should be *stable* under change of a base reference context; [ii] The covering sieves should be *transitive*.

The notion of covering sieves generalizes topologically Whitehead's *covering relations* formulated in terms of *abstractive sets* as discussed in chapter 5. From a physical perspective, it is instructive to think of covering sieves as generalized topological measures of localization in the continuum. Philosophically, they express the *generic morphology of the mereotopological internal relations constituting the process-theoretic scheme of extensive connection in the continuum.* The operation assigning to each reference context of the base category a collection of covering sieves satisfying the closure conditions, stated previously, gives rise to the notion of a Grothendieck topology on the base category of reference contexts.

In application of these notions to quantum theory, we think of the base category of reference contexts as a category of Boolean algebraic contexts related to typical quantum measurement situations.[23] It is a physical fact that any quantum measurement situation involves the preparation of a Boolean algebraic reference context (conceived locally in a proper way) with respect to which a measurement value can be obtained operationally by the use of some measurement device. In this setting a fundamental admonition is given in the Kochen-Specker theorem,[24] according to which the complete description of a quantum mechanical system is impossible to produce by way of a single system of Boolean devices extended globally. By contrast, in every concrete quantum measurement context, the set of logical events that can be actualized in *that* context forms a Boolean algebra. This fact motivates the assertion that *a Boolean algebra within the partially ordered set of quantum logical events plays the role of a local logical reference frame relative to which a measurement result is coordinatized.* Thus, in the words of Arthur Eddington:

> In Einstein's theory of relativity the observer is a man who sets out in quest of truth armed with a measuring-rod [a metrical concept]. In quantum theory he sets out with a sieve [a topological concept].[25]

The construction of a suitable Grothendieck topology on the base category of contexts is significant for the following reasons: First, it explicitly elucidates, with great precision, the conceptions of local and global in a categorical measurement environment, such that these conceptions become *detached from their restricted spatial connotation in terms of geometric point-spaces*—the latter being wholly absent in the case of quantum theory. Instead, these conceptions are depicted exclusively in terms of *relational observable information.* Secondly, it permits the *gluing of local observable information terms into global ones* by utilization of the notion of a sheaf for a suitable Grothendieck topology.

This is most easily demonstrated by first exploring the general notion of sieves, introduced earlier, and then proceeding to focus on the notion of covering sieves. In preface, it is instructive to note at this stage that in the usual definition of a localization scheme induced by a topological space X we use the *open neighborhoods* U of a point in X; such neighborhoods are actually inclusions $U \rightarrowtail X$. The neighborhoods U in topological spaces can be replaced by morphisms of reference contexts $C \to B$, *not* necessarily inclusions, and this can be done in any complete category. In effect, a covering of a topological space by open sets can be replaced by a new notion of covering, provided by a family of morphisms targeting some object in a categorical environment, and satisfying the above conditions [i] and [ii], which abstract the decisive features of a localization process.

A *B-sieve* with respect to a reference context B in the base category $\mathcal{B}$ is a family S of $\mathcal{B}$-morphisms targeting B, such that if $C \to B$ belongs to S and $D \to C$ is any $\mathcal{B}$-morphism, then the composite $D \to C \to B$ belongs to S. We may think algebraically of a B-sieve as a right *B-ideal.* We notice that in the case of $\mathcal{O}(X)$, since $\mathcal{O}(X)$-morphisms are inclusions of open loci, a right U-ideal is identical with a *downwards closed* U-subset.

We notice that if S is a *B-sieve* and $h: C \to B$ is any arrow to the reference context B, then: $h^*(S) = \{f \,/\, cod(f) = C, (h \circ f) \in S\}$ is a C-sieve, referred to as the *pullback* of S along h, or the *restriction of* S along h , where $cod(f)$ denotes the codomain of f. We notice that for a context B in $\mathcal{B}$, the set of all arrows into B is a B-sieve, called the *maximal sieve* on B, and denoted by, $t(B) := t_B$.

As we proceed to the next stage of development, the key conceptual issue to be settled is the following: How is it possible to *restrict* or *reduce* the set of all B-sieves for each reference context B in $\mathcal{B}$, denoted by $\mathbf{K}(B)$, such that each B-sieve of the restricted set can acquire the interpretation of a *covering* B-sieve with respect to a generalized covering system? Equivalently stated, we wish to impose the satisfaction of appropriate conditions on the set of B-sieves for each context B in $\mathcal{B}$, such that the subset of B-sieves obtained, denoted by $\mathbf{K}_\chi(B)$, implements the constitutive properties of a localization process. In this sense, the B-sieves of $\mathbf{K}_\chi(B)$ for each locus B in $\mathcal{B}$, to be thought as covering B-sieves, can be legitimately used for the implementation of localization processes. The appropriate conditions depicting the covering B-sieves from the set of all B-sieves, for each reference context B in $\mathcal{B}$, are the following:

[1] We interpret an arrow $C \to B$, where C,B are contexts in $\mathcal{B}$, as a *figure* of B, and thus we interpret B as an *extension* of C in $\mathcal{B}$. It is a natural requirement that the set of all figures of B should belong in $\mathbf{K}_\chi(B)$ for each context B in $\mathcal{B}$.

[2] The covering sieves should be *stable* under pullback operations. This requirement means, in particular, that the intersection of covering

sieves should also be a covering sieve, for each reference context *B*, in the base category $\mathcal{B}$.

[3] Finally, it would be desirable to impose: [i] a *transitivity* requirement on the specification of the covering sieves, such that, intuitively stated, covering sieves of figures of a context in covering sieves of this context, should also be covering sieves of the context themselves, and [ii] a requirement of *common refinement* of covering sieves.

If we take into account the above requirements we can define a *generalized covering system*, called a *Grothendieck topology*, in the environment of the base category $\mathcal{B}$ as follows:

A *Grothendieck topology* on $\mathcal{B}$ is an operation $\mathbf{J}$, which assigns to each reference context *B* in $\mathcal{B}$, a collection $\mathbf{J}(B)$ of *B*-sieves, called covering *B*-sieves, such that the following three conditions are satisfied:

[1] For every reference context *B* in $\mathcal{B}$, the maximal *B*-sieve $\{g : cod(g) = B\}$ belongs to $\mathbf{J}(B)$ (*maximality condition*).

[2] If *S* belongs to $\mathbf{J}(B)$ and $h: C \to B$ is a figure of *B*, then $h^*(S) = \{f : C \to B, (h \circ f) \in S\}$ belongs to $\mathbf{J}(C)$ (*stability condition*).

[3] If *S* belongs to $\mathbf{J}(B)$, and if for each figure $h: C_h \to B$ in *S*, there is a sieve R_h belonging to $\mathbf{J}(C_h)$, then the set of all composites $h \circ g$, with $h \in S$, and $g \in R_h$, belongs to $\mathbf{J}(B)$ (*transitivity condition*).

As a consequence of the definition and conditions above, we can easily verify that any two *B*-covering sieves have a common refinement, that is: if *S*,*R* belong to $\mathbf{J}(B)$, then $S \cap R$ belongs to $\mathbf{J}(B)$.

As a first application we may consider the partially ordered set of open subsets of a topological space *X*, viewed as the *base category of open reference contexts*, $\mathcal{O}(X)$. Then we specify that *S* is a covering U-sieve if and only if *U* is contained in the union of open sets in *S*. The above specification fulfills the requirements of covering sieves posed above, and consequently, defines a Grothendieck topology on $\mathcal{O}(X)$.

A category of reference contexts $\mathcal{B}$ together with a Grothendieck topology $\mathbf{J}$, is called a *site*, denoted by $(\mathcal{B}, \mathbf{J})$. The notion of a site generalizes the notion of a topological measurement state space in categorical terms and constitutes the *relational variable topological background* for the effectuation of a localization scheme for *filtering* or *coarse-graining* or *grading* the information contained in a global structure of partially ordered physical events. This filtering occurs via localizing structural frames determined by a homologous operational physical procedure of measurement.

As will be demonstrated throughout the remainder of Part II, this conceptual framework is uniquely suited to the coherent and consistent description of localization processes in quantum theory. The basic idea is to associate the functioning of a localization scheme in a global quantum event continuum with covering sieves of local Boolean contexts of quantum measurement. Before applying the machinery of sites to quantum theory, however, it is instructive to discuss in detail the important, simplest case of a localization scheme induced by a topology on a space.

Notes

1. Roland Omnès, *The Interpretation of Quantum Mechanics* (Princeton: Princeton University Press, 1994).
2. H. Poincare, *The Foundations of Science: Science and Hypothesis, The Value of Science and Methods* (New York: The Science Press, 1921), 52.
3. Alfred North Whitehead, *Process and Reality: An Essay in Cosmology*, Corrected Edition, ed. D. Griffin and D. Sherburne (New York: Free Press, 1978), 63.
4. Whitehead, *Process and Reality*, 288.
5. Whitehead, *Process and Reality*, 66.
6. Whitehead, *Process and Reality*, 67.
7. Whitehead, *Process and Reality*, 288.
8. S. Awodey, *Category Theory*, Oxford Logic Guides (Oxford: Oxford University Press, 2006).
9. Elias Zafiris, "Probing Quantum Structure Through Boolean Localization Systems," *International Journal of Theoretical Physics* 39, no. 12 (2000); Elias Zafiris, "Boolean Coverings of Quantum Observable Structure: A Setting for an Abstract Differential Geometric Mechanism," *Journal of Geometry and*

Physics 50, no. 99 (2004); Elias Zafiris, "Interpreting Observables in a Quantum World from the Categorial Standpoint," *International Journal of Theoretical Physics* 43, no. 1 (2004).

10. J. L. Bell, "From Absolute to Local Mathematics," *Synthese* 69 (1986); J. L. Bell, "Categories, Toposes and Sets," *Synthese* 51, no. 3 (1982).

11. We note that an interesting analysis of the general notion of duality from a process-theoretic standpoint has been presented in T. E. Eastman, "Duality without Dualism" in *Physics and Whitehead*, ed. T. Eastman and H. Keeton (Albany: State University of New York Press, 2004).

12. Whitehead, *Process and Reality*, 291.

13. Whitehead, *Process and Reality*, 292.

14. Awodey, *Category Theory*; S. Mac Lane and I. Moerdijk, *Sheaves in Geometry and Logic* (New York: Springer-Verlag, 1992).

15. Guerino Mazzola, "Towards Big Science: Geometric Logic of Music and Its Technology" (paper presented at KlangArt-Kongress 1995, Universität Osnabrück).

16. Mac Lane and Moerdijk, *Sheaves in Geometry and Logic*.

17. N. Rescher, *Studies in Leibniz's Cosmology* (Berlin: Ontos-Verlag, 2006), 86.

18. Whitehead, *Process and Reality*, 288.

19. Whitehead, *Process and Reality*, 289.

20. Mac Lane and Moerdijk, *Sheaves in Geometry and Logic*.

21. Whitehead, *Process and Reality*, 66.

22. Mac Lane and Moerdijk, *Sheaves in Geometry and Logic*.

23. Elias Zafiris, "Probing Quantum Structure Through Boolean Localization Systems," *International Journal of Theoretical Physics* 39, no. 12 (2000); Elias Zafiris, "Boolean Coverings of Quantum Observable Structure: A Setting for an Abstract Differential Geometric Mechanism," *Journal of Geometry and Physics* 50, no. 99 (2004); Elias Zafiris, "Interpreting Observables in a Quantum World from the Categorial Standpoint," *International Journal of Theoretical Physics* 43, no. 1 (2004).

24. Simon Kochen and Ernst Specker, "The problem of hidden variables in quantum mechanics," *Journal of Mathematics and Mechanics* 17 (1967): 59-87.

25. A. S. Eddington, *The Theory of Groups* in *The World of Mathematics*, Vol. 3, ed. J. R. Newman (New York: Dover, 2003).

CHAPTER 7

Sheaves of Germs: The Topological Case

7.1 Localization over a Topological Space

The topological case refers to a localization scheme on a global partial order of physical events. These events are defined as taking place over a base category of reference contexts $\mathcal{O}(X)$, consisting of open sets U of a topological measurement space X, the arrows between them being inclusions. In this case, the reference contexts of the operational environment employed for observation are all the open sets U of X, *partially ordered by inclusion.* Equivalently, the open set inclusions $U \rightarrowtail X$ are considered as varying base reference frames of open loci over which a global partial order of events fibers. We may use the suggestive term 'local observer' to refer to an event-registering device associated with a reference context U of the base category $\mathcal{O}(X)$. Note that the meaning of 'local' here is understood with respect to the topology of X, such that a 'local observer' in a measurement situation taking place over a reference context U individuates events by means of local real-valued observables, being continuous maps $s : U \to \mathbb{R}$. Thus, the 'local observers' do *not* have a global 'perception' of continuous functions $f : X \to \mathbb{R}$, but rather always register events as localized over the associated reference contexts in terms of local observables. Of course, appropriate conditions are further needed for gluing their findings together; as will be explained presently, these conditions are the necessary and sufficient conditions for a topological sheaf-theoretic structure.

7.2 Uniform Localization of Observables

In order to clarify the functioning of a *topological localization process*, we will describe in detail the important case of localization of a commutative, unital $\mathbb{R}$-algebra of observables of some system over a base localizing category $\mathcal{O}(X)$. Since observables are thought of as *global functions* on the $\mathbb{R}$-coordinatized state-space of this system, the process of localization forces the replacement of the algebra of global observables $\mathcal{A}$ by an algebraic structure which will give us *all local and global functional information* together. All these functional elements should *interlock compatibly* together in an appropriate manner, which is characterized by the property that the combined structure respects the extension from local to global, as well as the restriction from global to local implied by the localization process. This structure is precisely formalized by the notion of a *sheaf of germs* of a commutative, unital $\mathbb{R}$-algebra of observables, denoted by $\mathbf{A}$, which incorporates all compatible local and global information together.

Let us first introduce precisely the general notion of a sheaf on a topological space, and then examine its applicability in the current situation. For this purpose, we consider the category of open sets $\mathcal{O}(X)$ in a topological space, partially ordered by inclusion. If $\mathcal{O}(X)^{op}$ is the opposite category of $\mathcal{O}(X)$, and *Sets* denotes the category of sets, we define:

A *presheaf* of sets on $\mathcal{O}(X)$ is a contravariant set-valued functor on $\mathcal{O}(X)$, denoted by $\mathbf{P}:\mathcal{O}(X)^{op} \to Sets$. For each base open set U of $\mathcal{O}(X)$, $\mathbf{P}(U)$ is a set, and for each arrow $F:V \to U$, $\mathbf{P}(F):\mathbf{P}(U) \to \mathbf{P}(V)$ is a set-function. If $\mathbf{P}$ is a presheaf on $\mathcal{O}(X)$ and $p \in \mathbf{P}(U)$, the value $\mathbf{P}(F)(p)$ for an arrow $F:V \to U$ in $\mathcal{O}(X)$ is called the *restriction* of p along F and is denoted by $\mathbf{P}(F)(p) := p \cdot F$. A presheaf $\mathbf{P}$ may be understood as a *right action* of $\mathrm{O}(X)$ on a set. This set is partitioned into sorts parameterized by the objects of $\mathrm{O}(X)$, and has the following property: If $F : V \to U$ is an inclusion arrow in $\mathrm{O}(X)$ and p is an element of $\mathbf{P}$ of sort U, then $p \cdot F$ is specified as an element of $\mathbf{P}$ of sort V. Such an action $\mathbf{P}$ is referred to as an $\mathrm{O}(X)$-*variable set*. A variable set of this form is entirely determined by its category of elements.

The *category of elements of a presheaf* $\mathbf{P}$, denoted by $\int(\mathbf{P}, \mathrm{O}(X))$, is described as follows: The objects of $\int(\mathbf{P}, \mathrm{O}(X))$ are all pairs (U, p), with U in $\mathrm{O}(X)$ and $p \in \mathbf{P}(U)$. The arrows of $\int(\mathbf{P}, \mathrm{O}(X))$, that is, $(U', p') \to (U, p)$, are those morphisms $Z : U' \to U$ in $\mathrm{O}(X)$, such that $p' = \mathbf{P}(Z)(p) := p \cdot Z$. Notice that the arrows in $\int(\mathbf{P}, \mathrm{O}(X))$ are those morphisms $Z : U' \to U$ in the base category $\mathrm{O}(X)$ that pull a chosen element $p \in \mathbf{P}(U)$ back into $p' \in \mathbf{P}(U')$.

The category of elements $\int(\mathbf{P}, \mathrm{O}(X))$ of a presheaf $\mathbf{P}$, together with the projection functor $\int_{\mathbf{P}} : \int(\mathbf{P}, \mathrm{O}(X)) \to \mathrm{O}(X)$, is called the *split discrete fibration* induced by $\mathbf{P}$[1], where $\mathrm{O}(X)$ is the *base category* of the fibration. We note that the fibers are categories in which the only arrows are identity arrows. If U is an open reference context of $\mathrm{O}(X)$, the inverse image of U under $\int_{\mathbf{P}}$ is simply the set $\mathbf{P}(U)$, although its elements are written as pairs so as to form a *disjoint union*.

From a physical viewpoint, the purpose of introducing the notion of a presheaf $\mathbf{P}$ on $\mathrm{O}(X)$ is the following: We identify an element of $\mathbf{P}$ of sort U, that is $p \in \mathbf{P}(U)$, with a *local observable*, which can be observed by means of a measurement procedure over the reference context U, being an open set of a topological space X. This identification forces the interrelations of local observables, over all reference contexts of the base category $\mathrm{O}(X)$, to fulfill the requirements of a uniform and homologous fibred categorical structure. The latter are to be understood according to the following two requirements: First, the reference contexts used for observational purposes, together with their structural morphisms, should form a mathematical category. The notion of uniformity implies that for any two local observables, being amenable to a measurement procedure over the same open domain of measurement U, the structure of all reference contexts that relate to the first cannot be distinguished, in any possible way, from the structure of contexts relating to the second. According to this, all the localized observables within any particular reference context should be *uniformly equivalent* to each other. The split discrete fibration induced by $\mathbf{P}$, where $O(X)$ is the base category of the fibration, provides a well-defined notion of *a uniform homologous fibred structure* of local observables in the following two ways: [1] By the arrows specification defined in the category of elements of $\mathbf{P}$, any local observable p, determined over the reference context U, is homologically related with any other local observable p' over the reference context U', and so on, by variation over all the reference contexts of the base category; [2] All the local observables p of $\mathbf{P}$, of the same sort U, viz., determined over the same reference context U, are uniformly equivalent to each other; this is because all the arrows in $\int(\mathbf{P},O(X))$ are induced by lifting arrows from the base category $O(X)$, being formed by partially ordering the reference contexts. We conclude that the topological localization process is consistent with the physical requirement of uniformity.

7.3 Gluing Sections and Local-to-Global Compatibility

The next crucial step of the construction aims at the satisfaction of a second physical requirement: Since we have assumed the existence of reference contexts (open observational domains) locally, according to the operational requirements of a corresponding physical procedure of measurement, the information gathered about local observables in different measurement situations should be collated together by appropriate means. Mathematically, this requirement is implemented via the completion of the *presheaf* $\mathbf{P}$, or equivalently, sheafification of $\mathbf{P}$.[2]

A *sheaf* is characterized as a presheaf $\mathbf{P}$ that satisfies the following condition: If $U = \bigcup_a U_a$, U_a in $\mathcal{O}(X)$, and elements $p_a \in \mathbf{P}(U_a)$, $a \in I$:index set, are such that for arbitrary a, $b \in I$, it holds:

$$p_a \mid U_{ab} = p_b \mid U_{ab}$$

where $U_{ab} := U_a \cap U_b$, and the symbol $\mid$ denotes the operation of *restriction* on the corresponding open reference context, then there exists a *unique* element $p \in \mathbf{P}(U)$, such that $p \mid U_a = p_a$ for each a in I. Then an element of $\mathbf{P}(U)$ is called a *section* of the sheaf $\mathbf{P}$ over the open context U. The sheaf condition means that sections can be *glued together uniquely* over the reference contexts of the base category $\mathcal{O}(X)$. In particular, the sheaf-theoretic qualification of a uniform and homologous fibred structure of observables, as above, also makes the latter *coherent* in terms of *local-global compatibility* of the information content it carries, under the inverse operations of restriction and extension.

Thus, we form the following conclusion: The structure of a sheaf arises by imposing the following two requirements upon the uniform and homologous fibred structure of elements of the corresponding presheaf: [i] *Compatibility of observable information under restriction from the global to the local level,* and [ii] *Compatibility of observable information under extension from the local to the global level.* According to the first of the above requirements, a sheaf constitutes a *separated presheaf* of local observables over a global topological space, meaning that two observables are conceived as being identical globally if and only if they are identical locally. In turn, according to the second requirement, locally compatible observables can be glued together within some global observable, which is also uniquely defined because of the first requirement.

Furthermore, it is obvious that each set of sort U, $\mathbf{P}(U)$, can be endowed with the structure of an $\mathbb{R}$- *algebra* under point-wise sum, product, and scalar multiplication, denoted correspondingly by $\mathbf{A}(U)$; in that case, the morphisms $\mathbf{A}(U) \to \mathbf{A}(V)$ stand for $\mathbb{R}$-linear morphisms of $\mathbb{R}$-algebras.

7.4 Contextuality: Germs and Stalks of Observable Sheaves

Now we shall introduce the physically important notions of *stalks* and *germs* of a sheaf. For this purpose it is necessary to explain the construction of the *inductive limit* (categorical colimit) of sets (or $\mathbb{R}$-algebras) of observables $\mathbf{A}(U)$, denoted by $Colim[\mathbf{A}(U)]$. Let us consider that x is a *point* of the topological measurement space X. Moreover, let K be a set consisting of open subsets of X, containing x, such that the following condition holds: For any two open reference contexts U, V, containing x, there exists an open reference context $W \in K$, *contained in the intersection* $U \bigcap V$. We may say that K constitutes a *basis* for the system of open reference contexts around x. We form the *disjoint union* of all $\mathbf{A}(U)$, denoted by:

$$\mathbf{D}(x) := \coprod_{U \in K} \mathbf{A}(U)$$

Then, we can define an *equivalence relation* in $\mathbf{D}(x)$ by requiring that $p \sim q$ for $p \in \mathbf{A}(U)$, $q \in \mathbf{A}(V)$, provided that they have the *same restriction to a smaller open set* contained in K. Then we define:

$$Colim_K[\mathbf{A}(U)] := \mathbf{D}(x)/\sim_K$$

Notice that if we denote the inclusion mapping of V into U by:

$$i_{V,U} : V \rightarrowtail U$$

and also the restriction morphism of sets from U to V by:

$$\rho_{U,V} : \mathbf{A}(U) \to \mathbf{A}(V)$$

we can introduce well-defined notions of addition and multiplication on the set $Colim_K[\mathbf{A}(U)]$ making it an $\mathbb{R}$-algebra.

Now, if we consider that K and Λ are two bases for the system of open reference contexts around $x \in X$, we can show that there are canonical isomorphisms between $Colim_K[\mathbf{A}(U)]$ and $Colim_\Lambda[\mathbf{A}(U)]$. In particular, we may take all the open subsets of X containing x: Indeed, we consider first the case when K is arbitrary and Λ is the set of all open subsets containing x. Then $\Lambda \supset K$ induces a morphism:

$$Colim_K[\mathbf{A}(U)] \to Colim_\Lambda[\mathbf{A}(U)]$$

which is an isomorphism, because whenever V is an open subset containing x, there exists an open subset U in K contained in V. Since we can repeat that procedure for all bases of the system of open sets domains around $x \in X$, the initial claim follows immediately.

Then, the *stalk* of $\mathbf{A}$ at the point $x \in X$, denoted by $\mathbf{A}_x$, is defined as the *inductive limit* of sets (or $\mathbb{R}$-algebras) of observables $\mathbf{A}(U)$:

$$Colim_K[\mathbf{A}(U)] := \coprod_{U \in K} \mathbf{A}(U) / \sim_K$$

where K is a basis for the system of open reference contexts around x, and $\sim_K$ denotes the equivalence relation of restriction within an open set in K. Note that this definition is *independent* of the chosen basis K. For an open reference context W containing the point x, we obtain an morphism of $\mathbf{A}(W)$ into the stalk at the point x:

$$i_{W,x} : \mathbf{A}(W) \to \mathbf{A}_x$$

For an element $p \in \mathbf{A}(W)$ its image:

$$i_{W,x}(p) := p_x = germ_x p$$

is called the *germ* of p at the point x. The fibred structure that corresponds to a sheaf of sets (or $\mathbb{R}$-algebras) of observables $\mathbf{A}$ is a *topological bundle* defined by the *continuous* mapping $\varphi: A \to X$, where:

$$A = \coprod_{x \in X} \mathbf{A}_x$$

$$\varphi^{-1}(x) = \mathbf{A}_x = Colim_{\{x \in U\}}[\mathbf{A}(U)]$$

The mapping φ is locally a homeomorphism of topological spaces.[3] The topology in A is defined as follows: for each $p \in \mathbf{A}(U)$, the set $\{p_x, x \in U\}$ is open, and moreover, an arbitrary open set is a union of sets of this form. In the physical state of affairs, we have identified an element of $\mathbf{A}$ of sort U that is a local section of $\mathbf{A}$, with a local observable p, which can be observed via a measurement procedure over the reference context U. Then the equivalence relation used in the definition of the stalk $\mathbf{A}_x$ at the point $x \in X$ is interpreted as follows: Two local observables $p \in \mathbf{A}(U)$ and $q \in \mathbf{A}(V)$ *induce the same contextual information* at x in X provided that they have the *same restriction* to a smaller open context contained in the basis K. Then the stalk $\mathbf{A}_x$ is the *set containing all contextual information* at x, that is, the set of all *equivalence classes*. Moreover, the image of a local observable $p \in \mathbf{A}(U)$ at the stalk $\mathbf{A}_x$, that is, the equivalence class of this local observable p, is precisely the germ of p at the point x.

Next, if we consider a local observable $p \in \mathbf{A}(U)$, it *determines a function:*

$$\dot{p}: x \mapsto germ_x p$$

whose domain is the open locus U and its codomain is the stalk $\mathbf{A}_x$ for each $x \in U$. Obviously, we may consider instead the disjoint union $A = \coprod_{x \in X} \mathbf{A}_x$ as the codomain of the function $\dot{p}$. In this sense, every local observable $p \in \mathbf{A}(U)$ gives rise to some *partial function:*

$$\dot{p}:U \to A$$

which is defined on the open reference context $U \subset X$. Hence, all local observables $p \in \mathbf{A}(U)$ admit a *functional representation*, established by means of the following correspondence:

$$\Delta(U): p \mapsto \dot{p}$$

Stated equivalently, each local observable $p \in \mathbf{A}(U)$ can be legitimately considered as a partial function:

$$\dot{p}:U \to A$$

defined over the open reference locus *U*, the value of which, at a point $x \in U$, viz., $germ_x p$, is the *contextual observable information induced at* x by the local observable p. Furthermore, such a partial function $\dot{p}:U \to A$ is identified with a *cross section of the topological bundle of germs*, defined by the continuous mapping $\varphi : A \to X$, such that:

$$\varphi^{-1}(x) = \mathbf{A}_x = Colim_{\{x \in U\}}[\mathbf{A}(U)]$$

Notice that the mapping φ is *locally* a homeomorphism of topological spaces, and thus the bundle is a topological *display bundle* (called equivalently an *étale bundle*). Now, as a consequence of the *equivalence* between sheaves on a topological space *X* and display topological bundles over *X*, every *sheaf can be considered as a sheaf of cross sections of the equivalent display topological bundle.*[4]

7.5 Completion and Functionalization: Display Bundles

It is also instructive to notice that the previous arguments can help us to understand the process of *completion* (or *sheafification*, or *germification*) of a presheaf.[5] For this purpose, we realize that the notions of germ, stalk and display bundle make sense for a general presheaf. More precisely, the germ at a point stands for an equivalence class of elements of the presheaf corresponding to open reference contexts around that point, under the equivalence relation of having the same germ. The stalk over this point is the set of all germs at this point. The display bundle is the disjoint union of all stalks.

The first crucial observation is that by the definition of a topology on the display bundle, as described previously, it is legitimate to consider *continuous sections* of the display bundle. Stated equivalently, this procedure amounts to transforming the elements of the presheaf into partial continuous functions (continuous sections) valued into the display space. Hence, we manage to *functionalize* the initial presheaf, by defining a new presheaf, called the *presheaf of sections of the initial presheaf* as follows: *It is the presheaf, which associates to each open locus of the base topological space the set of continuous sections from that open locus into the display space*. Now, there is an obvious morphism from the initial presheaf to its presheaf of sections—a morphism that maps each element of the category of elements of the initial presheaf to the continuous section, thus sending each point in an open reference context of the base space to the germ of this element at that point.

The second crucial observation is that *the associated functionalized presheaf* of sections of a presheaf is actually also *localized* (locally determined), viz., it is a sheaf, identified as the *sheaf of cross sections* of the corresponding topological display bundle. Thus, the latter sheaf is called the *sheaf associated with the initial presheaf*. Moreover, the process of completion of a presheaf into the sheaf of cross sections of the corresponding topological display bundle is *functorial*, meaning that there exists a functor sending each presheaf to its completion, viz., to its associated sheaf of sections, called the *sheafification functor*. As a corollary, we conclude that *a presheaf is a sheaf, viz., a complete presheaf, if and only if the morphism to its associated functional presheaf of sections is an isomorphism*. Thus, the process of *completion* of a presheaf is equivalent to the combined processes of *functionalization and localization* of its elements. Consequently, the associated sheaf of sections of a corresponding presheaf contains, by its construction, the totality of *local contextual information* compatible with that given in the initial presheaf via its restriction property. In this sense, the sheaf constitutes the completion of the presheaf.

7.6 Events: From Set-Theoretic to Sheaf-Theoretic Semantics

Now, let us consider a sheaf of $\mathbb{R}$-algebras of local observables, identified as a sheaf of real-valued continuous cross-sections of the corresponding display bundle. Then, the set of germs of all these sections at a point, viz., the stalk at this point, is also an $\mathbb{R}$-algebra. Most important, the stalk at this point is a *local* $\mathbb{R}$-algebra, meaning that it has a *unique maximal ideal*. In turn, this maximal ideal consists of all germs vanishing at the considered point. The quotient of the stalk by this maximal ideal is isomorphic to the field of real numbers. Equivalently, this means that the morphism evaluating a germ of the stalk at a point to the real numbers (viz., providing a real value at the corresponding non-vanishing equivalence class of sections at the considered base point) is a surjective morphism of $\mathbb{R}$-algebras having for a kernel the maximal ideal of the stalk at this point:

$$\mathbf{ev}_x : \mathbf{A}_x \to \mathbf{A}_x / \mu_x \cong \mathbb{R}$$

$$germ_x p \mapsto \mathbf{ev}_x(germ_x p) = p(x)$$

Thus, *the evaluation morphism of a germ* of the stalk at a point of the base space is an $\mathbb{R}$-valued measurement of this observable germ, interpreted as an *observed event* of the corresponding system, and subsequently encoded by means of an $\mathbb{R}$-state of its topological state-space.

Proceeding to the next stage of the development of this framework, the sheaf of germs of real-valued continuous functions on a topological space X is an object in the functor category of sheaves $\mathbf{Sh}(X)$ on varying reference contexts U, being open sets of X, partially ordered by inclusion. The morphisms in $\mathbf{Sh}(X)$ are all natural transformations between sheaves.[6] It is instructive to notice that a sheaf makes sense only if the *base category of reference contexts* is specified, which is equivalent to the determination of a *topology* on the space X.

The functor category of sheaves $\mathbf{Sh}(X)$ provides an exemplary case of a construct known as *topos.*[7] A topos can be conceived as a *local mathematical framework* corresponding to a generalized model of set theory, or as a *generalized algebraic space*, corresponding to a categorical universe of variable observable information sets over the multiplicity of the reference contexts of the base category. We recall that formally a topos is a category, which has a terminal object, pullbacks, exponentials, and a subobject classifier, which is understood as an object of generalized truth values. The particular significance of the sheaf of real-valued continuous functions on X is due to the following isomorphism: *The sheaf of germs of continuous real-valued functions on X is isomorphic to the object of Dedekind real numbers in the topos of sheaves* $\mathbf{Sh}(X)$.[8] This isomorphism validates the physical intuition of considering a local observable as a *continuously variable real number* over its context of definition.

The transition in the semantics of the 'physical event continuum' from the topos of $\mathbf{Sets}$ to the topos of sheaves $\mathbf{Sh}(X)$ is an instance of the basic idea of *sheaf-theoretic localization of physical observability* referring to the interpretation of observed events. We initially notice that in the former case observables are identified with (continuous) functions determined completely by their values at points. In the latter case observables are identified with local continuous sections of the display space determined completely by their germs.

In order to analyze in more detail the transition in the semantics, we note that in the former case a continuous function from a base topological space *X* to the topological space R can be considered as a continuous section from *X* to the product space $X \times \mathrm{R}$. This product space is isomorphic set-theoretically to a space having a copy of R at each point, being the inverse image of the projection from $X \times \mathrm{R}$ to the base *X*. The value that is taken at a point is the value taken by the function. Thus, this type of modeling of an observable is only appropriate in capturing its *point properties*.

In contradistinction, in the latter case we strive to capture *local properties* of observables. This is because in the sheaf-theoretic local environment, we associate *not the value that a section takes at a point of the base space but its germ*. In this sense, instead of the product total space $X \times \mathrm{R}$, we have the topological display space, such that the inverse image of each point of the base space is not a copy of R, but the *stalk* at that point. This essentially means that the *transition of semantics* from the topos of sets **Sets** to the topos of sheaves $\mathbf{Sh}(X)$ amounts to a shift of focus from point-wise descriptions of observables to local descriptions of observables.

Obviously the topological display space is a much richer and larger space than the rigid space $X \times \mathrm{R}$, since the display space provides information about the *local behavior* of observables around each point of the base space in terms of germs, instead of merely point-wise behavior of observables in terms of their values in the real numbers. Thus, *observed events are not determined by the values of continuous functions (observables) at points of the base space, but by the evaluation morphisms of observable germs at those points*, according to our previous remarks.

We conclude that the meaning of the idea of sheaf-theoretic localization of physical observability referring to the interpretation of observed events, effectuated by the transition from the topos of sets **Sets** to the topos of sheaves $\mathbf{Sh}(X)$, amounts to a *topological relativization* with respect to the *local behavior* of physical observables as opposed to their point behavior.

Let us now further elaborate the sheaf-theoretic semantics of observed events from a process-theoretic standpoint. We argue that each event is a *novel actualized fact* and respects the selection constraints of the local context with respect to which it can be individuated. Thus events are essentially individuated and actualized with respect to local reference contexts U in $\mathcal{O}(X)$. It is crucial to notice that an event is not specified sheaf-theoretically by the evaluation of an observable, but rather by the evaluation of a corresponding observable germ; that is, an event is specified in a global sense by the *equivalence class of all compatible* local reference contexts conceived in all different local levels nested within each other with respect to which observable information can be glued.

Thus an observable germ is precisely the semantic information carrier of a *seed of extension* in the ontological formation of the continuum; it is, in other words, the germ of extensive connection from the local to the global via the topological gluing procedure. Notice also that the notion of a germ is *not* defined in abstraction from its local context, since its very definition makes sense only with respect to such a local context, where local sections agree. Moreover in the conception of a germ, a local context is *never* considered apart from the local information content which is contextualized. It is thus useful to think of germs as representing the *internal constitution of actual occasions* in their process of *concrescence*, since they integrate compatible information at higher and higher local levels (reference contexts) by the formation of equivalence classes. In this way, it is via the concept of the germ that an actual occasion's objective global relevance can be analyzed—that is, via the semantic transition from the local to the global.

In Whiteheadian philosophy, an actual occasion is considered the starting point of its own process of concrescence, and this process is analyzable 'genetically.' In terms of the sheaf theoretic framework presented herein, the genetic division constitutive of an actual occasion is perfectly represented via the concept of the germ. Thus an actual occasion determines its germ of local observables or local sections. This, in turn, involves the determination of an appropriate local context with respect to which the germ is initially defined and eventually evaluated or actualized. The actualization or 'satisfaction' of the actual occasion's process of concrescence, in other words, can be understood in the context of quantum measurement as the evaluation / actualization of this germ as a quantum event.

7.7 Covering Sieves as Representable Functors

Before explaining the applicability of sheaf-theoretic localization of physical observability to the concept of a 'quantum event continuum,' it is important to first revisit and clarify the notions of sieves and covering sieves introduced in Chatper 6 as they pertain to the topological localization process discussed above. A U-sieve with respect to a reference context U in $\mathcal{O}(X)$, is defined as a family S of $\mathcal{O}(X)$-morphisms with codomain U, such that if $V \to U$ belongs to S and $D \to V$ is any $\mathcal{O}(X)$-morphism, then the *nested composite inclusions* (*internal relations* among local reference contexts) $D \to V \to U$ belongs to S. Now, since all $\mathcal{O}(X)$-morphisms are inclusions, a U-sieve with respect to U is actually a *downwards closed* U-subset.

If we consider the contravariant representable functor of U in $\mathcal{O}(X)$, viz., *the network of all internal relations* that the reference context U has with all the other contexts in $\mathcal{O}(X)$, denoted by $\mathbf{y}[U] := Hom_{\mathcal{O}(X)}(-,U)$, then it is easy to realize that a U-sieve is *equivalent* to a subfunctor $\mathbf{S} \rightarrowtail \mathbf{y}[U]$ in $Sets^{\mathcal{O}(X)^{op}}$.

In detail, given a U-sieve S, we define:

$$\mathbf{S}(V) = \{g \,/\, g : V \to U, g \in S\} \subseteq \mathbf{y}[U](V)$$

This definition yields a functor $\mathbf{S}$ in $Sets^{\mathcal{O}(X)^{op}}$, which is obviously a subfunctor of $\mathbf{y}[U]$. Conversely, given a subfunctor $\mathbf{S} \rightarrowtail \mathbf{y}[U]$ in $Sets^{\mathcal{O}(X)^{op}}$, the set:

$$S = \{g \,/\, g : V \to U, g \in \mathbf{S}(V)\}$$

for some reference context V in $\mathcal{O}(X)$, is a U-sieve. Thus, we conclude:

$$\langle\ U\text{-sieve: } S\ \rangle = \langle\ \text{Subfunctor of } \mathbf{y}[U] : \mathbf{S} \rightarrowtail \mathbf{y}[U]\ \rangle$$

We notice that if S is a U-sieve and $h : V \to U$ is any inclusion to the reference context U, then:

$$h^*(S) = \{f \,/\, cod(f) = V, (h \circ f) \in S\}$$

is a V-sieve, called the *pullback* of S along h. Consequently, we may define a presheaf functor $\mathbf{K}$ in $Sets^{\mathcal{O}(X)^{op}}$, such that its action on reference contexts U in $\mathcal{O}(X)$, is given by:

$$\mathbf{K}(U) = \{S \,/\, S : U - sieve\}$$

and on arrows $h : V \to U$, by $h^*(-) : \mathbf{K}(U) \to \mathbf{K}(V)$, given by:

$$h^*(S) = \{f \,/\, cod(f) = V, (h \circ f) \in S\}$$

We notice that for a reference context U in $\mathcal{O}(X)$, the set of all inclusion arrows into *U*, called the *maximal sieve* on *U*, and denoted by $t(U) := t_U$, is a U-sieve. Moreover, if we consider the partially ordered set of open subsets of a topological space *X*, viewed as the base category of open reference contexts, $\mathcal{O}(X)$ as above, then a U-sieve S (where *U* is an open reference context) is a covering U-sieve if and only if *U* is contained in the union of open sets in *S*. This fulfills the requirements of covering sieves, and consequently, defines a *Grothendieck topology J* on

$\mathcal{O}(X)$. A Grothendieck topology J exists as a presheaf functor $\mathbf{K}_\chi$ in $Sets^{\mathcal{O}(X)^{op}}$, such that by acting on reference contexts U in $\mathcal{O}(X)$, J gives the set of all covering U-sieves, denoted by $\mathbf{K}_\chi(U)$; whereas by acting on figures $h: V \to U$, it gives a morphism $h^*(-): \mathbf{K}_\chi(U) \to \mathbf{K}_\chi(V)$, expressed as:

$$h^*(S) = \{f \,/\, cod(f) = V, (h \circ f) \in S\}\text{, for } S \in \mathbf{K}_\chi(U).$$

Notes

1. S. Mac Lane and I. Moerdijk, *Sheaves in Geometry and Logic* (New York: Springer-Verlag, 1992).
2. Mac Lane and Moerdijk, *Sheaves in Geometry and Logic*. See also: A. Mallios, *Geometry of Vector Sheaves: An Axiomatic Approach to Differential Geometry*, Vol. 1 (Kluwer Academic Publishers, Dordrecht, 1998).
3. To each a in A there exists an open set V, with $a \in V \subseteq A$, such that $\varphi(V)$ is open in X and $\varphi | V$ is a homeomorphism (continuous function that has a continuous inverse) $V \to \varphi(V)$.
4. Mac Lane and Moerdijk, *Sheaves in Geometry and Logic*. See also: Mallios, *Geometry of Vector Sheaves*, and A. Mallios, "On Localizing Topological Algebras," *Contemporary Mathematics* 341, no. 79 (2004).
5. Mallios, "On Localizing Topological Algebras."
6. Mac Lane and Moerdijk, *Sheaves in Geometry and Logic*.
7. S. Awodey, *Category Theory*, Oxford Logic Guides (Oxford: Oxford University Press, 2006). See also: Mac Lane and Moerdijk, *Sheaves in Geometry and Logic*.
8. Mac Lane and Moerdijk, *Sheaves in Geometry and Logic*.

CHAPTER 8

Sheaves of Boolean Germs: The Quantum Topological Case

8.1 Quantum Logical Event Algebras and Observables

The quantum case refers to a localization scheme on a global partial order of quantum physical events L taking place over a base category of reference contexts $\mathcal{B}$. This base category consists of *Boolean algebras B* associated with the operational procedure of quantum measurements, the arrows between them being Boolean algebra homomorphisms. In this case, the reference contexts of the operational environment employed for observation are all the Boolean algebras B, whereas their morphisms are not inclusions but *all* Boolean homomorphisms. Equivalently, the arrows $B \to L$ are considered as varying base reference frames of Boolean algebras over which a global partial order of quantum events L fibers.

Consider now the algebra L of *logical events* or *propositions* associated with the measurement of a quantum system. A *quantum logical event algebra* L in $\mathcal{L}$ is a partially ordered set of quantum logical events, endowed with a maximal element 1, and with an operation of orthocomplementation $[-]^{*}: L \to L$, which satisfy, for all $l \in L$, the following conditions:

[a] $l \leq 1$

[b] $l^{**} = l$

[c] $l \vee l^{*} = 1$

[d] $l \leq l' \Rightarrow l'^{*} \leq l^{*}$

[e] $l \perp l' \Rightarrow l \vee l' \in L$

[f] for $l, l' \in L$, $l \leq l'$ implies that l and l' are compatible, where $0 := 1^{*}$, $l \perp l' := l \leq l'^{*}$, and the operations of meet $\wedge$ and join $\vee$ are defined as usual. Recall that $l, l' \in L$ are *compatible* if the sublattice generated by $\{l, l^{*}, l', l'^{*}\}$ is a *Boolean algebra*, namely, if it is a *Boolean sublattice*. In order to have a well defined theory of observables over L, we also impose a σ-*completeness* condition, namely that the *join of countable families* of pairwise orthogonal events exists. It is important to notice here that a quantum logical event algebra L is *not* a Boolean algebra. It is technically called an *orthomodular* σ-*orthoposet*.

We can now construct a locally small cocomplete category, denoted by $\mathcal{L}$, which is called the category of quantum logical event algebras. The objects of $\mathcal{L}$ are the quantum logical event algebras L, whereas the arrows of $\mathcal{L}$ are quantum algebraic homomorphisms, defined as follows:

A *quantum logical algebraic homomorphism* in $\mathcal{L}$ is a morphism $H : K \rightarrow L$, which satisfies, for all $k \in K$, the following conditions:

[a] $H(1) = 1$

[b] $H(k^{*}) = [H(k)]^{*}$

[c] $k \leq k' \Rightarrow H(k) \leq H(k')$

[d] $k \perp k' \Rightarrow H(k \vee k') \leq H(k) \vee H(k')$

A *quantum observable* Ξ is defined to be an algebraic morphism from the *Borel algebra* of the real line $Bor(\mathbb{R})$ to the quantum event algebra L,

$$\Xi : Bor(\mathbb{R}) \rightarrow L$$

such that:

[a] $\Xi(\varnothing) = 0$, $\Xi(\mathbf{R}) = 1$

[b] $E \cap F = \varnothing \Rightarrow \Xi(E) \perp \Xi(F)$, for $E, F \in Bor(\mathbb{R})$

[c] $\Xi(\bigcup_n E_n) = \vee_n \Xi(E_n)$, where $E_1, E_2, \ldots$
is a sequence of mutually disjoint Borel sets of the real line.

Addition and multiplication over $\mathbb{R}$ induce on the set of quantum observables the structure of a *partial commutative algebra* over $\mathbb{R}$. In most of the cases the stronger assumption of a non-commutative algebra of quantum observables is adopted.

Thus, we may construct a category, denoted by $\mathcal{O}_Q$, which is called the *category of quantum observables*. Its objects are quantum observables $\Xi : Bor(\mathbb{R}) \rightarrow L$ and its arrows $\Xi \rightarrow \Theta$ are commutative triangles, or equivalently the quantum algebraic homomorphisms $H : L \rightarrow K$ in $\mathcal{L}$, preserving by definition the join of countable families of pairwise orthogonal events, such that $\Theta = H \circ \Xi$ in the following diagram is again a quantum observable.

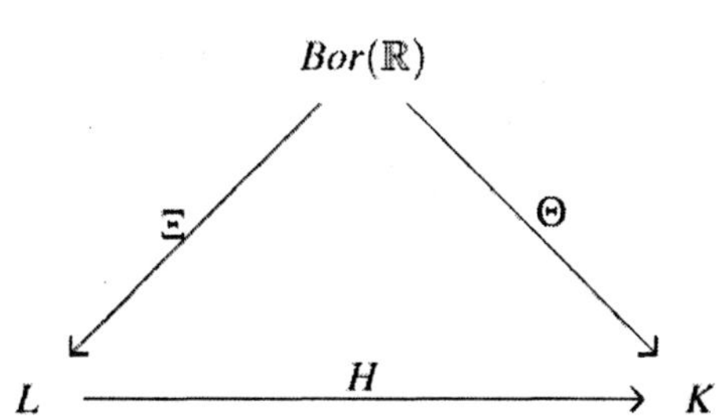

Correspondingly, we may construct a category, denoted by $\mathcal{O}_B$, which is called the *category of Boolean observables*. Its objects are the Boolean observables $\xi : Bor(\mathbb{R}) \to B$ and its arrows are the Boolean algebraic homomorphisms $h : B \to C$ in $\mathcal{B}$, such that $\theta = h \circ \xi$ in the following diagram is again a Boolean observable.

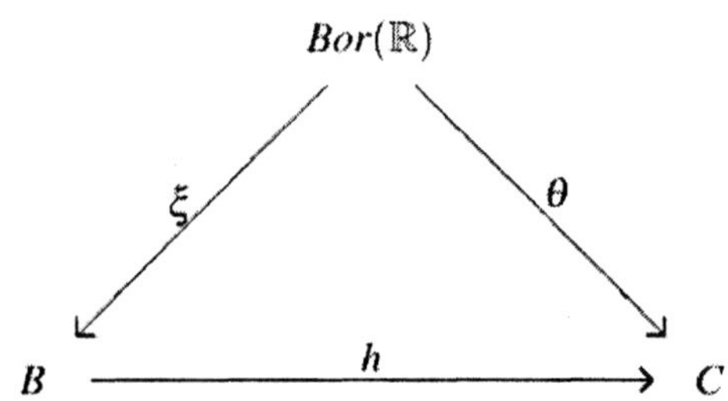

If L is isomorphic with the orthocomplemented lattice of *orthogonal projections* on a complex Hilbert space, then from von Neumann's *spectral theorem* the quantum observables are in bijective correspondence with the hypermaximal Hermitian operators on the Hilbert space.[1] It is important to note, here, that the original quantum logical formulation of quantum theory depends in an essential way on the *identification of propositions with projection operators on a complex Hilbert space*. A non-classical, *non-Boolean* logical structure is thereby induced which has its origins in quantum theory. More accurately, the Hilbert space quantum logic has been initially axiomatized as a complete, atomic, orthomodular lattice.[2] Equivalently, it could be cast as isomorphic to the *partial Boolean algebra of closed subspaces* of the Hilbert space associated with the quantum system, or alternatively the *partial Boolean algebra of projection operators* of the system.[3] By contrast, the propositional logic of classical mechanics is *Boolean logic*; that is, for the propositions of classical mechanics, the class of models over which validity and associated semantic notions are defined is the class of Boolean logic structures.

In this setting a fundamental result is provided by the Kochen-Specker theorem, according to which *there are no two-valued homomorphisms on a quantum logical event algebra L* globally. Thus, a quantum event algebra *cannot be embedded* into a Boolean algebra. Despite this admonition, however, and as stressed in chapter 2, in every concrete quantum measurement context, the set of logical events that can be actualized in this context *always* forms a Boolean event algebra. This fact motivates the assertion that *a Boolean algebra within the lattice of quantum logical events plays the topological role of a local logical reference frame* relative to which a measurement result is always coordinatized.

Likewise, together with a logical event structure, there always exists a corresponding *probabilistic structure*, defined by means of convex sets of measures on that logic. In this sense, the probabilistic structure of a classical system is described by convex sets of probability measures on the Boolean algebra of events of this system; whereas the probabilistic structure of a quantum system is described by convex sets of probability measures on the quantum logical event structure of that system.[4] More accurately, in the case of quantum systems, if the quantum event algebra is denoted by L, each *quantum probability measure*, or *quantum probabilistic state*, is defined by a mapping:

$$p : L \to [0,1]$$

such that the following conditions are satisfied: $p(1) = 1$ and $p(x \vee y) = p(x) + p(y)$, if $x \perp y$, where x, $y \in L$. In the Hilbert space formulation of quantum theory, L denotes the Hilbert space quantum event algebra, whereas a quantum state is defined by the Hilbert space inner product:

$$\langle \varphi, \mathrm{P}x\varphi \rangle$$

where $x \in L$, φ is a normalized vector in the Hilbert space, and $\mathrm{P}x$ is the orthogonal projection operator corresponding to $x \in L$.

8.2 The Topological Significance of Boolean Contexts

The set-theoretic axiomatizations of quantum event algebras *hides the intrinsic topological significance* of Boolean localizing contexts in the formation of these structures. On the other hand, the operational procedures followed in typical quantum measurement situations are based explicitly in the preparation and employment of appropriate *Boolean contexts.* That is to say, even if according to the Kochen-Specker theorem there *does not exist a global two-valued truth-functional* assignment pertaining to the *global event structure* of a quantum system, there *always* exists a *local two-valued truth-functional* assignment with respect to a *complete Boolean algebra of projection operators* on the Hilbert space of a quantum system, identified with a *complete Boolean subalgebra* of the global non-Boolean event algebra of a quantum system. More precisely, each *self-adjoint operator* representing an observable has *associated with it* a Boolean subalgebra which is identified with the Boolean algebra of projection operators belonging to its *spectral decomposition.* Hence, given a set of observables of a quantum system, there always exists a complete Boolean algebra of projection operators, viz., a *local Boolean subalgebra* of the global non-Boolean event algebra of a quantum system with respect to which a local two-valued truth-functional assignment is meaningful, if and only if the given observables are *simultaneously measurable.* Consequently, the possibility of local two-valued truth-functional assignments of the global non-Boolean event algebra of a quantum system points to the conclusion that *complete Boolean algebras play the topological role of local Boolean logical frames for contextual true/false value assignments.*

Because of this, we propose a shift in the semantics of quantum events from the set-theoretic level to the sheaf-theoretic, topological level. In particular, the epistemic path proposed in this volume implements the intuitively clear idea of probing the structure of a quantum algebra of events in terms of *localizing Boolean contexts*, admitting an unquestionably operational interpretation. In this regard, the quantum-theoretic localization process we describe involves the following: First, the introduction of a *localization scheme* consisting of (local or partial information) Boolean algebras *completely covering* a quantum algebra of quantum events by means of coordinatizing morphisms. These morphisms from the Boolean localizing reference contexts capture, in essence, separate, *complementary potential relations*, thus providing a structural decomposition or *classification* of a quantum event algebra in the descriptive terms of local Boolean reference contexts. These contexts thus provide potentiality windows for the actualization of concrete events via the evaluation or measurement of the observables contextualized. Second, the coherent realization of the quantum-theoretic localization process demands the satisfaction of *partial compatibility* between *overlapping* local Boolean covers. This guarantees an efficient *gluing procedure* between different coordinatizations of a quantum algebra of events via the measurement of observables. Third, the qualitative content incorporated in local Boolean reference contexts should be *integrated globally*, such that it respects the requirement of compatibility under both *extension* from the local level to the global level, and *restriction* from the global level to the local level.

8.3 Logical Internal Relation: Coherent Local-to-Global Ontological Formation

The above three constituent characteristics capture the essence of the sheaf-theoretic semantics of a quantum event algebra induced by a localization scheme of Boolean reference contexts. Thus, every *local observable* (*potential fact*) is *internally related*, via the *congruence* and *compatibility* conditions induced by the localization scheme, to a global quantum event structure of actualities. We will now proceed to explain how these relations are expressed in terms of *covering sieves* of quantum event algebras.

The *mutually implicative internal relation* between the local Boolean quantum level and the global, partially Boolean quantum level, is formulated mathematically via a sheaf-theoretic semantics in terms of *covering sieves of a quantum event algebra*. This mereotopological, internal relational structure is the mechanism by which local, intra-contextual logical implication is both restricted and extended globally, inter-contextually. Before defining precisely this concept of covering sieves of a quantum event algebra, it is important to stress the consequences referring to the interpretation of a 'quantum event continuum' via its sheaf-theoretic conceptualization. As emphasized earlier, *a quantum event continuum is not given as a completed, already extant set-theoretic structure; rather, it is a mereotopological process of continuous ontological formation that unfolds via the punctuated actualization of potential facts with respect to their local Boolean contexts*. Most significantly, this process of formation is *coherent* in the sense of respecting the rules of transition from the local or partial to the global and conversely. Moreover, the actualization of any potential fact is an *objective augmentation* of the already formed global event structure. The functional dependence inherent in sheaf-theoretic localization thus *relativizes* the presupposed rigid relations between quantum events with respect to *variable* local Boolean contexts *conditioning* the actualization of events. In this way, each *new* actualized event creates a *novelly integrated whole*, and this is due to the fact that its relational integration to the previously established totality is possible only via gluing and compatibility. It is elucidative to quote Alfred North Whitehead in this respect:

> The ultimate metaphysical principle is the advance from disjunction to conjunction, creating a novel entity other than the entities given in disjunction. The novel entity is at once the togetherness of the "many" which it finds, and also it is one among the disjunctive "many" which it leaves; it is a novel entity, disjunctively among the many entities which

> it synthesizes. The many become one and are increased by one. In their natures, entities are disjunctively "many" in process of passage to conjunctive unity . . . Thus the production of "novel togetherness" is the ultimate notion embodied in the term "concrescence."[5]

Thus, when Whitehead says that the "many become one and are increased by one" this can be interpreted from a sheaf-theoretic perspective as follows: The many become one by compatible interconnection or *gluing* of their relational information content (encompassing the information of the already actualized events in local partially-compatible contexts), and this process is generative of a novel quantum event; thus, the many are increased by one, because the newly integrated totality now includes this novel event. Together with the compatible local context of this event's actualization, the novel totality, in turn, imposes further compatibility constraints on new local contexts (specified by the requirements of the conjunctive gluing procedure) for the integration of new observable information, and so on.

The defining conditions characterizing a sheaf incorporate the formal requirements of this process, wherein every local contextual actualization of a potential fact is internally related to a unique global totality of already actualized facts (events). It is necessary to emphasize that the sheaf-theoretic localization of quantum observability is thus conceived as an *ontological process* and not as a mere epistemic restructuring of a fixed, perpetually extant global totality of quantum events. In this sense it provides a conceptual avenue for a further category-theoretic investigation of the already established relevance and suitability of Whitehead's process philosophy for the interpretation of quantum mechanics.[6] For example, a global quantum event algebra, although hypothesized to exist in a way corresponding to reality, is *not* given as a completed whole initially, but has to be *formed via actualization* of local potential facts with respect to Boolean contexts adding precisely factual content to the global. Furthermore, the *bidirectional form* of consistent internal relation depicted via sheaf-theoretic localization reflects the fact that every potential quantum event, actualized with respect to a local

Boolean context, is always logically correlated (by means of an equivalence relation) to any other potential quantum event actualized with respect to any other compatible local Boolean context. *Globally, this means that quantum observables are not determined by their measured values, but by their Boolean observable germs, viz., equivalence classes of compatible Boolean observables forming an inductive limit.* Concomitantly, global quantum events are determined by evaluations of Boolean observable germs.

8.4 Functor of Boolean Reference Frames

A *Boolean shaping functor* or functor of *local Boolean coefficients* of a quantum categorical event structure, $\mathbf{M}: \mathcal{B} \rightarrow \mathcal{L}$, assigns to Boolean event algebras in $\mathcal{B}$ (constitutive of a model coordinatizing category) the underlying quantum event algebras from $\mathcal{L}$, and likewise assigns to Boolean homomorphisms the underlying quantum algebraic homomorphisms. Throughout the remainder of the discussion, for reasons of simplicity in the notation, the Boolean shaping functor will not appear explicitly in the formulas, although it will be always implicitly assumed.

If $\mathcal{B}^{op}$ is the opposite category of $\mathcal{B}$, then $Sets^{\mathcal{B}^{op}}$ denotes the functor category of *presheaves* on Boolean event algebras, with objects all functors $\mathbf{P}: \mathcal{B}^{op} \rightarrow Sets$, and morphisms all natural transformations between such functors. Each object $\mathbf{P}$ in this category is a contravariant set-valued functor on $\mathcal{B}$, called a presheaf on $\mathcal{B}$. For each Boolean event algebra B of $\mathcal{B}$, $\mathbf{P}(\mathrm{B})$ is a set, and for each arrow $f: C \rightarrow B$, $\mathbf{P}(f): \mathbf{P}(B) \rightarrow \mathbf{P}(C)$ is a set-function. If $\mathbf{P}$ is a presheaf on $\mathcal{B}$ and $p \in \mathbf{P}(B)$, the value $\mathbf{P}(f)(p)$ for an arrow $f: C \rightarrow B$ in $\mathcal{B}$ is called the *restriction* of x along f and is denoted by $\mathbf{P}(f)(p) = p \cdot f$.

The *category of elements* of a presheaf $\mathbf{P}$, denoted by $\int(\mathbf{P}, \mathcal{B})$ has as its objects all pairs (B, p), and morphisms $(B', p') \rightarrow (B, p)$ are those morphisms $u: B' \rightarrow B$ of $\mathcal{B}$ for which $p \cdot u = p'$.

Projection on the second coordinate of $\int(\mathbf{P},\mathcal{B})$, defines a functor $\int_{\mathbf{P}} : \int(\mathbf{P},\mathcal{B}) \to \mathcal{B}$. $\int(\mathbf{P},\mathcal{B})$ together with the projection functor $\int_{\mathbf{P}}$ is defined as the *split discrete fibration* induced by $\mathbf{P}$, where $\mathcal{B}$ is the base category of the fibration as in the diagram below.

$$\begin{array}{ccc} \int(\mathbf{P},\mathcal{B}) & & \\ \Big\downarrow{\scriptstyle \int_{\mathbf{P}}} & & \\ \mathcal{B} & \xrightarrow{\mathbf{P}} & \mathbf{Sets} \end{array}$$

We note that the fibers are categories in which the only arrows are identity arrows. If B is an object of $\mathcal{B}$, the inverse image under $\int_{\mathbf{P}}$ of B is simply the set $\mathbf{P}(B)$, although its elements are written as pairs so as to form a disjoint union.

The *representation* or *realization* functor of the category of quantum event algebras $\mathcal{L}$ into the category of presheaves of Boolean event algebras $Sets^{\mathcal{B}^{op}}$ is defined by:

$$\mathbf{R} : \mathcal{L} \to Sets^{\mathcal{B}^{op}}$$

The *functor of Boolean reference frames* of a quantum event algebra L in $\mathcal{L}$ is the *image* of the representation functor $\mathbf{R}$, evaluated at L, into the category of presheaves of Boolean events algebras $Sets^{\mathcal{B}^{op}}$, that is:

$$\mathbf{R}(L) := \mathbf{R}_L : \mathcal{B}^{op} \to Sets$$

Notice that the representation functor of $\mathcal{L}$ is completely determined by the action of the functor of Boolean reference frames, for each quantum event algebra L in $\mathcal{L}$, on the objects and arrows of the category of Boolean event algebras $\mathcal{B}$, specified as follows: Its action on an object B in $\mathcal{B}$ is given by

$$\mathbf{R}(L)(B) := \mathbf{R}_L(B) = Hom_{\mathcal{L}}(B, L)$$

whereas its action on a morphism $x : D \to B$ in $\mathcal{B}$, for $v : B \to L$ is given by

$$\mathbf{R}(L)(x) : Hom_{\mathcal{L}}(B, L) \to Hom_{\mathcal{L}}(D, L)$$

$$\mathbf{R}(L)(x)(v) = v \circ x$$

It is important to stress that the functor of Boolean reference frames of a quantum event algebra L in $\mathcal{L}$, is a *presheaf* $\mathbf{R}(L) := \mathbf{R}_L : \mathcal{B}^{op} \to Sets$. Thus, we can legitimately consider the category of elements $\int(\mathbf{R}_L, \mathcal{B})$, together with the projection functor $\int_{\mathbf{R}_L} : \int(\mathbf{R}_L, \mathcal{B}) \to \mathcal{B}$, viz., the split discrete fibration induced by the functor of Boolean frames of L, where $\mathcal{B}$ is the base category of Boolean reference contexts of the fibration. Hence, the functor of Boolean frames of a quantum event algebra induces a *uniform and homologous fibered representation of quantum events in terms of Boolean reference frames* for the measurement of observables.

Thus, in mathematical terms, the quantum localization process in the descriptive terms of Boolean reference frames is understood by means of an *action* of the category of (local) Boolean event algebras on a set-theoretic global structure of quantum events. A global quantum event algebra is then *partitioned* into sorts parameterized by the *Boolean frames* of the base category of local Boolean contexts. In this way, the functioning of a Boolean localization scheme on a global structure of quantum events is represented by means of a *fibered construct*, understood geometrically, as a locally variable set over the base category of local Boolean reference contexts. The notion of functional dependence incorporated in this action forces *the partially ordered structure of quantum events to fiber over the base category of local Boolean reference contexts*.

It is important to emphasize that the functor of Boolean reference frames of a quantum event algebra L, viz., *the network of relationships* that L has with all Boolean contexts B, formalized as a presheaf, incorporates the physical requirement of *uniformity* among events. By this we mean that for any two quantum events observed over the same Boolean domain of measurement, the structure of all Boolean logical reference frames that relate to the first cannot be distinguished in any possible way from the structure of Boolean frames relating to the second. According to the principle of uniformity, in other words, all the localized quantum events within any particular Boolean reference context should be *uniformly equivalent* to each other. The compatibility of the Boolean localization process of a quantum event algebra with the requirement of uniformity entails that the partial order of relations in a global structure of quantum events is induced by lifting an appropriate family of morphisms from the base category of local Boolean reference contexts to the fibers. Equivalently, *the partial order of relations among quantum events is induced by corresponding relations among their localizing Boolean reference contexts.*

8.5 Functorial Boolean-Quantum Internal Relation

The network of relationships defined by a quantum event algebra with Boolean logical frames, formalized categorically by the notion of a presheaf functor of Boolean frames inducing a Boolean localization scheme, is the semantic information carrier of a *mutually implicative internal relation* between the local Boolean level and the global quantum level. This *dipolar* mutually implicative internal relation is formulated in the category theoretic syntax in terms of a pair of adjoint functors between the category of presheaves of Boolean event algebras and the category of quantum event algebras, thus forming a categorical *adjunction*. This categorical adjunction formalizes the process-theoretic operation of the category of *transmutation* by relating internally and bidirectionally the local Boolean and global quantum levels of the event structure. More precisely, we formulate the following proposition:[7]

There exists a pair of adjoint functors $\mathbf{L} \dashv \mathbf{R}$ *as follows:*

$$\mathbf{L} : Sets^{\mathcal{B}^{op}} \rightleftarrows \mathcal{L} : \mathbf{R}$$

The Boolean frames-quantum adjunction consists of the functors $\mathbf{L}$ and $\mathbf{R}$, called left and right adjoints with respect to each other respectively, as well as the natural bijection:

$$Nat(\mathbf{P}, \mathbf{R}(L)) \cong Hom_{\mathcal{L}}(\mathbf{LP}, L)$$

The above bijective correspondence, interpreted functorially, says that the *Boolean realization functor* of a quantum categorical event structure $\mathcal{L}$, realized for each quantum event algebra L in $\mathcal{L}$ by its *functor of Boolean frames*, viz., by

$$\mathbf{R}(L) : B \mapsto Hom_{\mathcal{L}}(B, L)$$

has a *left adjoint* functor $\mathbf{L} : Sets^{\mathcal{B}^{op}} \to \mathcal{L}$, which is defined for each presheaf of Boolean algebras $\mathbf{P}$ in $Sets^{\mathcal{B}^{op}}$ as the colimit (inductive limit):

$$\mathbf{L(P)} = Colim\{\textstyle\int(\mathbf{P}, \mathcal{B}) \to \mathcal{B} \to \mathcal{L}\}$$

Thus, the following diagram commutes:

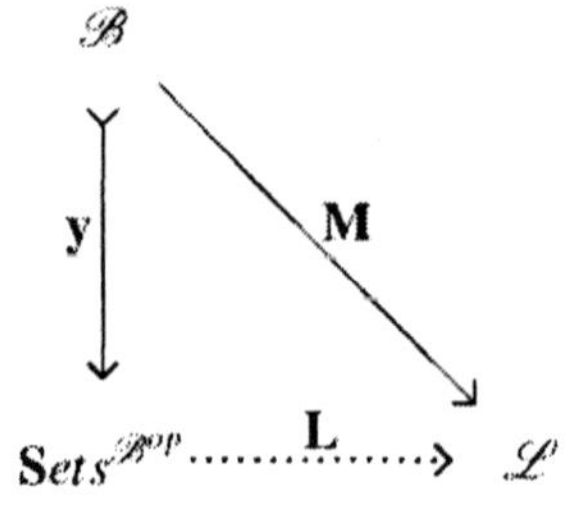

The physical meaning of the adjunction between presheaves of Boolean logical frames and quantum event algebras is made transparent if we consider that the mutually inverse pair of internally related adjoint functors formalizes the *mutually implicative processes of encoding and decoding information* relevant to the structural form of their domain and codomain categories. If we think of $Sets^{\mathcal{B}^{op}}$ as the categorical universe of variable local Boolean frames modeled in $Sets$, and of $\mathcal{L}$ as the categorical universe of quantum event structures, then the functor $\mathbf{L}: Sets^{\mathcal{B}^{op}} \rightarrow \mathcal{L}$ signifies a *translational code* of information from the level of locally Boolean events to the level of global quantum events; and likewise, the Boolean realization functor $\mathbf{R}: \mathcal{L} \rightarrow Sets^{\mathcal{B}^{op}}$ signifies a translational code in the inverse direction.

In general, the content of the information cannot remain completely invariant with respect to translation from one categorical universe to another and conversely. However, there remain two alternatives for a variable set over local Boolean algebras $\mathbf{P}$ to exchange information with a quantum algebra L: *either* the content of information is exchanged in *quantum terms* with the inductive limit in the category of elements of $\mathbf{P}$ translating, represented as the quantum morphism $\mathbf{LP} \rightarrow L$; *or* the content of information is exchanged in *Boolean terms* with the functor of Boolean frames of L translating, represented correspondingly as the natural transformation $\mathbf{P} \rightarrow \mathbf{R}(L)$.

In the first case, from the perspective of L, information is being received in quantum terms while in the second, from the perspective of $\mathbf{P}$, information is being sent in Boolean terms. The natural bijection then corresponds to the assertion that these *two distinct ways of communicating are equivalent*. Thus, the fact that these two functors are adjoint expresses a relation of variation, regulated by *two poles*, with respect to the meaning of the information related to observation. We claim that the totality of the content of information included in quantum observable structures remains *invariant under Boolean encodings*, corresponding to local Boolean logical frames, *if and only if* the adjunctive correspondence can be appropriately restricted to an *equivalence* of the functorially correlated categories. This task can be accomplished by defining an appropriate Grothendieck topology on the base category of Boolean frames, which essentially paves the way for the understanding of a quantum event algebra as a *sheaf* of local Boolean algebras over an appropriately specified covering system of the former. This categorical equivalence is interpreted physically in terms of a *transmuted information invariance property* between the Boolean and the quantum species of structure, referring to the *bidirectional internal relation* between variable local Boolean algebras and global quantum ones.

For any presheaf $\mathbf{P}$ in $Sets^{\mathcal{B}^{op}}$, the *unit* natural transformation of the Boolean frames-quantum adjunction is defined as follows:

$$\delta_{\mathbf{P}} : \mathbf{P} \to \mathbf{RLP}$$

Likewise, for each quantum event algebra L in $\mathcal{L}$ the *counit* natural transformation is defined as follows:

$$\varepsilon_L : \mathbf{LR}(L) \to L$$

The representation of a quantum event algebra L in $\mathcal{L}$, in terms of the functor of Boolean frames $\mathbf{R}(L)$ of L, is *full and faithful*, if and only if the counit is a *quantum algebraic isomorphism*. In turn, the counit is a quantum algebraic isomorphism if and only if the right adjoint functor is full and faithful. In the latter case we characterize the Boolean shaping functor $\mathbf{M} : \mathcal{B} \to \mathcal{L}$ as a *proper* or *dense shaping functor*. It can be shown that the Boolean realization functor is full and faithful if it corresponds to a *functor of Boolean localizations* of L for an appropriate *Grothendieck topology*.[8] Thus, the counit is an isomorphism if it is restricted to some *Boolean localization system* of L.

From the above, we deduce that the representation of a quantum event algebra L in $\mathcal{L}$, in terms of the functor of Boolean frames $\mathbf{R}(L)$ of L, is full and faithful, if the Boolean frames-quantum adjunction is *restricted* to a Boolean localization system of L. As a corollary we obtain that a quantum categorical event structure L is a *reflection* of the topos of presheaves $Sets^{B^{op}}$.

8.6 Quantum Localization: Sheaf of Boolean Reference Frames

According to the defining requirements of a localization scheme, the established fibred representation of quantum events, realized for each quantum event algebra by its functor of Boolean frames, should be also coherent under extension from the local to the global level and conversely. Thus, we should define an appropriate Grothendieck topology **J** on the base category of Boolean contexts B, such that: [i] The Boolean reference frames acquire the semantics of *local frames* with respect to that Grothendieck topology **J** on B, and [ii] The functor of Boolean frames for a quantum event algebra L in L becomes a *sheaf on the site* (B,**J**) for that **J**. For this purpose, we use the machinery of covering sieves of Boolean frames, defined as follows:

A B-sieve S on a Boolean reference context B in B is called a *covering sieve* of B, if all the arrows $s: C \to B$ belonging to the sieve S, taken together, form an *epimorphic family* in L. This requirement may be, equivalently, expressed in terms of a map:

$$G_S : \coprod_{(s:C \to B) \in S} C \to B$$

being an *epimorphism* in L. It can be proved that the specification of covering sieves on Boolean contexts B in B, in terms of epimorphic families of arrows in L, *does indeed define a categorical topological covering system (Grothendieck topology)* **J** on B.[9] Thus, the function **J**, which assigns to each Boolean reference context B in B, a collection $\mathbf{J}(B)$ of covering B-sieves, being epimorphic families of arrows in L, constitutes a Grothendieck topology **J** on B. Then, we can show that the *functor of Boolean reference frames* $\mathbf{R}_L$ is a *sheaf for* *the Grothendieck topology* $\mathbf{J}$, defined by covering sieves of epimorphic families on the category of Boolean reference contexts, as above.[10] Hence, the Boolean reference frames $B \to L$ acquire the *semantics of local Boolean frames* with respect to the Grothendieck topology $\mathbf{J}$, defined in terms of covering sieves of epimorphic families of arrows in $\mathcal{L}$. Moreover, since the functor of Boolean frames for a quantum event algebra L becomes a *sheaf* restricted on the site $(\mathcal{B}, \mathbf{J})$ for that $\mathbf{J}$, the split discrete fibration $\int \mathbf{R}_L : \int(\mathbf{R}_L, \mathcal{B}) \to \mathcal{B}$, induced by the *sheaf of Boolean frames* of L on the site $(\mathcal{B}, \mathbf{J})$, forces a uniform, homologous, and also coherent *fibered representation of quantum events in terms of local Boolean reference frames*.

More concretely, the definition of the Grothendieck topology $\mathbf{J}$ on the base category of Boolean reference contexts $\mathcal{B}$, induces the functionality of a quantum localization scheme, formulated in terms of covering sieves on a quantum event algebra L as follows: We say that a sieve on a quantum event algebra L defines a covering sieve by Boolean reference frames, such that the domains of the frames belonging to the sieve are Boolean contexts in $\mathcal{B}$, if all these morphisms in the sieve collectively define an *epimorphism* (that is, collectively *cover* it completely):

$$T : \coprod_{(E \in [\mathcal{B}]_0, \psi_E : E \to L)} E \to L$$

where $[\mathcal{B}]_0$ denotes the set of Boolean reference contexts of the base category $\mathcal{B}$. From the physical point of view, *covering sieves of a quantum event algebra by local Boolean reference frames are interpreted as Boolean localization systems of a quantum event algebra.*[11] It is instructive to think of covering sieves of a quantum event algebra L as *subfunctors* of the functor of Boolean frames of L, viz., of the functor $\mathbf{R}(L) := \mathbf{R}_L : \mathcal{B}^{op} \to \mathcal{S}ets$. These subfunctors of $\mathbf{R}(L)$, that is, the covering sieves of L, are *partially ordered* by inclusion. The partial order relation among covering sieves of a quantum event algebra L is interpreted as an operation of *coarse-graining* among Boolean localization systems.

8.7 Boolean Localization Systems and Boolean Germs

A Boolean localization system forces a transition in the semantics of quantum events from the set-theoretic to the topological sheaf-theoretic level. This is due to the following proposition:[12]

A covering sieve of a quantum event algebra L *by local Boolean reference frames, that is, a Boolean localization system of* L, *induces an isomorphism:*

$$L \cong \mathbf{Colim}\{\textstyle\int(\mathbf{R}_L, \mathcal{B}) \to \mathcal{B}\}$$

where, $\mathbf{R}_L$ denotes a sheaf of Boolean frames of L on the site $(\mathcal{B}, \mathbf{J})$, for $\mathbf{J}$ the Grothendieck topology of epimorphic families, and $\int(\mathbf{R}_L, \mathcal{B})$ is the associated fibration induced by the category of Boolean frames of L.

Thus, a quantum event algebra L is represented *isomorphically* by the *inductive limit taken in the category of local Boolean reference frames of the sheaf* $\mathbf{R}_L$ with respect to a *Boolean localization system*. Let us now explain the meaning of this *inductive limit* (colimit) construction. For this purpose we consider the set:

$$\mathbf{Y}(\mathbf{R}_L) = \{(\psi_B, q) / (\psi_B : B \to L), q \in B\}$$

Note that the morphisms $\psi_B : B \to L$ denote Boolean reference frames of L, encoded as elements in the category of Boolean frames of L, viz., $\int(\mathbf{R}_L, \mathcal{B})$. We notice that if there exists a Boolean morphism $u : B' \to B$ such that: $u(q') = q$ and $\psi_B \cdot u = \psi_{B'}$ by restriction along $u : B' \to B$, then we may define a *transitive* and *reflexive* relation $\mathfrak{R}$ on the set

$\mathbf{Y}(\mathbf{R}_L)$. Of course the inverse also holds true. Then we obtain the following:

$$(\psi_B \cdot u, q)\Re(\psi_B, u(q'))$$

for any $u: B' \to B$ in the category $\mathcal{B}$. The next step is to make this relation also *symmetric* by postulating that for ζ, η in $\mathbf{Y}(\mathbf{R}_L)$, where ζ, η denote pairs in the above set, we have:

$$\zeta \sim \eta$$

if and only if $\zeta \Re \eta$ or $\eta \Re \zeta$. Finally by considering a sequence ξ_1, ξ_2, . . . , ξ_k of elements of the set $\mathbf{Y}(\mathbf{R}_L)$ and also ζ, η such that there exists a path of Boolean transition arrows:

$$\zeta \sim \xi_1 \sim \xi_2 \sim \ldots \sim \xi_{k-1} \sim \xi_k \sim \eta$$

We may define an equivalence relation on the set $\mathbf{Y}(\mathbf{R}_L)$ as follows:

$$\zeta \rhd\lhd \eta := \zeta \sim \xi_1 \sim \xi_2 \sim \ldots \sim \xi_{k-1} \sim \xi_k \sim \eta$$

Then for each $\zeta = (\psi_B, q) \in \mathbf{Y}(\mathbf{R}_L)$ we define the Boolean germ (equivalence class) at ζ as follows:

$$Q_\zeta = \{\iota \in \mathbf{Y}(\mathbf{R}_L) : \zeta \rhd\lhd \iota\}$$

We finally define:

$$\mathbf{Y}(\mathbf{R}_L)/\rhd\lhd := \{Q_\zeta : \zeta = (\psi_B, q) \in \mathbf{Y}(\mathbf{R}_L)\}$$

and use the notation $Q_\zeta = \|(\psi_B, q)\|$, where $\|(\psi_B, q)\|$ denotes the *Boolean germ* at $\zeta = (\psi_B, q)$. The *quotient set* $\mathbf{Y}(\mathbf{R}_L)/\rhd\lhd$ defines the *inductive limit* in the category of local Boolean reference frames of the sheaf $\mathbf{R}_L$, that is,

$$\mathbf{Y}(\mathbf{R}_L)/\rhd\lhd = \mathbf{Colim}\{\int(\mathbf{R}_L,\mathcal{B}) \to \mathcal{B}\}$$

by noticing that it is naturally endowed with a quantum event algebra structure as follows:

[1] The orthocomplementation is defined as:
$Q_\zeta^* = \| (\psi_B, q) \|^* = \| (\psi_B, q^*) \|$.

[2] The unit element is defined as: $\mathbf{1} = \| (\psi_B, 1) \|$.

[3] The partial order structure on the quotient set $\mathbf{Y}(\mathbf{R}_L)/\rhd\lhd$ is defined as: $\| (\psi_B, q) \| \preceq \| (\psi_C, r) \|$ if and only if $d_1 \preceq d_2$ where we have made the following identifications: $\| (\psi_B, q) \| = \| (\psi_D, d_1) \|$ and $\| (\psi_C, r) \| = \| (\psi_D, d_2) \|$, with d_1, $d_2 \in D$ according to the pullback diagram of event algebras:

$$\begin{array}{ccc} D & \xrightarrow{\ \beta\ } & B \\ \gamma\downarrow & & \downarrow \\ C & \longrightarrow & L \end{array}$$

such that $\beta(d_1) = q$, $\gamma(d_2) = r$.

8.8 Gluing Isomorphism of Local Boolean Reference Frames

In order to understand the above pullback diagram we have to remember that $\mathbf{R}_L$ is a *sheaf of Boolean frames* of L on the *site* $(\mathcal{B}, \mathbf{J})$, for $\mathbf{J}$ the Grothendieck topology of *epimorphic families*. The fact that

$\mathbf{R}_L$ is a sheaf of Boolean frames of L means the following: Let us consider two Boolean reference frames $\psi_B : B \to L$ and $\psi_{B'} : B' \to L$ of L. Then the *pullback* (*categorical overlap*) of the Boolean reference frames ψ_B and $\psi_{B'}$ with common codomain a quantum event algebra L, consists of the event algebra $B \times_L B'$ and two arrows $\psi_{BB'}$ and $\psi_{B'B}$, called *projections*, as shown in the following diagram:

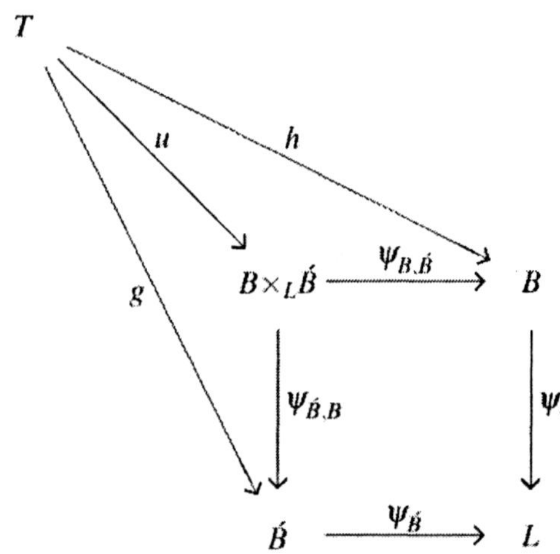

The square *commutes* and for any event algebra T and arrows h and g that make the outer square commute, there is a unique $u : T \to B \times_L B'$ that makes the whole diagram commute. Now, if ψ_B and $\psi_{B'}$ are injective morphisms, then their pullback is isomorphic with their intersection $B \cap B'$.

The fact that the functor $\mathbf{R}_L$ of Boolean frames of L is a sheaf, means that there exists a *gluing isomorphism* of Boolean reference frames on their overlap, defined as follows:

$$\Omega_{B,B'} : \psi_{B'B}(B \times_L B') \to \psi_{BB'}(B \times_L B')$$

$$\Omega_{B,B'} = \psi_{BB'} \circ \psi_{B'B}{}^{-1}$$

The gluing isomorphism $\Omega_{B,B'}$ means that the Boolean reference frames of L, that is, $\psi_{B'B}(B \times_L B')$ and $\psi_{BB'}(B \times_L B')$, *cover the same part of a quantum event algebra* L in a *compatible way*. Furthermore, an immediate consequence of the previous definition is the satisfaction of the following *cocycle conditions*:

$$\Omega_{B,B} = id_B$$
$$\Omega_{B,B'} \circ \Omega_{B',B''} = \Omega_{B,B''} \qquad \textit{if} \quad B \times_L B' \times_L B'' \neq 0$$

$$\Omega_{B,B'} = \Omega^{-1}{}_{B',B} \qquad \textit{if} \quad B \times_L B' \neq 0$$

We conclude that the gluing isomorphism between two Boolean reference frames $\Omega_{B,B'}$ in the sheaf of Boolean frames $\mathbf{R}_L$ assures that $\psi_{B'B}(B \times_L B')$ and $\psi_{BB'}(B \times_L B')$ are being *glued and thus cover* the same part of a quantum event algebra L in a compatible way.

Consequently, we state the following conclusion: *A quantum event algebra* L *can be represented isomorphically by the inductive limit taken in the category of elements of the sheaf functor of Boolean reference frames of* L. Hence, a quantum event algebra can be represented by means of a *sheaf of Boolean germs*, with respect to the topology $\mathbf{J}$. Globally, this means that quantum observables are *not* determined by their measured values, but they *are* determined by their *Boolean observable germs*, viz., equivalence classes of partially compatible Boolean observables. Concomitantly, quantum events are specified by *evaluations* of their corresponding Boolean germs with respect to Boolean localization systems. Thus, there exists a uniform, homologous and coherent fibered representation of quantum events with respect to local Boolean reference frames forming Boolean localization systems of a quantum event algebra.

We conclude that every local observable (potential fact) is *internally related*, via the congruence and compatibility relations induced by the correlate Boolean localization system, to a global quantum event structure of actualities. Conversely, a global quantum event continuum is *not* given as a completed a priori set-theoretic structure; rather, it is constituted, in a *continuous* topological process of ontological formation, by *actualizations of potential facts* with respect to local Boolean contexts. These actualizations are expressed categorically by *evaluations of Boolean germs*. Thus, given the the functional role of Boolean germs as internally constitutive of becoming actual occasions, this sheaf-theoretic process of continuous formation is fully coherent in the sense of respecting the rules of *extension* from the local to the global, and *restriction* of the local by the global.

Notes

1. V. S. Varadarajan, *Geometry of Quantum Theory* (New York: Springer-Verlag, 1985).
2. G. Birkhoff and J. von Neumann, "The logic of quantum mechanics," *Annals of Mathematics* 37 (1936).
3. Simon Kochen and Ernst Specker, "The problem of hidden variables in quantum mechanics," *Journal of Mathematics and Mechanics* 17 (1967): 59-87.
4. Elias Zafiris, "Sheaf-Theoretic Representation of Quantum Measure Algebras," *Journal of Mathematical Physics* 47, 092103 (2006).
5. Alfred North Whitehead, *Process and Reality: An Essay in Cosmology*, Corrected Edition, ed. D. Griffin and D. Sherburne (New York: Free Press, 1978), 21.
6. M. Epperson, *Quantum mechanics and the philosophy of Alfred North Whitehead*, American philosophy series, no. 14 (New York: Fordham University Press, 2004); M. Epperson, "Quantum Mechanics and Relational Realism: Logical Causality and Wavefunction Collapse," *Process Studies* 38, no. 2 (2009).
7. Elias Zafiris, "Boolean Coverings of Quantum Observable Structure: A Setting for an Abstract Differential Geometric Mechanism," *Journal of Geometry and Physics* 50, no. 99 (2004); Elias Zafiris, "Quantum Event Structures from the perspective of Grothendieck Topoi," *Foundations of Physics* 34, no. 7 (2004); Elias Zafiris, "Interpreting Observables in a Quantum World from the Categorial Standpoint," *International Journal of Theoretical Physics* 43, no. 1 (2004); Elias Zafiris, "Sheaf-Theoretic Representation of Quantum Measure Algebras," *Journal of Mathematical Physics* 47, 092103 (2006).

8. See note 7.

9. Zafiris, "Quantum Event Structures"; Elias Zafiris, "Generalized Topological Covering Systems on Quantum Events Structures," *Journal of Physics A: Mathematical and General* 39 (2006).

10. See note 9.

11. Zafiris, "Interpreting Observables."

12. See note 9.

CHAPTER 9

Functorial Entanglement and Logical Classification

9.1 Composite Systems and Entanglement

In the Hilbert space formulation of quantum theory, *entanglement* refers to the phenomenon displayed by a *composite* quantum system according to which the behavior and properties of the latter, consisting of several subsystems, is *not reducible* to the properties of these individual subsystems. This phenomenon is manifest in the particular type of *correlations* found in the *joint probability distributions* of events associated with the subsystems constitutive of a composite quantum system. More precisely, in the Hilbert space formulation, a single quantum system is represented by a complex Hilbert space of states, whereas a composite quantum system is represented by the *tensor product* of the Hilbert spaces of the states of its subsystems over the complex numbers. In this formulation, it is the tensor product decomposition structure of the Hilbert space of states of the composite system which accounts for the appearance of entanglement phenomena among subsystems.

Let us briefly recall that a *state* of a quantum system is represented by a positive, Hermitian (self-adjoint) operator of trace 1 called a *density operator* (or statistical operator). The density operator is represented as a linear sum of orthogonal, one-dimensional projection operators:

$$\rho = \Sigma_i \lambda_i P_i$$

where the weight coefficients λ_i are positive numbers summing up to unity, that is $\Sigma_i \lambda_i = 1$. In the above linear sum, if only a single coefficient λ is different from zero, then the state is called a *pure state*—that is, $\rho = \lambda P$. Equivalently, a pure state can be characterized by a unit state vector Ψ belonging to the one-dimensional subspace of the Hilbert space (*ray*) into which P projects. Thus, a non-pure state, called a *mixed state*, is considered as a weighted linear sum (called a convex combination) of pure states (or one-dimensional projection operators) where the total weight is one. Note that this is *different* from a superposition of pure states with complex coefficients, which is also a pure state according to the *superposition principle* of quantum theory. Moreover, a mixed state can be decomposed in many different ways in the form:

$$\rho = \Sigma_i \gamma_i \rho_i$$

where the coefficients γ_i are positive weights and the ρ_i are other density operators (which may correspond to pure states).

The particular form of the density operator expressed by $\rho = \Sigma_i \lambda_i P_i$ is called the *spectral decomposition* of the density operator in terms of *one-dimensional projectors*. In the Hilbert space formulation, each Hermitian operator has associated with it a *complete Boolean algebra* identified as a Boolean algebra of projection operators belonging to its spectral decomposition. Hence, given a set of observables of a quantum system, there always exists a complete Boolean algebra of projection operators, viz., a local Boolean subalgebra of the global non-Boolean event algebra of a quantum system with respect to which a local two-valued truth-functional assignment is meaningful, if and only if the given observables can be *simultaneously measurable*. As we have already explained, the possibility of local two-valued truth-functional assignments of the global non-Boolean event algebra of a quantum system points to the assumption that complete Boolean algebras play the role of *local Boolean logical* frames for *contextual* true/false value assignments. Thus, as introduced in chapters 2 and 3, an observational procedure is represented by a Boolean algebra of orthogonal projectors, where each one of these projectors corresponds to one of the possible results of the observation. For example, the probability for the result ε in state ρ is given by:

$$\mu_\varepsilon = Tr(\rho P_\varepsilon)$$

where the numbers (probabilities) μ_ε are obtained by the *trace* of the operator in the parentheses, whereas their positivity is induced by the positivity of ρ. Now, if we assume that the observed or measured values associated with the results ε are real numbers, represented by the coefficients ζ_ε, then the Hermitian operator

$$Z = \Sigma_\varepsilon \zeta_\varepsilon P_\varepsilon$$

is identified as the observable associated with this measurement procedure.

It is important to keep in mind that the form of decomposition of a density operator in terms of a *convex combination* of one-dimensional, orthogonal projection operators is *not unique* but *depends on* the local Boolean logical frame of projectors associated with some observational procedure. Thus, the weight coefficients

$$\lambda_i = Tr(\rho P_i)$$

where P_i are one-dimensional projectors, are interpreted as *conditional probabilities* with respect to the local Boolean frame that P_i belongs to. Notice that since P_i may belong to different overlapping local Boolean frames, the coefficients λ_i are interpreted as conditional probabilities with respect to a Boolean localization system of compatible overlapping local Boolean frames. Thus, λ_i is the conditional probability of an equivalence class (Boolean germ) conditioned by the corresponding Boolean localization system. We conclude that the density operator of a quantum system provides a description of states in which all possible decompositions are, in a well-defined sense, *implicitly* present at once, albeit *potentially*.

As introduced in chapters 2 and 3, in a composite quantum system consisting of two subsystems, the Hilbert space of states is represented by the tensor product of the Hilbert spaces of the states of the two subsystems over the complex numbers, denoted by:

$$H = H_1 \otimes H_2$$

The tensor product of H_1 and H_2 considered as linear spaces over the complex numbers is the *unique* object in the category of complex linear spaces, which satisfies the following *universal property*: There exists a bilinear morphism $\tau : H_1 \times H_2 \to H_1 \otimes H_2$ such that given any linear space S and any bilinear morphism $f : H_1 \times H_2 \to S$ there exists a unique linear morphism $g : H_1 \otimes H_2 \to S$ such that the latter bilinear morphism admits the factorization $f = g \circ \tau$. Consequently, factorization via the tensor product of two complex linear spaces H_1, H_2, viz., $H_1 \otimes H_2$, should be understood as the *universal way to linearize any bilinear morphism* form $H_1 \times H_2$ to any other complex linear space S. Note that, since the tensor product construction is defined by a universal mapping property, it is unique up to unique isomorphism.

In the Hilbert space $H = H_1 \otimes H_2$ there exists a special class of state vectors, called *product states*, which have the product form

$$\Psi = \Psi^{[1]} \otimes \Psi^{[2]}$$

corresponding to the cartesian product of the state vectors of the two subsystems. But the tensor product Hilbert space $H = H_1 \otimes H_2$

additionally contains all *linear combinations* of such product states. More precisely, if we choose *orthonormal bases* $\Psi^{[1]}{}_a$ and $\Psi^{[2]}{}_b$ in H_1 and H_2 correspondingly, a general vector state of the tensor product Hilbert space $H = H_1 \otimes H_2$ is written in the following form:

$$\Psi = \sum_{ab} \eta_{ab} \Psi^{[1]}{}_a \otimes \Psi^{[2]}{}_b$$

where the state vector Ψ represents a pure state of the total system composed of subsystems [1] and [2].

We observe that in general, a pure state of this total system is *not reduced to the product* of the state vectors of its two subsystems. Thus, it constitutes a correlated or *entangled state*, meaning that each subsystem does *not* possess a separable state within the composite system. Conceptually, this means that each one of the subsystems of the composite system does *not* have an individual, separable and definite state independent of the state of the composite system, and most significantly, this is the case *irrespective* of the spatial distance between the subsystems. This form of correlation, in other words, is *not* of any dynamical nature and constitutes a consequence of the universal property of the tensor product Hilbert space $H = H_1 \otimes H_2$ as a means of linearizing the bilinear morphisms from the cartesian product $H_1 \times H_2$. As introduced in chapter 2.8, quantum entanglement in this regard is well exemplified in experimental EPR correlations, and these will be further analyzed in section 9.4 of the present chapter.

The essential aspect of entanglement phenomena is that the behavior of the whole system is not reducible to the behavior of its constitutive subsystems—i.e., the whole is more than the sum of its parts;[1] more specifically, the parts do not assume an individuation or localization independent of the whole. Thus there exists a *mutually implicative bidirectional relation* between the parts and the whole. The observant reader may have already realized that the general formulation of entanglement phenomena lies at the heart of the proposed sheaf-theoretic approach; and by contrast, the particular case of these phenomena as explicated by the nature of the tensor product Hilbert space of states $H = H_1 \otimes H_2$ constitutes only an *instance* of this general framework of *functorial entanglement*, which will be explained later.

At this stage, it is instructive to examine the possibility of assigning a notion of *partial state* to each of the subsystems [1] and [2], even though neither of these possesses an individual, separable state. This notion of partial state would thus encompass all the *statistical information* about H_1 that the density operator of the composite system ρ_{12} incorporates.

In order to be able to define such a notion of partial state for each of the subsystems [1] and [2] it is necessary to consider *local actions* of each one of [1] and [2]. By a local action of subsystem [1], for example, we mean a measurement which can be performed by an observable of subsystem [1]. This is an operation which is represented by Hermitian operators of the form $A^{[1]} \otimes 1^{[2]}$, where $A^{[1]}$ is a Hermitian operator in H_1 corresponding to the chosen observable. Notice that this is equivalent to employing a *local Boolean frame* consisting of projection operators belonging to the spectral decomposition of the chosen observable. Similarly, a local action of subsystem [2] is represented by Hermitian operators of the form $1^{[1]} \otimes B^{[2]}$. We stress the fact that none of the subsystems [1] and [2] have access to all of the observables (Boolean frames) of the composite system. More concretely, the algebras of observables of the subsystems can be obtained by the operation of *restriction* or *localization* of the algebra of observables of the composite system to each one of them.

Thus, the individual subsystems subsume the state ρ_{12} of the composite system *only partially*. We define the partial state of subsystem [1] as the *reduced density operator* $\rho_1 = Tr_{[2]}\rho_{12}$ obtained by *partial tracing* over subsystem [2] (and analogously for subsystem [2]) by the following requirements:

$$Tr(\rho_1 A^{[1]}) = Tr(\rho_{12}(A^{[1]} \otimes 1^{[2]}))$$

$$Tr(\rho_2 B^{[2]}) = Tr(\rho_{12}(1^{[1]} \otimes B^{[2]}))$$

Thus, for all observables $A^{[1]}$ of subsystem [1], and all observables $B^{[2]}$ of subsystem [2], the reduced density operators ρ_1 and ρ_2 correspondingly constitute restrictions of ρ_{12} to the respective subsystems. We note that *many* different states ρ_{12} of the composite system may have the same restrictions on the algebras of observables of the two subsystems. Hence, from the point of view of subsystems [1] and [2], many different states of the composite system have *identical restrictions* to each subsystem. According to the above requirements, the reduced density operator ρ_1, for example, reproduces the same statistical distribution for an event caused by a local action of subsystem [1] as ρ_{12} does, in the sense that we could either apply the operation $A^{[1]} \otimes 1^{[2]}$ on the composite system (thus leaving subsystem [2] unaffected) or apply the operation $A^{[1]}$ directly on subsystem [1]. Note that the assumption of considering an event caused by a local action of subsystem [1] leaving subsystem [2] unaffected respects the requirement of *Einstein locality* in spacetime if the two subsystems are sufficiently separated; in other words, the probability of an event at subsystem [1] is independent of subsystem [2] if the regions associated with these subsystems are *spacelike separated.*

However, it is important to realize that the reduced density operators ρ_1 and ρ_2 are *not* sufficient to determine the *probabilities of pairs* of correlated events between the two subsystems. These pairs of correlated events do not have any dynamical origin but are implied by the entanglement of the states of the composite system if we consider *compatible local actions* of the subsystems—i.e., measurements which can be performed by *compatible observables* of subsystems [1] and [2]. Equivalently, correlations between events constitutive of the subsystems can be observed with coincidence measurements performed between compatible local Boolean frames within some Boolean localization system of the composite system—i.e., Boolean frames representing the local measurement contexts of the compatible, correlated observables.

The condition of local Boolean frame compatibility between observables of the subsystems [1] and [2] means that the reduced density operators ρ_1 and ρ_2 constitute restrictions of some pure state of the composite system only if their *eigenvalues* are identical with respect to these compatible Boolean frames. Equivalently, a vector state of the tensor product Hilbert space $H = H_1 \otimes H_2$ reflecting the condition of local Boolean frames compatibility, is written in the following form:

$$\Psi = \sum_j \eta_j \Psi^{[1]}{}_j \otimes \Psi^{[2]}{}_j$$

where the state vector Ψ represents a pure state of the composite system; $\Psi^{[1]}{}_j$, $\Psi^{[2]}{}_j$ are orthonormal bases of the Hilbert spaces H_1 and H_2 of the subsystems [1] and [2], respectively, corresponding to the spectral decompositions of ρ_1 and ρ_2 with respect to compatible Boolean frames of [1] and [2] (or compatible local actions of the two subsystems); and η_j are the identical eigenvalues of [1] and [2] with respect to these bases.

In the physical state of affairs the correlated pairs of events usually refer to some *conserved* physical quantity like charge, energy, momentum or spin orientation of the subsystems (corresponding to some specified observable of the combined system) and persist *irrespective* of the metrical distance between the subsystems. It is important for the understanding of these entanglement correlations to emphasize the significance of an *event* ontology, together with the crucial role that the *compatibility between local Boolean frames* (with respect to which events occur by measurement of corresponding observables) plays, in Boolean localization systems. This is the case because entanglement correlations *cannot be reduced* to correlations between assumed pre-existing states assigned to the subsystems before the occurrence of the events (with respect to their corresponding local Boolean frames). This conclusion reaffirms the significance of Whitehead's event ontology in the relational realist understanding of entanglement correlations in quantum theory.

9.2 Functorial Entanglement

The ontological significance of events in interpreting entanglement phenomena forces a careful re-thinking of the notion of a subsystem of a composite system as exemplified by the concept of a partial state, or reduced density operator pertaining to its description. Since a partial state of a subsystem is used for the evaluation of probabilities of the events referring to this subsystem exclusively, irrespective of any correlations of these events with events referring to any other subsystem, it becomes clear that:

[i] The notion of a subsystem is *not fundamental* and retains meaning only insofar as a *complete Boolean frame* is designated. This frame corresponds to the measurement of some observable via the registration of some corresponding fact interpreted as an observed event. Hence, the notion of a subsystem before the existence of some observed events remains only as a *potentiality* expressed by its partial state description. This potentiality, under the designation of some Boolean frame, acquires the interpretation of a *probability function* for the evaluation of event-probabilities when the subsystem is definable as a reference linkage among observed events referring to the corresponding observable.

[ii] The separation of a composite system into subsystems is an *idealization*; it does not correspond to a 'partition' of the system into subsystems in terms of the density operators by which the subsystems are described. Rather, it is properly understood as the *algebraic operation of restriction or localization of the algebra of observables* of the composite system into appropriate *subalgebras* of observables corresponding to *potential subsystems* which can be realized only after the designation of local Boolean frames. Intuitively, these subalgebras represent only the observables subsumed by the potential subsystems, distinguished in this way only after the generation of concrete measurement events. Furthermore, this observables-induced distinguishability of subsystems within a composite system, for example of the subsystems [1] and [2] according to the preceding, is effectuated by considering observables of

the form $A^{[1]} \otimes 1^{[2]}$ and $1^{[1]} \otimes B^{[2]}$ within the algebra of observables of the total system.

[iii] The observables-induced distinguishability of potential subsystems within a composite system allows for an understanding of entanglement correlations between these subsystems under the condition of *compatibility* between their corresponding local Boolean frames within the Boolean localization system of a composite system. For example, in the case of two potential subsystems, the condition of compatibility means that the reduced density operators ρ_1 and ρ_2 constitute restrictions of some pure state of the composite system only if their eigenvalues are identical with respect to these compatible Boolean frames.

Reflecting on the above, we conclude that the notion of entanglement or non-separability pertaining to the description of a composite quantum system in terms of its potential parts—and the description of the parts in terms of the system as a whole—implies an *intrinsic relativity* with respect to the *compatible local Boolean frames distinguishing the subsystems* and their correspondingly compatible observables. In this way, it becomes clear that the same pure state of a total system may entail completely different entanglement correlations depending on the *choice* of different sets of local Boolean frames. Thus, it is important to formulate the notions of entanglement and non-separability *functorially*. This means that we need a functorial framework to properly express entanglement correlations among potential subsystems of a composite system which are *observable-induced* and *do not* depend on artificial choices of subsystems. For this purpose the *sheaf-theoretic uniform fibration* partitioning a general quantum event algebra into observable sorts, indexed by local Boolean frames and thus forming Boolean localization systems, is indispensable. As the discussion proceeds, we will show that the sheaf-theoretic *inductive limit* construction of a quantum event algebra with respect to some Boolean localization system can be expressed in terms of a *functorial tensor product decomposition*, providing a general *functorial formulation of entanglement phenomena* from the *sheaf-theoretic* perspective of our approach.

It was shown in chapter 8.7 that a general quantum event algebra L is represented *isomorphically* by the inductive limit taken in the category of local Boolean frames of the sheaf $\mathbf{R}_L$ with respect to a Boolean localization system:

$$L \cong Colim\{\int(\mathbf{R}_L, \mathcal{B}) \to \mathcal{B}\} := Colim\mathbf{R}_L = \mathbf{LR}_L$$

The functorial tensor product decomposition is obtained by expressing categorically the above colimit (inductive limit) as a *coequalizer of a coproduct.* For this purpose, we consider the colimit of any functor $\mathbf{X}: I \to \mathcal{L}$ from some index category $\mathbf{I}$ to $\mathcal{L}$. Let $\mu_i : \mathbf{X}(i) \to \coprod_i \mathbf{X}(i)$, $i \in I$, be the injections into the coproduct. A morphism from this coproduct, $\chi : \coprod_i \mathbf{X}(i) \to \mathcal{L}$, is determined *uniquely by the set of its components* $\chi_i = \chi\mu_i$. These components χ_i form a cocone over $\mathbf{X}$ to the quantum event algebra vertex L only when for all arrows $v : i \to j$ of the index category I the following conditions are satisfied:

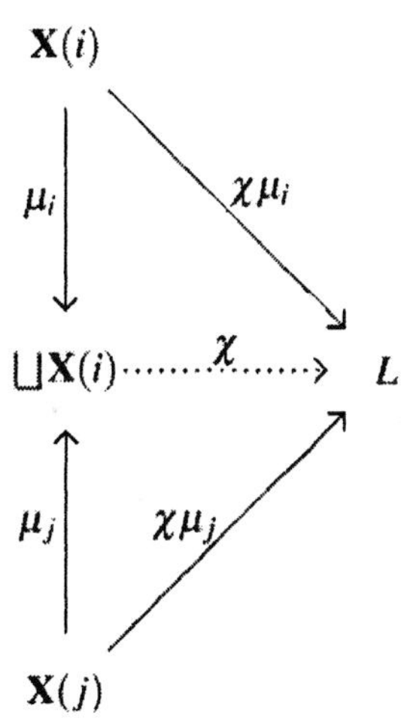

$$(\chi\mu_j)\mathbf{X}(v) = \chi\mu_i$$

So we consider all $\mathbf{X}(domv)$ for all arrows v with its injections μ_v and obtain their coproduct $\coprod_{v:i \to j} \mathbf{X}(domv)$. Next we construct two arrows ζ

and η, defined in terms of the injections μ_v and μ_i, for each $v: i \to j$ by the conditions:

$$\zeta \mu_v = \mu_i$$

$$\eta \mu_v = \mu_j \mathbf{X}(v)$$

together with their coequalizer χ:

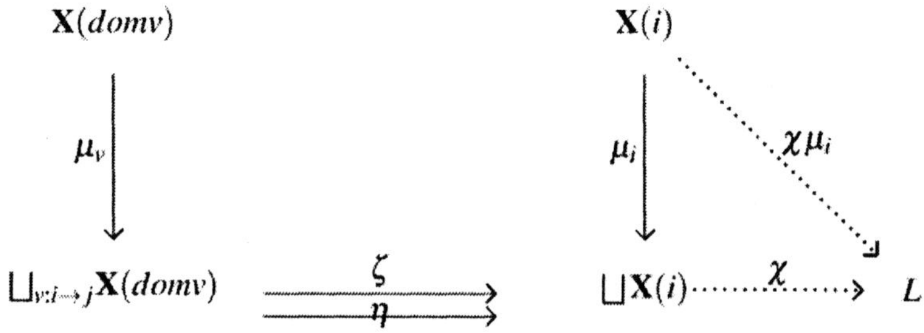

The *coequalizer condition* $\chi\zeta = \chi\eta$ informs us that the arrows $\chi\mu_i$ form a *cocone* over $\mathbf{X}$ to the quantum event algebra vertex $\mathcal{L}$. We further note that since χ is the coequalizer of the arrows ζ and η this cocone is the *colimiting cocone* for the functor $\mathbf{X}: I \to \mathcal{L}$ from some index category I to $\mathcal{L}$. Hence the *colimit of the functor* $\mathbf{X}$ *can be constructed as a coequalizer of coproduct* according to the following diagram:

$$\bigsqcup_{v:i\to j} \mathbf{X}(domv) \underset{\eta}{\overset{\zeta}{\rightrightarrows}} \bigsqcup \mathbf{X}(i) \xrightarrow{\ \chi\ } Colim\mathbf{X}$$

In our case the index category is the category of elements of the presheaf of Boolean algebras $\mathbf{P}$, that is $\int(\mathbf{P}, \mathcal{B})$, and the functor $\mathbf{M} \circ \int(\mathbf{P}, \mathcal{B})$ plays the role of the functor $\mathbf{X}: I \to \mathcal{L}$. In the diagram above, the second coproduct is over all the objects (B,p) with $p \in \mathbf{P}(B)$ of the category of elements, while the first coproduct is over all the maps $v:(B',p') \to (B,p)$ of that category, so that $v: B' \to B$ and the condition $p \cdot v = p'$ is satisfied. We conclude that the colimit $\mathbf{L}_{\mathbf{M}}(P)$ can be equivalently presented as the following coequalizer of coproduct:

$$\coprod_{v:\acute{B}\to B} \mathbf{M}(\acute{B}) \;\underset{\eta}{\overset{\zeta}{\rightrightarrows}}\; \coprod_{(B,p)} \mathbf{M}(B) \xrightarrow{\chi} \mathbf{P} \otimes_{\mathcal{B}} \mathbf{M}$$

The preceding is referred to as the *functorial tensor product decomposition* of the colimit (inductive limit) in the category of elements of $\mathbf{P}$ induced by the shaping functor of local Boolean frames $\mathbf{M}$ of a quantum categorical event structure. In order to understand the induced tensor product decomposition semantics in functorial terms, we ignore for the moment the quantum algebraic structure of the category $\mathcal{L}$, and we simply take $\mathcal{L} = Sets$. Then the coproduct $\coprod_p \mathbf{M}(B)$ is a coproduct of sets, which is equivalent to the product $\mathbf{P}(B) \times \mathbf{M}(B)$ for $B \in \mathcal{B}$. The coequalizer is thus the definition of the tensor product $\mathbf{P} \otimes_{\mathcal{B}} \mathbf{M}$ of the set valued functors:

$$\mathbf{P}: \mathcal{B}^{op} \to Sets, \qquad \mathbf{M}: \mathcal{B} \to Sets$$

$$\coprod_{B,\acute{B}} \mathbf{P}(B) \times Hom(\acute{B},B) \times \mathbf{M}(\acute{B}) \;\underset{\eta}{\overset{\zeta}{\rightrightarrows}}\; \coprod_{B} \mathbf{P}(B) \times \mathbf{M}(B) \xrightarrow{\chi} \mathbf{P} \otimes_{\mathcal{B}} \mathbf{M}$$

According to the above diagram for elements $p \in \mathbf{P}(B)$, $v: B' \to B$ and $q' \in \mathbf{M}(B')$ the following equations hold, symmetric in $\mathbf{P}$ and $\mathbf{M}$:

$$\zeta(p,v,q') = (p \cdot v, q'), \qquad \eta(p,v,q') = (p, v(q'))$$

Hence the elements of the set $\mathbf{P} \otimes_{\mathcal{B}} \mathbf{M}$ are all of the form $\chi(p,q)$, which can be equivalently written as:

$$\chi(p,q) = p \otimes q, \quad p \in \mathbf{P}(B), q \in \mathbf{M}(B)$$

Thus if we take into account the definitions of ζ and η above, we obtain:

$$p \cdot v \otimes q' = p \otimes v(q'), \quad p \in \mathbf{P}(B), q' \in \mathbf{M}(B'), v : B' \to B$$

We conclude that the set $\mathbf{P} \otimes_B \mathbf{M}$ is actually the *quotient* of the set $\coprod_B \mathbf{P}(B) \times \mathbf{M}(B)$ by the *equivalence relation* generated by the above equations. Furthermore if we define the arrows

$$k_B : \mathbf{P} \otimes_B \mathbf{M} \to L, \qquad l_B : \mathbf{P}(B) \to Hom_{\mathcal{L}}(\mathbf{M}(B), L)$$

they are related under the Boolean frames-quantum adjunction by:

$$k_B(p,q) = l_B(p)(q), \qquad B \in \mathcal{B}, p \in \mathbf{P}(B), q \in \mathbf{M}(B)$$

Here we consider k as a function on $\coprod_B \mathbf{P}(B) \times \mathbf{M}(B)$ with components $k_B : \mathbf{P}(B) \times \mathbf{M}(B) \to L$ satisfying

$$k_{B'}(p \cdot v, q') = k_B(p, v(q'))$$

in agreement with the equivalence relation defined above.

Now we replace the category *Sets* by the category of quantum event algebras $\mathcal{L}$ under study. The element q in the set $\mathbf{M}(B)$ is replaced by a generalized element $q : C \to \mathbf{M}(B)$ from some object C of $\mathcal{L}$. Then we consider k as a function $\coprod_{(B,p)} \mathbf{M}(B) \to L$ with components $k_{(B,p)} : \mathbf{M}(B) \to L$ for each $p \in \mathbf{P}(B)$, which for all arrows $v : B' \to B$ satisfy:

$$k_{(B', p \cdot v)} = k_{(B,p)} \circ \mathbf{M}(v)$$

Then the condition defining the bijection holding by virtue of the Boolean frames-quantum adjunction is given by:

$$k_{(B,p)} \circ q = l_B(p) \circ q : C \to L$$

This argument, being natural in the object C, is determined by setting $C = \mathbf{M}(B)$ with q being the identity map. Hence the bijection takes the form $k_{(B,p)} = l_B(p)$.

In particular, if $\mathbf{P}$ is the sheaf of local Boolean frames $\mathbf{R}_L$ with respect to a Boolean localization system of L, such that locally:

$$L \cong Colim\{\int(\mathbf{R}_L, \mathrm{B}) \to \mathrm{B}\} := Colim \mathbf{R}_L$$

then we obtain:

$$\coprod_{v:\acute{B}\to B} \mathbf{M}(\acute{B}) \underset{\eta}{\overset{\zeta}{\rightrightarrows}} \coprod_{(B,\psi_B)} \mathbf{M}(B) \xrightarrow{\chi} \mathbf{R}(L) \otimes_{\mathscr{B}} \mathbf{M}$$

According to the diagram above for elements $\psi_B \in \mathbf{R}(L)(B)$, $v: B' \to B$ and $\xi' \in \mathbf{M}(B')$, the following equations hold:

$$\zeta(\psi_B, v, \xi') = (\psi_B \cdot v, \xi'), \qquad \eta(\psi, v, \xi') = (\psi_B, v(\xi'))$$

symmetric in $\mathbf{R}(L)$ and $\mathbf{M}$. Hence the elements of $\mathbf{R}(L) \otimes_{\mathrm{B}} \mathbf{M}$ are all of the form $\chi(\psi_B, \xi)$. This element can be written as:

$$\chi(\psi_B, \xi) = \psi_B \otimes \xi, \quad \psi_B \in \mathbf{R}(L)(B), \xi \in \mathbf{M}(B)$$

Thus if we take into account the definitions of ζ and η above, we obtain:

$$\psi_B \cdot v \otimes \xi' = \psi_B \otimes v(\xi')$$

We conclude that $\mathbf{R}(L) \otimes_{\mathrm{B}} \mathbf{M}$ is actually the quotient of $\coprod_{(B,\psi_B)} \mathbf{M}(B)$ by the smallest equivalence relation generated by the above equations. Moreover, if there exists a Boolean algebra D in B and Boolean homomorphisms $w: D \to B$, $h: D \to B'$, such that: $w(\bar{\xi}) = \xi$, $h(\bar{\xi}) = \xi'$, $\psi_B \cdot w = \psi_D$, $\psi_{B'} \cdot h = \psi_D$ and thus $\psi_B \cdot w = \psi_{B'} \cdot h$

where

$$\xi \in \mathbf{M}(B), \xi' \in \mathbf{M}(B'), \overline{\xi} \in \mathbf{M}(D), \psi_B \in \mathbf{R}(L)(B), \psi_{B'} \in \mathbf{R}(L)(B'), \psi_D \in \mathbf{R}(L)(D)$$

then the identification equations take the general symmetric form:

$$\psi_B \otimes \xi = \psi_{B'} \otimes \xi'$$

If we denote by $l_Q(B)$ the ideal generated by the equivalence relation, corresponding to the above identification equations, for each B in $\mathcal{B}$, we conclude that locally—i.e., with respect to a Boolean localization system—an element of $\mathbf{R}(L) \otimes_B \mathbf{M}$ can be written in the form of a *Boolean germ*:

$$\psi_B \otimes \xi = (\psi_B, \xi) + l_Q(B)$$

Thus, a *quantum event algebra admits a sheaf-theoretic representation so that it can be described locally—that is, over a particular cover in a Boolean localization system*, as follows:

$$\mathbf{R}(L)(B) \otimes_B \mathbf{M}(B) = [\mathbf{R}(L)(B) \times \mathbf{M}(B)] \; modulo \; [l_Q(B)]$$

Furthermore, we mention that precisely analogous conclusions hold if we consider the categories of quantum observables and quantum probabilistic states defined over a quantum event algebra according to the following commutative diagrams:

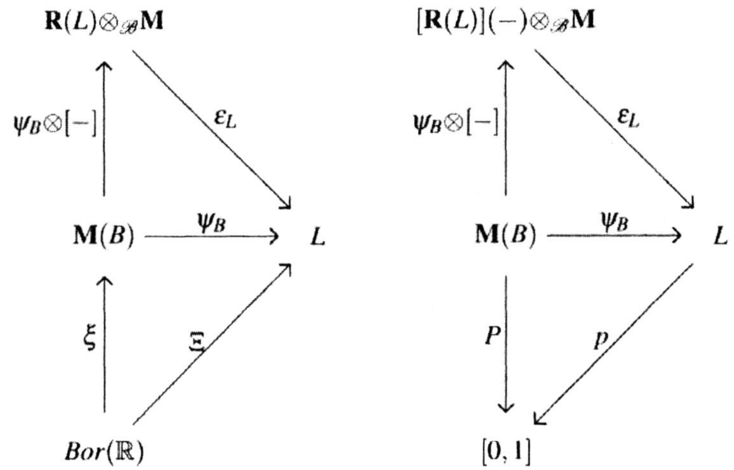

9.3 Interpretation of Quantum Events

If we focus our attention on a Boolean localization system of compatible overlapping local Boolean frames, we can accurately specify the information encoded in the *ideal* $l_Q(B)$ in $\mathbf{R}(L)(B) \otimes_{\mathrm{B}} \mathbf{M}(B)$, where $\psi_B : \mathbf{M}(B) \to L$, and B in B, stands for a local Boolean frame belonging to this system. More concretely, the ideal $l_Q(B)$ contains *information about all the other local Boolean frames in the localization system* which are compatible in pullback diagrams over $L \cong \mathbf{R}(L) \otimes_{\mathrm{B}} \mathbf{M}$ with the specified one. This is clear if we inspect the gluing isomorphism $\Omega_{B,B'} = \psi_{BB'} \circ \psi_{B'B}{}^{-1}$ and notice that its existence guarantees the satisfaction of the relations needed for the validity of the identification equations: $\psi_B \otimes \xi = \psi_{B'} \otimes \xi'$ in a Boolean localization system. Thus, the information encoded in the ideal $l_Q(B)$ essentially refers to all other local Boolean frames that are compatible with some specified one in a Boolean localization system. This is a *uniquely peculiar characteristic of the local behavior of quantum event algebras in comparison to classical event algebras*. Remarkably, each observable-induced local Boolean frame in a covering sieve (Boolean localization system) of a global quantum event algebra incorporates local observable information (in the form of a *Boolean germ*) about *all other local Boolean frames compatible with it in the same covering sieve*. This is a crucial observation and pertains to discussions of alleged 'quantum *non-local causality*' characterizing the supposedly 'paradoxical' behavior of quantum systems. As discussed earlier in chapters 2 and 3, from our perspective, the improper characterization of non-local correlations as 'paradoxical' stems from two factors: The first factor has to do with the failure to acknowledge that entanglement correlations between potential subsystems are *observable or Boolean-frame induced, presupposing the existence of events*. The second factor stems from the misconceived attempt to *assimilate the potential relations, described via topologically interlocking local Boolean frames constitutive of a Boolean localization system, to actual causal relations*—i.e., in terms of a non-local, metrical, efficient causal mechanism of information propagation between spacetime points.

The *Boolean germ relation*—that is, the equivalence relation of local Boolean frames belonging to a Boolean localization system of a quantum event algebra—provides the following insights for the interpretation of events in a quantum world:

The first is that the Boolean germ relation, which is observable induced, establishes a *criterion of equivalence* among experimental measurement contexts / observational procedures. Thus, in a Boolean localization system of L using local Boolean frames $(B,\psi_B : \mathbf{M}(B) \to L)$ and Boolean coefficients $b \in B$ we can form their equivalence classes, which modulo the compatibility conditions on overlaps, will represent a *single quantum event* in L. The fact that two different local Boolean frames $(B,\psi_B : \mathbf{M}(B) \to L)$ and $(C,\psi_C : \mathbf{M}(C) \to L)$ in a Boolean localization system are compatible, reflects the possibility of using *two distinct experimental contexts* of compatible observables, or equivalently setting up two different measurement procedures of compatible observables, and hence obtaining two outcome events quantified by the Boolean propositions $b \in B$ and $c \in C$. Now, the equivalence relation between these two local Boolean frames means that the corresponding observational procedures are in fact *equivalent*, and thus they record the same quantum event.

The second insight is provided by the application of the *Stone representation theorem* for Boolean algebras.[2] This theorem establishes a *duality* (categorical equivalence) between the category of *Boolean event algebras* and the category of *compact totally disconnected Hausdorff topological spaces* (Stone spaces). In the finite case this duality restricts to an equivalence of the category of finite Boolean algebras and the opposite of the category of finite sets. In this way, each Boolean algebra B has an associated topological space $S(B)$, called its *Stone space*. The points of the Stone space $S(B)$ are the *ultrafilters* of the Boolean algebra B, or equivalently the homomorphisms from B to the 2-element Boolean algebra. A basis of the topology of $S(B)$ consists of sets of the form $\{x \in S(B) \mid b \in x\}$ where $b \in B$. Conversely, given any topological space X, the collection of subsets of X that are compact open or equivalently clopen, that is, both closed and open, is a Boolean algebra.

Now, the assignment of a Boolean algebra to its Stone space is *functorial*, that is, it corresponds to the object part of a contravariant functor from the category of Boolean algebras to the category of Stone spaces

$$Hom_{\mathcal{B}}(-,2) : \mathcal{B}^{op} \to \mathcal{T}$$

where $\mathcal{B}$, $\mathcal{T}$ are the categories of Boolean algebras and Stone spaces correspondingly. The functor $Hom_{\mathcal{B}}(-,2)$ sends a Boolean algebra B to the Stone space $S(B) := X_B$, and a Boolean algebra homomorphism $F : B \to A$ to the continuous function $Hom_{\mathcal{B}}(F,2)$, whose action is defined by precomposition with F, viz., it sends $H : A \to 2$ of X_A to $F \circ H : B \to 2$ of X_B. Now, there exists an inverse functor

$$Hom_{\mathcal{T}}(-,2) : \mathcal{T} \to \mathcal{B}^{op}$$

which sends a Stone space S to the set of continuous functions from S to the Stone space 2, endowed with the structure of a Boolean algebra, and further identified as the Boolean algebra of its clopen sets. Moreover, it sends a continuous function of Stone spaces $f : S_1 \rightarrow S_2$ to the homomorphism of Boolean algebras $Hom_{\mathcal{T}}(f,2)$, whose action is defined by postcomposition with f correspondingly. Notice that the *duality* between the category of Boolean algebras and the category of Stone spaces has been based on a *dual interpretation of the two-element set* 2, viz., as the two-element true/false Boolean algebra or as a two-point Stone space.

According to the above duality, it is legitimate to replace Boolean algebras by *fields of subsets of a sample space*, obtaining in physical grounds an operational instantiation of the notion of observed event. Hence, if we replace each Boolean algebra B in $\mathcal{B}$ by its set-theoretic representation $[\Sigma, B_\Sigma]$, consisting of a local sample space Σ and its local field of subsets B_Σ, it is possible to define local sample space *covers* $(B_\Sigma, \psi_{B_\Sigma} : \mathbf{M}(B_\Sigma) \rightarrow L)$ and corresponding localization systems for a quantum event algebra L in $\mathcal{L}$. Again from local sample space covers $(B_\Sigma, \psi_{B_\Sigma} : \mathbf{M}(B_\Sigma) \rightarrow L)$ we may form their *equivalence classes*, which, modulo the conditions for compatibility on overlaps, will represent a single quantum event in L. Under these circumstances we may interpret the equivalence classes of local sample space covers $\psi_{B_\Sigma} \otimes a$, $a \in \mathbf{M}(B_\Sigma)$ as the *statistical experimental actualizations* of the quantum events in $\mathcal{L}$. Moreover, the local sample space covers $(B_\Sigma, \psi_{B_\Sigma} : \mathbf{M}(B_\Sigma) \rightarrow L)$ and $(C_\Sigma, \psi_{C_\Sigma} : \mathbf{M}(C_\Sigma) \rightarrow L)$ are compatible in a localization system for L, if and only if for some $(D_\Sigma, \psi_{D_\Sigma} : \mathbf{M}(D_\Sigma) \rightarrow L)$ in the localization system, and $a \in \mathbf{M}(B_\Sigma)$, $b \in \mathbf{M}(C_\Sigma)$, $c, d \in \mathbf{M}(D_\Sigma)$ the following conditions hold:

$$\psi_{B_\Sigma} \otimes a = \psi_{D_\Sigma} \otimes c$$

$$\psi_{C_\Sigma} \otimes b = \psi_{D_\Sigma} \otimes d$$

The pullback compatibility condition (topological gluing isomorphism), which is in injective correspondence with the one in $\mathcal{L}$ since it holds in a localization system, is interpreted in the measurement context as denoting that two local sample space representations of a quantum event are compatible if and only if their corresponding measurements are equivalent to measurements conducted in experimental contexts of compatible observables. Thus, the equivalence relation between these two local sample spaces means that the corresponding observational procedures are in fact *equivalent*, and thus they record the same quantum event.

We conclude that it is precisely this internal equivalence relation between compatible observable-induced Boolean frames in a localization system, obtained by the functorial tensor product representation of the inductive limit (in the category of local Boolean frames of the sheaf $\mathbf{R}_L$*), which permits an indirect logical inference from one Boolean frame to another after some event has actualized with respect to one of them.*

9.4 EPR Correlations

As a particular application of the functorial scheme of sheaf-theoretic entanglement we have developed so far, it is instructive to include a brief discussion of *EPR correlations*, which were introduced in chapter 2.8. The EPR correlations are due to the phenomenon of *entanglement* or *non-separability* pertaining to the description of a quantum system with reference to *two potential subsystems* emerging by the depiction of certain compatible local Boolean frames distinguishing them and corresponding to compatible observables. For example, in Bohm's version of the EPR correlations, these refer to correlated pairs of events associated with two compatible observable-induced distinguished subsystems (*spin in some designated direction*) and are due to *total spin conservation*, persisting *irrespective* of the metrical distance between the subsystems.

The hallmark of EPR correlations is that they *cannot* be reduced to correlations between assumed, pre-existing states assigned to the subsystems before the occurrence of measurement events (with respect to their corresponding compatible observable-induced local Boolean frames). Thus, our understanding of EPR correlations is based on an *event ontology* and their explanation stems from the *sheaf-theoretic compatibility relation* between local Boolean frames (with respect to which events occur by measurement of corresponding compatible observables) in Boolean localization systems. These Boolean localization systems are *indexed by the axis of the observable spin*, since spin observables having different spin axes are not compatible.

Notice that the two *observable-induced subsystems* constitutive of an EPR-pair *do not* assume an individuation or localization independent of the total system. More precisely, for example, the *singlet pure state* of the total system is not reducible to a product which would be assigned to the state vectors of the two subsystems. Thus, it is an entangled state, which means that each subsystem does not possess a separable state independent of the state of the total system. Thus, there exists a *mutually implicative bidirectional relation* between the observable-induced subsystems and the total system. This relation is manifest via *correlation events* generated by coincidence spin measurements in designated directions (induced by the actions of corresponding detectors):

$$\Psi = (1/\sqrt{2})\{\Psi^{[1]}{}_{[+]} \otimes \Psi^{[2]}{}_{[-]} - \Psi^{[1]}{}_{[-]} \otimes \Psi^{[2]}{}_{[+]}\}$$

A natural question is whether it is possible to write the singlet state above as a kind of *composition of 'potential parts'* which become *observable-induced*, thus reflecting their *non-separability*. We denote the 'potential parts' by $\widehat{\Psi_1}$ and $\widehat{\Psi_2}$ correspondingly:

$$\widehat{\Psi_1} = (1/\sqrt{2})\{\widehat{\Psi^{[1]}{}_{[+]}} \otimes \Psi^{[2]}{}_{[-]} - \widehat{\Psi^{[1]}{}_{[-]}} \otimes \Psi^{[2]}{}_{[+]}\}$$

$$\widehat{\Psi_2} = (1/\sqrt{2})\{\Psi^{[1]}{}_{[+]} \otimes \widehat{\Psi^{[2]}{}_{[-]}} - \Psi^{[1]}{}_{[-]} \otimes \widehat{\Psi^{[2]}{}_{[+]}}\}$$

$$\Psi = \widehat{\Psi_1}\, e\, \widehat{\Psi_2}$$

Notice that the 'potential parts' $\widehat{\Psi_1}$ and $\widehat{\Psi_2}$ are *not-separable* from the entangled singlet state Ψ, since they *do not* constitute tensor factors of Ψ. As we have shown, the notion of 'potential parts' in our scheme, to be interpreted as states of potential subsystems, emerges by the operation of restriction or localization of the algebra of observables of the total system to these subsystems. Now, the *observable-induced distinguishability* of these potential subsystems by events pertaining to $\widehat{\Psi_1}$ or $\widehat{\Psi_2}$, for instance measurement of spin in some direction, allows an understanding of correlated pairs of events between the subsystems under the *condition of compatibility* between their corresponding local Boolean frames (i.e., spin measurement in the same direction) within a Boolean localization system of a composite system (indexed by the direction of spin measurement). The *condition of compatibility* means that given the reduced density operators ρ_1 and ρ_2 of the potential subsystems, they constitute *restrictions of the singlet state* of the total system, and thus pertain to $\widehat{\Psi_1}$ and $\widehat{\Psi_2}$, only if their eigenvalues are identical (*perfectly anti-correlated*) with respect to these compatible Boolean frames of spin measurement in a designated direction as an implementation of the *total spin conservation* principle. Notice that the singlet state is characterized by *rotational invariance* so the above argument is valid *independent* of the choice of direction of spin measurement *modulo* the compatibility of the local Boolean frames of the two subsystems (corresponding to the spin observable in each one of them in the designated direction).

For example, let us consider the spin observable in the *z*-direction $\mathbf{S}_z^{[1]}$ pertaining to the 'potential part' $\widehat{\Psi_1}$, and thus indexing it by the direction z (we note that the spin operators can be represented using the standard Pauli matrices). The measurement of the observable $\mathbf{S}_z^{[1]}$ induces a local action of subsystem [1] (distinguished by the subalgebra of observables of the total system 'visible' to it), which under the assumption that it leaves subsystem [2] unaffected (and thus respects Einstein locality) has the form $\mathbf{S}_z^{[1]} \otimes 1^{[2]}$. This is equivalent to employing a *local Boolean frame* consisting of projection operators belonging to the spectral decomposition of $\mathbf{S}_z^{[1]} \otimes 1^{[2]}$, which we denote by B. Analogously, we may consider the spin observable in the z-direction $\mathbf{S}_z^{[2]}$ pertaining to the 'potential part' $\widetilde{\Psi_2}$ and thus indexing it by z. Symmetrically, the measurement of the observable $\mathbf{S}_z^{[2]}$ induces a local action of subsystem [2] of the form $1^{[1]} \otimes \mathbf{S}_z^{[2]}$, which is equivalent to employing a corresponding local Boolean frame, denoted by B'. Then, we define the *partial state* of subsystem [1] induced by the local action $\mathbf{S}_z^{[1]}$ of subsystem [1], as the reduced density operator $\rho_1 = Tr_{[2]}\rho_{12}$ (where ρ_{12} denotes the density operator of the total system corresponding to the singlet state) obtained by partial tracing over subsystem [2] (and analogously for subsystem [2]) by:

$$Tr(\rho_1 \mathbf{S}_z^{[1]}) = Tr(\rho_{12}(\mathbf{S}_z^{[1]} \otimes 1^{[2]}))$$

$$Tr(\rho_2 \mathbf{S}_z^{[2]}) = Tr(\rho_{12}(1^{[1]} \otimes \mathbf{S}_z^{[2]}))$$

Thus, we obtain:

$$\rho_1 = (1/2)(\Pi^z{}_{\Psi^{[1]}{}_{[+]}} + \Pi^z{}_{\Psi^{[1]}{}_{[-]}})$$

$$\rho_2 = (1/2)(\Pi^z{}_{\Psi^{[2]}{}_{[+]}} + \Pi^z{}_{\Psi^{[2]}{}_{[-]}})$$

where $\Pi^z{}_{\Psi^{[1]}{}_{[+]}}$ is the *projection operator* of the local Boolean frame in the direction z for subsystem $[1]$, which projects in the $[+]$ orientation and similarly for the others.

We note that many different density operators of the total system of the form ρ_{12} may have the same restrictions on each one of the two subsystems. Hence, from the point of view of subsystems [1] and [2], many different states of the total system have identical restrictions to each one of them and it is *not* possible to be distinguished by them. For example the state of the total system:

$$\Psi = (1/\sqrt{2})\{\Psi^{[1]}{}_{[+]} \otimes \Psi^{[2]}{}_{[-]} + \Psi^{[1]}{}_{[-]} \otimes \Psi^{[2]}{}_{[+]}\}$$

gives exactly the *same partial states* ρ_1 and ρ_2 induced by corresponding observables as the singlet state does. This is to be expected since the reduced density operator ρ_1, for example, gives the *statistical distribution* for an event referring to a measurement of $\mathbf{S}_z{}^{[1]}$ leaving subsystem [2] unaffected. However, the reduced density operators ρ_1 and ρ_2 are *not* sufficient to determine the *probabilities of pairs of correlated events* between the two subsystems. These pairs of events can be observed only by *coincidence measurements* performed between compatible local Boolean frames, i.e., measurement of spin in z direction for subsystems [1] and [2] within some Boolean localization system of the total system corresponding to these compatible measurements. The compatibility condition with respect to local Boolean frames [1] and [2] means that the reduced density operators ρ_1 and ρ_2 constitute restrictions of the singlet pure state of the total system, characterized by total spin zero in direction z, *only if* their eigenvalues are identical (equal with $1/2$) with respect to these compatible Boolean frames, such that in the direction z the total spin orientation *conservation* is satisfied (perfect anti-correlation).

Taking into account our previous analysis we realize that there exists an *internal equivalence relation* between the compatible observables induced by $\mathbf{S}_z{}^{[1]} \otimes 1^{[2]}$ and $1^{[1]} \otimes \mathbf{S}_z{}^{[2]}$ for the Boolean frames of subsystems [1] and [2], correspondingly, in the Boolean localization system formed in the z direction for the total system. This is obtained by the *functorial tensor product representation* of the inductive limit (in the category of local Boolean frames of the sheaf $\mathbf{R}_L$ localized in the z direction, where L is the quantum event algebra of the total system) which permits an *indirect logical inference* from one Boolean frame B to the other compatible one B' after some detection event has been registered with respect to one of them. These Boolean frames are thus *topologically amalgamated together* such that in one equivalence class (Boolean germ) of the *inductive limit* the projector $\Pi^z{}_{\Psi[1]_{[+]}}$ of the Boolean frame B is glued together with the projector $\Pi^z{}_{\Psi[2]_{[-]}}$ of B' and in another the projector $\Pi^z{}_{\Psi[1]_{[-]}}$ of B is glued together with the projector $\Pi^{z}{}_{\Psi[2]_{[+]}}$ of B'. The Boolean localization system in the z direction is depicted in the following diagram:

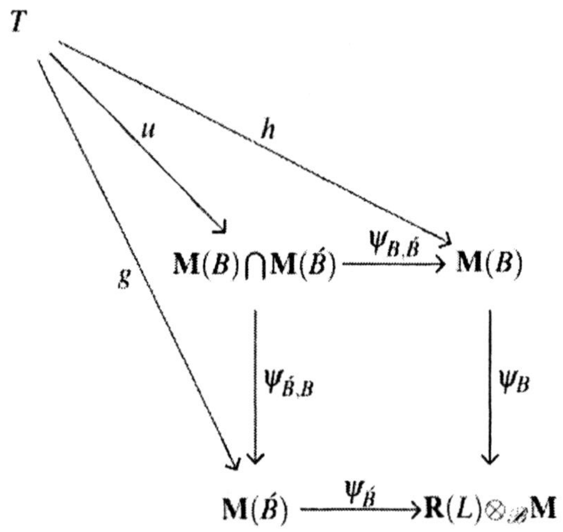

Where this commutative diagram is considered in $\mathcal{L}$, $\mathbf{M}(B)$ and $\mathbf{M}(B')$ are the images of the Boolean frames B and B' in L (the quantum event algebra of the total system) under the action of the shaping functor $\mathbf{M}$; and $\mathbf{R}(L)\otimes_{B}\mathbf{M}$ is the inductive limit consisting of Boolean germs in the sheaf of Boolean frames $\mathbf{R}(L)$, being locally (with respect to this Boolean localization system) isomorphic to L. The gluing morphism of compatible Boolean frames on their overlap is given by:

$$\Omega_{B,B'} = \psi_{BB'} \circ \psi_{B'B}{}^{-1}$$

The corresponding Boolean germs are defined by the equivalence relation generated by the following identification equations:

$$\psi_{B} \otimes \Pi^{z}{}_{\Psi[1]_{[+]}} = \psi_{B'} \otimes \Pi^{z}{}_{\Psi[2]_{[-]}}$$

$$\psi_{B} \otimes \Pi^{z}{}_{\Psi[1]_{[-]}} = \psi_{B'} \otimes \Pi^{z}{}_{\Psi[2]_{[+]}}$$

If we denote by l_Q the *ideal* generated by the equivalence relation, creating *Boolean germs* as above, we conclude that locally—that is, with respect to this Boolean localization system—the quantum event algebra L over each of the Boolean frames B and B' of this system takes the form:

$$Q(B) := \mathbf{R}(L)(B) \otimes_B \mathbf{M}(B) = [\mathbf{R}(L)(B) \times \mathbf{M}(B)] modulo [l_Q(B)]$$

and similarly over $\mathbf{M}(B')$. Notice that the ideal $l_Q(B)$, referring to subsystem [1] from the perspective of the Boolean localization system of the total system, contains potential information about the other compatible local Boolean frame of subsystem [2] from the particular way that these Boolean frames are glued together in this Boolean localization system. Thus, it is precisely $Q(B)$, the quantum event algebra restricted to subsystem [1] from the *global* perspective of the total system, which *explains* the entanglement EPR correlations implicated in the singlet state of total spin zero of the total system—or equivalently, of the total system's 'potential part' $\widetilde{\Psi_1}$ (and likewise for subsystem [2]) together with the associated quantum event semantics.

9.5 Logical Classification of Quantum Events

As a logical state of affairs, observed events in quantum theory correspond to evaluations of observables with respect to some *Boolean logical binary schemata* (*yes/no experiments* subsumed by measuring devices), interpreted as *property-indicating facts*. Note that this is meaningful from a topological perspective only if it respects the normative constraints of the *relational decomposition* of a quantum event structure with respect to local Boolean frames, as well as the rules of compatibility under relational extension from the local to the global and conversely. The functionality of local Boolean frames is crucial for a relational realist interpretation of quantum theory. For this purpose it is necessary to study their functional role with respect to the logical structure and coherence of quantum theory.

We have already deduced that the representation of a quantum event algebra L in $\mathcal{L}$, in terms of the functor of Boolean frames $\mathbf{R}(L)$ of L, is full and faithful, if the Boolean frames-quantum adjunction is restricted to a system of Boolean localizations of L. As a corollary we have obtained that a quantum categorical event structure $\mathcal{L}$ is a reflection of the topos of presheaves $Sets^{\mathcal{B}^{op}}$. Thus, there exists a subobject functor for a quantum categorical event structure $\mathcal{L}$ equipped with Boolean localization systems.

The *subobject functor* of the category of quantum event algebras $\mathcal{L}$ is defined as follows:

$$\mathbf{Sub} : \mathcal{L}^{op} \to Sets$$

The subobject functor of $\mathcal{L}$ is a contravariant functor by pulling back. Composition of this functor with a proper Boolean modeling functor defines a presheaf functor in $Sets^{\mathcal{B}^{op}}$, called the *Boolean frames modeled subobject functor* of $\mathcal{L}$ as follows:

$$\mathbf{Sub} \circ \mathbf{M} : \mathcal{B}^{op} \to \mathcal{L}^{op} \to Sets$$

In a compact notation we obtain:

$$\Upsilon_{\mathbf{M}} := \Upsilon(\mathbf{M}(-)) := \mathbf{Sub} \circ \mathbf{M} : \mathcal{B}^{op} \to Sets$$

such that:

$$\mathcal{B}^{op} \ni B \mapsto \{[Dom(m) \rightarrowtail \mathbf{M}(B)]\} \in \mathbf{Sets}$$

where the range denotes the set of subobjects of $\mathbf{M}(B)$, viz., the set of equivalence classes of monic quantum homomorphisms (quantum monomorphisms) m from $Dom(m)$ to $\mathbf{M}(B)$.

The set $\Upsilon_{\mathbf{M}}(B) = \Upsilon(\mathbf{M}(B))$ is defined as the *set of all subobjects* of $\mathbf{M}(B)$, for every B in $\mathcal{B}$, in the category $\mathcal{L}$. Notice that the set $\Upsilon(\mathbf{M}(B))$, for every B in $\mathcal{B}$, is a *partially ordered* set under inclusion of subobjects of $\mathbf{M}(B))$.

A natural question arising in this categorical setting is the following: *Is the subobject functor representable in $\mathcal{L}$ by means of a concrete quantum event algebra Ω in $\mathcal{L}$, which would play the role of a classifying object in $\mathcal{L}$?* The representation of the subobject functor in a quantum categorical event structure $\mathcal{L}$ via $\Upsilon_{\mathbf{M}}$ is significant because it allows one to interpret the concrete classifying object Ω as an object that contains the *information of equivalence classes of truth valuations* with respect to all compatible Boolean frames belonging to a Boolean localization system of a quantum event algebra L in $\mathcal{L}$. In that case $\mathcal{L}$ is endowed with a *subobject classifier*, defined as a universal monic quantum homomorphism,

$$\mathrm{T} := True : 1 \rightarrowtail \Omega$$

such that every subobject of L in $\mathcal{L}$, is uniquely a pullback of the universal monic T, according to the following diagram:

$$\begin{array}{ccc} K & \xrightarrow{\;!\;} & 1 \\ \Big\downarrow m & & \Big\downarrow T \\ L & \xrightarrow{\;\phi_m\;} & \Omega \end{array}$$

We can prove the following proposition:[3]

The functor $\Upsilon_{\mathbf{M}}$ is representable in $\mathcal{L}$ if and only if the evaluation of the unit of the Boolean frames-quantum adjunction at $\Upsilon_{\mathbf{M}}$, restricted to a *functor of Boolean localizations of L for every L in $\mathcal{L}$, is an isomorphism.*

The basic idea of the proof is the following: The counit of the Boolean frames-quantum adjunction, for each quantum event algebra L in $\mathcal{L}$, is given by:

$$\varepsilon_L : \mathbf{LR}(L) \to L$$

The counit evaluated at L, that is ε_L, is a quantum algebraic isomorphism if it is restricted to a Boolean localization system of L. On the other hand, we have seen that for any presheaf $\mathbf{P} \in Sets^{B^{op}}$, the unit is defined as follows:

$$\delta_{\mathbf{P}} : \mathbf{P} \to \mathbf{RLP}$$

It is easy to see that if we consider the Boolean frames modeled subobject functor $\Upsilon(\mathbf{M}(-))$ as $\mathbf{P} \in Sets^{B^{op}}$, we obtain the following natural transformation:

$$\delta_{\Upsilon(\mathbf{M}(-))} : \Upsilon(\mathbf{M}(-)) \to \mathbf{RL}\Upsilon(\mathbf{M}(-))$$

or equivalently:

$$\delta_{\Upsilon(\mathbf{M}(-))} : \Upsilon(\mathbf{M}(-)) \to Hom_{\mathcal{L}}(\mathbf{M}(-), \mathbf{L}\Upsilon(\mathbf{M}(-))$$

Hence, by inspecting the unit of the adjunction arrow $\delta_{\Upsilon(\mathbf{M}(-))}$ evaluated at $\Upsilon(\mathbf{M}(-))$, we conclude that the *Boolean frames modeled subobject functor becomes representable* in $\mathcal{L}$ if and only if, given that the counit ε_L for every L in $\mathcal{L}$ is an isomorphism, the unit $\delta_{\Upsilon(\mathbf{M}(-))}$ is also an isomorphism. Thus, $\Upsilon_{\mathbf{M}}$ becomes representable in $\mathcal{L}$ if and only if the unit $\delta_{\Upsilon(\mathbf{M}(-))}$, restricted to a functor of Boolean localizations of L, called the *localized unit* at $\Upsilon_{\mathbf{M}}$, is an isomorphism.

As a corollary we deduce that if the evaluation of the localized unit at $\Upsilon_{\mathbf{M}}$ is an isomorphism, then the *quantum classifying event algebra* is given by the inductive limit taken in the category of elements of $\Upsilon_{\mathbf{M}}$, according to:

$$\Omega := \mathbf{L}\Upsilon(\mathbf{M}(-)) = Colim\{\int(\Upsilon(\mathbf{M}(-)),\mathcal{B}) \to \mathcal{B} \to \mathcal{L}\}$$

Hence, the following diagram is a classifying pullback square in $\mathcal{L}$:

$$\begin{array}{ccc} Dom(\lambda) & \xrightarrow{\quad !\quad} & 1 \\ \Big\downarrow{\lambda} & & \Big\downarrow{T} \\ \mathbf{M}(B) & \xrightarrow{[\delta_{\Upsilon(\mathbf{M}(B))}]^{\lambda}} & \mathbf{L}\Upsilon(\mathbf{M}(-)) := \Omega \end{array}$$

From the preceding we conclude that if the evaluation of the localized unit at $\Upsilon_{\mathbf{M}}$ is an isomorphism, then the subobject functor $\mathbf{Sub} : \mathcal{L}^{op} \to Sets$ of $\mathcal{L}$ is representable in $\mathcal{L}$ by:

$$\Omega := \mathbf{L}\Upsilon(\mathbf{M}(-))$$

and thus, $\mathcal{L}$ is endowed with a subobject classifier defined by a universal monic quantum homomorphism, $\mathrm{T} := True : 1 \rightarrowtail \Omega$. In the following section, we will provide an *explicit representation of the classifying Boolean modeled quantum event algebra* by exploiting the *functorial tensor product decomposition of the inductive limit*, and we will apply this representation to the problem of measurement in quantum mechanics.

9.6 Truth Valuations in Boolean Localization Systems

The quantum classifying event algebra Ω is given by the inductive limit in the category of elements of the Boolean frames modeled subobject functor, that is:

$$\Omega := \mathbf{L}\Upsilon(\mathbf{M}(-)) = Colim\{\int(\Upsilon(\mathbf{M}(-)), \mathcal{B}) \to \mathcal{B} \to \mathcal{L}\}$$

where the category of elements of $\Upsilon(\mathbf{M}(-))$ is denoted by $\int(\Upsilon(\mathbf{M}(-)), \mathcal{B})$. Its objects are all pairs (B, λ), where λ is a subobject of $\mathbf{M}(B)$, that is, a monic quantum homomorphism in $\mathbf{M}(B)$. The morphisms of the category of elements of $\Upsilon(\mathbf{M}(-))$ are given by the arrows $(B', \lambda') \to (B, \lambda)$, namely, they are those morphisms $v : B' \to B$ of $\mathcal{B}$ for which $\lambda * v = \lambda'$, where $\lambda * v$ denotes the pullback of the subobject λ of $\mathbf{M}(B)$ along v, and λ' is a subobject of $\mathbf{M}(B')$.

The inductive limit in the category of elements of the Boolean frames modeled subobject functor can be equivalently represented as a coequalizer of the coproduct; that is, it can be expressed as a tensor product of functors analogous to the construction of section 9.2:

$$\coprod_{v:\acute{B} \to B} \mathbf{M}(\acute{B}) \underset{\eta}{\overset{\zeta}{\rightrightarrows}} \coprod_{(B,\lambda)} \mathbf{M}(B) \xrightarrow{\chi} \Upsilon(\mathbf{M}(-)) \otimes_{\mathcal{B}} \mathbf{M}$$

where the second coproduct is over all the pairs (B, λ) with $\lambda \in \Upsilon(\mathbf{M}(B))$ of the category of elements, while the first coproduct is over all the morphisms $v : (B', \lambda') \to (B, \lambda)$ of that category, so that $v : B' \to B$ and the condition $\lambda * u = \lambda'$ is satisfied.

First, we may interpret the above representation of the inductive limit (in the category of elements of the Boolean frames modeled subobject functor) in the category *Sets*. In this case, the coproduct $\coprod_{(B,\lambda)} \mathbf{M}(B)$ is a coproduct of sets, which is equivalent to the product $\Upsilon(\mathbf{M}(B)) \times \mathbf{M}(B)$

for B in $\mathcal{B}$. The coequalizer is thus the definition of the tensor product $\Upsilon(\mathbf{M}(-)) \otimes \mathbf{M}$ of the set valued functors:

$$\Upsilon(\mathbf{M}(-)) : \mathcal{B}^{op} \to Sets, \qquad \mathbf{M} : \mathcal{B} \to Sets$$

$$\coprod_{B,\acute{B}} \Upsilon(\mathbf{M}(B)) \times Hom(\acute{B}, B) \times \mathbf{M}(\acute{B}) \quad \overset{\zeta}{\underset{\eta}{\rightrightarrows}}$$

$$\overset{\zeta}{\underset{\eta}{\rightrightarrows}} \quad \coprod_{B} \Upsilon(\mathbf{M}(B)) \times \mathbf{M}(B) \xrightarrow{\chi} \quad \Upsilon(\mathbf{M}(-)) \otimes \mathbf{M}(B)$$

According to the preceding diagram for elements $\lambda \in \Upsilon(\mathbf{M}(B))$, $v : B' \to B$ and $q' \in \mathbf{M}(B')$ the following equations hold:

$$\zeta(\lambda, v, q') = (\lambda * v, q'), \qquad \eta(\lambda, v, q') = (\lambda, v(q'))$$

symmetric in $\Upsilon(\mathbf{M}(B))$ and $\mathbf{M}$. Hence the elements of the set $\Upsilon(\mathbf{M}(-)) \otimes_{B} \mathbf{M}$ are all of the form $\chi(\lambda, q)$. This element can be written as:

$$\chi(\lambda, q) = \lambda \otimes q, \quad \lambda \in \Upsilon(\mathbf{M}(B)), q \in \mathbf{M}(B)$$

Thus if we take into account the definitions of ζ and η above, we obtain:

$$[\lambda * v] \otimes q' = \lambda \otimes v(q'), \quad \lambda \in \Upsilon(\mathbf{M}(B), q' \in \mathbf{M}(B'), v : B' \to B$$

We conclude that the set $\Upsilon(\mathbf{M}(B)) \otimes_{B} \mathbf{M}$ is actually the quotient of the set $\coprod_{B} \Upsilon(\mathbf{M}(B)) \times \mathbf{M}(B)$ by the equivalence relation generated by the above equations. Furthermore, if we define $\lambda * v = \lambda'$, $v(q') = q$, where λ' is a subobject of $\mathbf{M}(B')$ and $q \in \mathbf{M}(B)$, we obtain the equation:

$$\lambda' \otimes q' = \lambda \otimes q$$

Next, since pullbacks exist in $\mathcal{L}$, we may consider the arrows $h:\mathbf{M}(D)\rightarrow\mathbf{M}(B)$ and $h':\mathbf{M}(D)\rightarrow\mathbf{M}(B')$ and the following pullback diagram in $\mathcal{L}$:

$$\begin{array}{ccc}
\mathbf{M}(D) & \xrightarrow{\quad h \quad} & \mathbf{M}(B) \\
\Big\downarrow{\scriptstyle h'} & & \Big\downarrow \\
\mathbf{M}(B') & \xrightarrow{\qquad} & L
\end{array}$$

such that the following relations are satisfied: $h(d)=q$, $h'(d)=q'$ and $\lambda * h=\lambda' * h'$. Then we obtain:

$$\lambda\otimes q=\lambda\otimes h(d)=[\lambda * h]\otimes d=[\lambda' * h']\otimes d=\lambda'\otimes h'(d)=\lambda'\otimes q'$$

We may further define:

$$\lambda * h=\lambda' * h'=\tau$$

where τ is a subobject of $\mathbf{M}(D)$. Then, it is obvious that:

$$\lambda\otimes q=\tau\otimes d$$

$$\lambda'\otimes q'=\tau\otimes d$$

It is then clear that we may define a partial order on the set $\Upsilon(\mathbf{M}(-))\otimes_B\mathbf{M}$ as follows:

$$\lambda\otimes b\leq\rho\otimes c$$

if and only if there exist quantum algebraic homomorphisms $\beta : \mathbf{M}(D) \to \mathbf{M}(B)$ and $\gamma : \mathbf{M}(D) \to \mathbf{M}(C)$, and some d_1, d_2 in $\mathbf{M}(D)$, such that: $\beta(d_1) = b$, $\gamma(d_2) = c$, and $\lambda * \beta = \rho * \gamma = \tau$. Thus we obtain:

$$\lambda \otimes b = \tau \otimes d_1$$

$$\rho \otimes c = \tau \otimes d_2$$

We conclude that:

$$\lambda \otimes b \leq \rho \otimes c$$

if and only if

$$\tau \otimes d_1 \leq \tau \otimes d_2 \Leftrightarrow d_1 \leq d_2$$

The set $\Upsilon(\mathbf{M}(-)) \otimes_{\mathcal{B}} \mathbf{M}$ may be further endowed with a maximal element which admits the following presentations:

$$\mathbf{1} = \tau \otimes 1 := \mathit{true} \qquad \forall \tau \in \Upsilon(\mathbf{M}(D))$$

$$\mathbf{1} = id_{\mathbf{M}(B)} \otimes b := \mathit{true} \qquad \forall b \in \mathbf{M}(B)$$

and an orthocomplementation operator:

$$[\tau \otimes d]^{\star} = \tau \otimes d^{\star}$$

Then it is easy to verify that the set $\Omega = \Upsilon(\mathbf{M}(-)) \otimes_{\mathcal{B}} \mathbf{M}$, endowed with the prescribed operations, is actually a *quantum event algebra* for every Boolean event algebra B in $\mathcal{B}$. Consequently, the elements of the quantum classifying algebra $\Omega = \Upsilon(\mathbf{M}(-)) \otimes_{\mathcal{B}} \mathbf{M}$ are equivalence classes (*Boolean germs of truth*) represented in tensor product form as follows:

$$[\delta_{\Upsilon(\mathbf{M}(B))}]^{\lambda}(b) := \Delta^{\lambda}(b) = \lambda \otimes b$$

where:

$$[\lambda * v] \otimes b' = \lambda \otimes v(b'), \quad \lambda \in \Upsilon(\mathbf{M}(B)), b' \in \mathbf{M}(B'), v : B' \to B, v(b') = b$$

and $[\delta_{\Upsilon(\mathbf{M}(B))}]^{\lambda} := \Delta^{\lambda}$ denotes a local Boolean cover of Ω in the Boolean localization system $[\delta_{\Upsilon(\mathbf{M}(-))}]^{(-)}$ of Ω using the unit isomorphism.

As a consequence, we obtain the following *criterion of truth valuations with respect to a Boolean localization system* of a quantum event algebra $\mathcal{L}$:

$$[\delta_{\Upsilon(\mathbf{M}(B))}]^{\lambda}(b) := \Delta^{\lambda}(b) = \lambda \otimes b = \mathit{true} \qquad \mathit{iff} \qquad b \in \mathit{Image}(\lambda)$$

Thus, we have shown explicitly that the category of quantum event algebras $\mathcal{L}$ endowed with a Boolean localization system has a *classifying object* Ω, which is defined by the inductive limit in the category of elements of the Boolean frames modeled subobject functor, viz., $\Omega = \Upsilon(\mathbf{M}(-)) \otimes_B \mathbf{M}$. The elements of the classifying quantum event algebra Ω, that is, the *Boolean germs of truth*, are equivalence classes represented in tensor product form as follows:

$$\Delta^{\lambda}(b) = \lambda \otimes b$$

where:

$$[\lambda * v] \otimes b' = \lambda \otimes v(b'), \quad \lambda \in \Upsilon(\mathbf{M}(B)), b' \in \mathbf{M}(B'), v : B' \to B, v(b') = b$$

and Δ^{λ} denotes a local Boolean cover of Ω in the Boolean localization system $\Delta^{(-)}$ of Ω using the unit isomorphism. Thus, we have proved that *truth valuations in quantum mechanics are localized with respect to equivalence classes of compatible Boolean frames belonging to a Boolean localization system of a quantum event algebra.* We require that a Boolean localization system of L in $\mathcal{L}$ is closed with respect to Boolean covers, viz., it contains all Boolean frames covering L and satisfying the compatibility conditions. In this way, the logical classifying Boolean germ *true* (equivalence class) used for the valuation of quantum propositions is characterized by:

$$\tau \otimes 1 := true \qquad \forall \tau \in \Upsilon(\mathbf{M}(D))$$

$$id_{\mathbf{M}(B)} \otimes b := true \qquad \forall b \in \mathbf{M}(B)$$

$$\lambda \otimes b = true \qquad iff \qquad b \in Image(\lambda), \quad \forall \lambda \in \Upsilon(\mathbf{M}(B))$$

9.7 Boolean Germ Classification of Measurement Events

The classifying quantum event algebra Ω in $\mathcal{L}$, induced by a Boolean localization system, plays a role similar to the role played by the Boolean algebra $2 := \{0,1\}$ in characterizing the logic of propositions referring to the behavior of classical systems. *Thus in the quantum case, subobjects of a quantum event algebra should be depicted in terms of characteristic functions taking values in* Ω. Let us explain the functionality of the logical classifying object Ω according to the criterion of truth valuations by the following diagram:

$$\begin{array}{ccc} Dom(l * e) & \longrightarrow & K \\ \downarrow {\scriptstyle l * e = \lambda} & & \downarrow {\scriptstyle l} \\ \mathbf{M}(B) & \xrightarrow{e} & L \end{array}$$

where $l : K \rightarrowtail L$ is a subobject of a quantum event algebra L, $e : \mathbf{M}(B) \to L$ is a Boolean cover of L in a Boolean localization system of L, and $l * e = \lambda$ is the pullback of l along e, that is, the subobject λ

of $\mathbf{M}(B)$. According to the truth-value criterion, the characteristic function of the subobject $l: K \rightarrowtail L$ of L is specified as an equivalence class of pullbacks of the subobject l along its restrictions on all Boolean covers in a Boolean localization system of L. For each Boolean cover $\mathbf{M}(B)$ of L, $(l * e) \otimes b = \lambda \otimes b = true$ if and only if b belongs to the $Image(\lambda)$, for all λ in $\Upsilon(\mathbf{M}(B))$. Thus, for each Boolean cover $\mathbf{M}(B)$ of L, the value $\mathbf{1} = true$ in Ω is assigned to all those b in $\mathbf{M}(B)$ belonging to the restriction of the subobject $l: K \rightarrowtail L$ of L with respect to the subobject λ of $\mathbf{M}(B)$, for all λ. In particular, if the Boolean covers are monomorphisms, each pullback is expressed as the intersection of the subobject l with the corresponding cover in the Boolean localization system.

Conceptually, every quantum event or proposition of a quantum event algebra L is contextualized with respect to all Boolean frames $\mathbf{M}(B)$ belonging to a Boolean localization system of L by means of the operation of restriction. In this way, a *quantum proposition refers to the germ (equivalence class) of all its restricted propositions with respect to all Boolean frames* $\mathbf{M}(B)$ belonging to a Boolean localization system of L. Note that all restricted propositions are *logically compatible* since they belong to a Boolean localization system of L. With respect to each such contextualization we obtain a *contextual truth valuation* of the restricted proposition associated with the corresponding frame $\mathbf{M}(B)$, specified by the rule $\lambda \otimes b = true$ if and only if b belongs to the $Image(\lambda)$, holding for every subobject λ of $\mathbf{M}(B)$, where b represents the restricted quantum proposition with respect to the Boolean frame $\mathbf{M}(B)$. We emphasize that all these contextual truth valuations of some quantum proposition are appropriately related to each other by the formation of equivalence classes *gluing together all its restrictions* with respect to all Boolean frames $\mathbf{M}(B)$ belonging to a Boolean localization system. This is achieved via the logical classifying 'Boolean germ *true*' specification constraints in Ω, thus yielding a complete inductive description of reality, defined by: $\tau \otimes 1 := true$ for all $\tau \in \Upsilon(\mathbf{M}(B))$ and $id_{\mathbf{M}(B)} \otimes b := true$ for all $b \in \mathbf{M}(B)$ and for all Boolean frame domains $\mathbf{M}(B)$ in a Boolean localization system.

Let us now apply the above truth valuation scheme to a typical measurement situation, such as the one occuring in the double-slit optics experiments discussed in chapters 2 and 4. The key measurement in these experiments is the measurement of a quantum system (e.g., a photon) passing through a slit, where a counter on the measuring apparatus will record passage through the slit by clicking (or not clicking if the photon does not pass through). If we denote a Boolean domain preparation context, which contains both the measuring apparatus as well as the system observed by $\mathbf{M}(B)$, then we may form the propositions:

$\langle c \rangle$:=[*counter clicks*]

$\langle d \rangle$:=[*system passes through the slit*]

$\langle \; c \Rightarrow d \; \rangle := \langle b \rangle :=$ [*counter clicks* $\Rightarrow$ *system passes through the slit*].

Notice that the proposition $\langle b \rangle$ is a *compound proposition* which refers to the *coupling* of the measuring apparatus together with the quantum system in the Boolean frame $\mathbf{M}(B)$. The proposition $\langle \; c \Rightarrow d \; \rangle := \langle b \rangle$ is assigned the value *true* in Ω, expressing a complete description of the state of affairs. More precisely, with respect to the Boolean logical frame of $\mathbf{M}(B)$, or more concretely with respect to $id_{\mathbf{M}(B)}$, we have that $id_{\mathbf{M}(B)} \otimes b = true$. We notice that the above truth valuation *does not* suffice in order to infer that the proposition $\langle d \rangle$ is true. In order to infer the above, we need to use the Boolean frame $\mathbf{M}(C)$ which contains only the measuring apparatus, which is a *subobject* of the Boolean frame $\mathbf{M}(B)$, viz., $\lambda : \mathbf{M}(C) \rightarrowtail \mathbf{M}(B)$. Then, we obtain that: $id_{\mathbf{M}(C)} \otimes c = true$ and $\lambda \otimes \xi = true$ if and only if ξ belongs to the *Image*(λ) in $\mathbf{M}(B)$. The Boolean contexts $\mathbf{M}(C)$ and $\mathbf{M}(B)$ induce the formation of a *Boolean localization system* for the evaluation of $\langle d \rangle$ and thus they are *glued* compatibly together. Hence, we deduce that $\lambda \otimes b = true$ because b belongs to the *Image*(λ) in $\mathbf{M}(B)$. In particular b in $\mathbf{M}(B)$ is the image of c in $\mathbf{M}(C) \rightarrowtail \mathbf{M}(B)$

and the truth of $\langle d \rangle$ is *indirectly* inferred by the measurement event through the *Boolean germ true* (equivalence class) defined by:

$$id_{\mathbf{M}(C)} \otimes c = \lambda \otimes b = id_{\mathbf{M}(B)} \otimes b = true$$

Notes

1. V. Karakostas, "Nonseparability, "Potentiality and the Context-Dependence of Quantum Objects," *Journal for General Philosophy of Science* 38 (2007).

2. P. T. Johnstone, *Stone Spaces* (Cambridge: Cambridge University Press, 1986).

3. Elias Zafiris, "Topos-Theoretic Classification of Quantum Events Structures in Terms of Boolean Reference Frames," *International Journal of Geometric Methods in Modern Physics* 3, no. 8 (2006).

CHAPTER 10

Quantum Localization in a Broader Conceptual Perspective

10.1 Boolean Covers and Physical Contexts

The association of a *local Boolean cover* in the mathematical descriptive language of sheaves, with a concrete *physical context* in the physical world, takes into account two fundamental distinctions inextricably connected with the quantum-theoretic formalism: The first of them refers to a distinction being made between an *event* and the *physical context* that constitutes a set of necessary and sufficient constraints for the occurrence of an event of the observed kind. To the event, there corresponds a formal descriptive proposition language. To the physical context there corresponds a context-description in a formal descriptive language assuming existence at the level of Boolean covers, followed by an appropriate terminology providing names for the characterization of the language of events occurring in that context. These latter descriptions can be said to belong to the constitutive level of the Boolean localization systems. The second refers to a distinction being made between *possessed* physical quantities, as those found in classical physics, and *dispositional* ones, as those found in quantum physics. The dispositional character of quantum observables is associated with the fact that they may only be specified via the measurement process, and more precisely, as *relationally* appearing with respect to theoretical or actual

Boolean preparatory environments, that is, Boolean logical frames. In the mathematical descriptive terminology this distinction is encoded in a *transition* from globally Boolean event structures to globally non-Boolean event structures, being *covered* by epimorphic families of local Boolean reference frames forming covering sieves. In this sense, creating a preparatory Boolean environment for a system to interact with a measuring device *does not* determine which event will take place, but it *does* determine the *kind* of event that will take place. It forces the outcome, whatever it is, to belong to a certain definite *Boolean algebra* of events for which the standard measurement conditions are invariant. Such a set of standard conditions for a definite kind of measurement is named a physical context and reflects a Boolean reference frame in the mathematical descriptive language.

10.2 Partial Congruence and Adjunction

From a general category-theoretic perspective the *relational realist* framework proposed for the interpretation of quantum event structures and their logical semantics is based on the existence of the Boolean frames-quantum adjunction. From this adjunction, characterized by means of the counit and unit natural transformations, we have constructed a *sheaf-theoretic representation* of quantum event algebras formulated with respect to Boolean localization systems, as well as a quantum subobject classifier, which plays the role of a *logical classifying object* in the quantum universe of discourse. In this way, the Boolean frames-quantum adjunction incorporates both the *representational semantics* of quantum event algebras as sheaves of local Boolean coefficients, and the *logical semantics* of the classifying Boolean germs concerning truth valuations with respect to Boolean localization systems.

Thus, from a physical viewpoint, the Boolean frames-quantum adjunction stands as a theoretical platform for *decoding* the global structural information contained in quantum algebras of events via *processes of localization* with respect to Boolean logical frames, realized as physical contexts for measurement of observables, and subsequent *processes of information classification* in terms of corresponding truth valuations. The functioning of this platform is based on the establishment of a *bidirectional* dependence between the Boolean and quantum structural levels being brought into *local* or *partial* congruence. Most significantly, this bidirectional dependence is implicated through the topos-theoretic universe of sheaves of sets over local Boolean frames of quantum events algebras forming Boolean localization systems, where these local Boolean frames play the role of local Boolean covers, effectuating in this way the specific form of the above partial congruence. Of course, the sheaf-theoretic requirements secure the *compatibility* of the locally Booleanized information in the overlapping regions of physical measurement contexts under extension from the local to the global and conversely.

Additionally the *global closure* of this bidirectional dependence is necessary to be constrained in order to obey certain conditions, such that its total constitutive information content, unfolded in the multitude of local Boolean reference frames, is both preserved and coherently organized in a logical manner. Remarkably, the necessary and sufficient conditions for both of these requirements, that is: [i] *preservation* of quantum information content in Boolean localization systems associated with compatible physical measurement contexts; and [ii] *logical classification* of quantum information by means of the classifying Boolean germ *true* (equivalence class) used for the valuation of quantum propositions, are supplied by the Boolean frames-quantum adjunctive correspondence itself, via the counit and unit natural transformations, respectively.

More concretely, regarding the first condition, we conclude that it is satisfied if the *counit* of the adjunction is an *isomorphism* for each quantum event algebra. In this case, there exists a full and faithful sheaf-theoretic representation of quantum events algebras in the descriptive terms of Boolean covering systems, characterized as Boolean localization systems of measurement. Regarding the second condition, we conclude that it is satisfied, given the *validity* of the first condition, if the *unit* of the adjunction is an *isomorphism* for the subobject functor. In this case respectively, the subobject functor is representable in the category of quantum event algebras by a *classifying object* in this category, called a *subobject classifier*. This classifying object, which is defined by an *inductive limit* operation, contains the information of equivalence classes of truth valuations with respect to all compatible Boolean frames belonging to a Boolean localization system of a quantum event algebra L in $\mathcal{L}$. Thus, it can be legitimately used for *truth valuations* of quantum propositions in exact correspondence with the use of the two-valued Boolean object, which is used for valuations of classical propositions. Moreover, the quantum subobject classifier provides the key logical device for the analysis of typical quantum measurement situations, providing a *criterion of truth* for a complete description of reality in the quantum regime with respect to a Boolean localization system.

10.3 Classical to Quantum: From Groups to Categories

It is instructive to draw an *analogy* between the notion of a reference frame as it is used in classical physics (presupposing the existence of a set-theoretic *spacetime* substratum) and the notion of a variable *Boolean localizing reference frame* in quantum physics. In classical or relativistic physics we always have to define first the notion of a reference frame (spatiotemporal reference frame) where we may describe a physical phenomenon or formulate a physical law. The consideration of a single reference frame is not enough. What we have to consider is a whole *class* of reference frames and their *transformations*. In classical theories the class of all reference frames together with their interrelations form a *group* of transformations. Then the physical laws involved should be *invariant* under the action of this group of transformations.

Now, a group of transformations is a very special kind of category having only one object. It is a natural requirement, in case we generalize to quantum theory for example, to obtain an analogous description of phenomena. For this purpose, we generalize from a spatiotemporal reference frame (*a metrical notion*) to an analogous localizing notion of reference frame (*a topological notion*), which admits a concrete physical interpretation and plays a similar role. This is the functionality of a Boolean reference frame in quantum theory. Thinking categorically means that we have to specify the transformations from a Boolean reference frame (corresponding to a quantum measurement situation) into another. Now the class of all Boolean reference frames and their transformations do not form a group but they form a *category*. So the quantum description of phenomena should be *invariant* under the action of the concrete category of all partially compatible Boolean reference frames. More concretely, the physical content of the topological sheaf-theoretic representation of quantum event algebras can be formulated in terms of an invariance principle. According to this, *the information content of a quantum event algebra is invariant under the category of gluing isomorphisms (technically called a groupoid) between overlapping local Boolean reference frames, along their intersections, in a Boolean localization system, preserving the quantum algebraic structure.* [1]

It is important to notice the obvious analogy to the metrical spacetime frame case. The requirement of *invariance* is the conceptual reason necessitating the introduction of the notion of a sheaf of Boolean logical frames, or equivalently the notion of a Boolean topological localization system of a quantum event algebra. We emphasize that the normative requirements for the localization of information in the quantum case *does not* refer to set-theoretic spacetime localization. Categorically thinking, localization has to do with the partiality of the information referring to a Boolean reference frame, interpreted in topological terms, and its consistent gluing effectuating a coherent *extension* or *induction* from the local to the global level.

It is instructive to note that in the special case of a group of transformations (that is the usual spacetime case with point-defined spatiotemporal reference frames) the notion of categorical localization is identical with the notion of spacetime localization. So, using the categorical framework we incorporate the classical spacetime localization case as a *very special case of the general localization notion* (the former is referring to a group of transformations, whereas the latter is referring to a category of transformations among the corresponding frames). Presheaves and sheaves (incorporating the gluing compatibility conditions) provide the conceptual and technical devices in order to express such a *generalization in the notion of localization*, which finds an important application in quantum theory (by introducing Boolean reference frames). Now the *gluing process* is a *vast generalization* of the usual process of requiring *inter-transformability* of our descriptions when we pass from some reference frame to another, if we take into account this more general localization idea. The benefit is that we obtain additionally a coherent *mereotopological* semantic methodology of extension from the local or partial into the global (considered in observable-theoretic terms as it is actually practiced via physical measurement) and conversely.

We claim that it is precisely the *sheaf-theoretic gluing process* which provides a rich and adequate mathematical model of Whitehead's scheme of *extensive connection*. The distinguishing feature of this claim is that the gluing process is of a purely *topological* nature and thus it captures the character of the relationship of extension idealized by Whitehead as being *prior to the metrical notions* of space and time:

> In this general description of the states of extension, nothing has been said about physical time or physical space, or of the more general notion of creative advance. These are notions which presuppose the more general relationship of extension. They express additional facts about the actual occasions. The extensiveness of space is really the spatialization of extension; and the extensiveness of time is really the temporalization of extension. Physical time expresses the reflection of genetic divisibility into coordinate divisibility. . .[2]

10.4 Topological Localization as Genetic Analysis

From a general philosophical standpoint, the general conceptual underpinning of the notion of sheaf-theoretic localization of observability is deeply related to the theoretical systems of Aristotle and Whitehead. According to Aristotle the *divisions* that we project into an observable continuum assume a *potential* existence, unless their *actuality* can be confirmed by property-indicating facts, viz., by experimental events. Thus, from an Aristotelian perspective, ontological *primacy* should be shifted from the level of states, to the level of observable and event information structures. This means that any complex systems theory (including quantum theory) should be viewed as a theory specifying a particular way of *correlating property-indicating events*, according to a schema of *potential division* of an observable continuum in terms of *local information carriers* (for instance Boolean reference frames), assuming its actuality *only* by the observed events and their correlations.

The process-theoretic system of Whitehead advocates this conception and poses additionally the natural condition that the correlations should conform to a *theory of congruence*, which should act as a normative requirement in order to make sense of the observed events and their correlations. Furthermore, a theory of congruence would manifest the potential division of an observable continuum as commonly presupposed in the study of natural phenomena, whereas its functional necessity and essential correctness would be affirmed only by means of the observed events (*actual occasions*).

From a logical perspective, observed events in quantum theory correspond to evaluations of potential facts (observables) with respect to some logical binary schemata (yes/no experiments subsumed by measuring devices), interpreted as property-indicating facts. Note that this is meaningful from a topological perspective only if it respects the normative constraints of the potential division of a global quantum event continuum with respect to local Boolean reference frames (preparation stage of a quantum measurement), as well as the rules of compatibility under extension from the local to the global and conversely. The *local-*

to-global bidirectional scheme corresponds to Whitehead's notion of *genetic analysis*, and is reflected in the modeling of the continuum related to the behavior of a quantum system via a *category of sheaves* of local information carriers (local Boolean reference frames). More precisely, the action of a presheaf corresponds to a potential division of the continuum according to the normative rules of a congruence theory of local Boolean observables, whereas the transformation of a presheaf into a sheaf (that is to a Boolean localization system) reflects the gluing and compatibility conditions determining the correlations of observed events under the transition from the local to the global, and conversely under the reduction of the global to the local.

Note that in full agreement with Whitehead, the genetic passage from phase to phase, viz., from each local Boolean observable frame to another, is *not* in physical time, meaning that the genetic process is *not* the temporal succession. *This is due to the dependence of each local phase in the genetic process on the global level, meaning that each such phase presupposes the global quantum level implicitly.* It is also important to notice that the conditioning of actualized events by such phases (considered as local Boolean reference frames) is indispensable for the existence of properties. In other words, every property is a possessed property, indicated by an actualized event, which is conditioned on the existence of a local phase, presupposing the global quantum level.

10.5 Boolean Localization and Decoherence

A significant application of the proposed Boolean localization scheme of quantum event structures concerns the elucidation of the phenomenon of quantum decoherence.[3] *Decoherence* is understood as the phenomenon according to which all the features of quantum observable behavior are suppressed, and instead a quantum system *behaves like* it can be described by classical probabilistic reasoning. Typical quantum features of this form constitute interference phases and the existence of off-diagonal elements in the density matrix description of a quantum system. Thus, decoherence is mathematically modeled either by the process of *diagonalization* of the density matrix with respect to a particular basis, called technically the *pointer basis*, or by the process of *minimization of interference phases* in a sufficiently *coarse-grained* resolution scale of quantum observable behavior. The essential aspect of all modeling approaches to decoherence is the explanation of the *emergence of classical behavior*, meaning the emergence of a classical statistical description referring to ensembles of equivalently prepared systems.

The first of the current modeling approaches to decoherence may be called *environment induced* decoherence, whereas the second may be called *coarse-graining* induced decoherence. These approaches are based on different conceptual foundations regarding the explanation of emergence of classical behavior. More specifically, environment-induced decoherence is an *extrinsic* approach in the sense that it requires the coupling action of an external to the system agent, called the *environment*, in order to explain the emergence of a classical statistical description with respect to some particular basis, that is the pointer basis. The crucial aspects of this approach may be summarized by stressing the *contingent* and extrinsic influence of some environment on a quantum system and the conception that the process of decoherence requires the passage of time, called *decoherence time*, within which energy and information are being dissipated to the environment. On the other side, coarse-graining induced decoherence is an *intrinsic* approach in the sense that classical behavior emerges intrinsically within the description of a quantum system by *coarsening* the scale of resolution of quantum observable behavior by means of *macroscopic* or gross or collective observables, called classical observables. Hence, decoherence is not conceived as a process requiring the coupling of a quantum system to some external environment, but as a process of resolving quantum observable behavior at sufficiently coarse-grained scales, expressed in terms of *macroscopic algebras of observables*, with respect to whom interference phases are being suppressed. The crucial aspects of the latter approach may be summarized by stressing the intrinsic character of emergence of a classical statistical description in terms of *commutative algebras of macroscopic observables* incorporated at sufficiently coarse-grained levels within a *global non-commutative algebra* of microscopic quantum observables.

We argue that the full implications of coarse-graining induced decoherence require a *topological* modeling approach based on the functional role of Boolean localization systems. More concretely, the phenomenon of decoherence should be understood as a consequence of covering the intrinsic structure of a global non-commutative (non-Boolean) algebra of quantum observables by covering sieves of local commutative (Boolean) algebras of macroscopic or classical observables corresponding to suitable coarse-grained resolution scales of quantum measurement and then localizing sheaf-theoretically at these scales.[4]

The crucial mathematical advantage of this topological approach to quantum decoherence is that *the phenomenon of decoherence can be understood as a process of topological localization of a global quantum (non-Boolean) algebra of observables at some local commutative (Boolean) algebra of classical observables associated with some macroscopic resolution scale*. Topologically speaking, this becomes possible via sieving the global information content of a quantum algebra of observables with respect to compatible covering families (covering sieves) of local commutative (Boolean) algebras of macroscopic observables assuming existence at various *coarse-grained* measurement levels. In this conceptual framework, it becomes possible to understand the notion of environment *not* as an external agent forcing the selection of some pointer basis with respect to whom a classical statistical or Boolean description emerges, but *internally as localizing* topologically a global algebra of fine-grained microscopic observables at some coarse-grained macroscopic level *sufficient* for emergence of classical behavior.

Most significantly, the topological approach not only models decoherence via the sheaf-theoretic mathematical technique of Boolean localization via coarse-graining sieves, but allows the conceptualization and modeling of *recoherence* from the local macroscopic to the global microscopic level of observable structure via sheaf-theoretic amalgamation of locally compatible observables, that is by gluing locally compatible observables together globally in the form of equivalence classes, which we have called *Boolean germs*.

The conceptual motivating precursors of the topological sheaf-theoretic modeling of coarse-graining induced decoherence are, on the physical side, the *consistent or decoherent histories* approach to quantum theory[5] and on the mathematical side, the *topological representation theorems* of algebras of observables by *sections of a sheaf* over a topological space, or a category endowed with a Grothendieck-type of a topology, tracing their roots to the famous *Stone and Gelfand topological representation theorems* of Boolean and commutative algebras, respectively.[6]

On the physical state of affairs, the consistent histories approach to quantum theory focuses on the formulation of an appropriate criterion, subsumed in the formulation of a *complex-valued functional* for pairs of histories, the vanishing of which specifies the condition of *consistency*, that is it decides which histories decohere. Typically, *decoherent sets* of histories contain coarse-grained histories. More precisely, in quantum mechanics, propositions about the attributes of a system at a fixed moment of time are represented by sets of projection operators. The projection operators P_α enable a partition of the possible alternatives α a system may exhibit at each moment of time. They are exhaustive and exclusive,

$$\sum_\alpha P_\alpha = 1, \qquad P_\alpha P_\beta = \delta_{\alpha\beta}\, P_\alpha$$

A projector is said to be *fine-grained* if it is of the form $|\alpha\rangle\langle\alpha|$, where $\{|\alpha\rangle\}$ are a complete set of states. Otherwise it is *coarse-grained*. A *homogenous quantum-mechanical history* is characterized by a string of time-dependent projections, $P^1_{\alpha_1}(t_1),\cdots P^n_{\alpha_n}(t_n)$, together with an initial state ρ. The time-dependent projections are related to the time-independent ones by:

$$P^k_{\alpha_k}(t_k) = e^{iH(t_k - t_0)} P^k_{\alpha_k} e^{-iH(t_k - t_0)}$$

where H is the Hamiltonian. The *candidate probability* for these homogeneous histories is:

$$p(\alpha_1, \alpha_2, \cdots \alpha_n) = \mathrm{Tr}\left(P^n_{\alpha_n}(t_n) \cdots P^1_{\alpha_1}(t_1) \rho P^1_{\alpha_1}(t_1) \cdots P^n_{\alpha_n}(t_n)\right)$$

It is straightforward to show that $p(\alpha_1, \alpha_2, \cdots \alpha_n)$ is both non-negative and normalized to unity when summed over $\alpha_1, \cdots \alpha_n$. However, it *does not* satisfy all the axioms of probability theory, and for that reason it is referred to as a candidate probability. For example, it does not satisfy the requirement of *additivity* on disjoint regions of sample space. More precisely, for each set of histories, one may construct coarser-grained histories by grouping the histories together. This may be achieved, for example, by *summing over* the projections at each moment of time, $\overline{P}_{\bar{\alpha}} = \sum_{\alpha \in \bar{\alpha}} P_\alpha$. The additivity requirement is then that the probabilities for each coarser-grained history should be the sum of the probabilities of the finer-grained histories of which it is comprised. Quantum-mechanical interference generally prevents this requirement from being satisfied. Histories of closed quantum systems *cannot* in general be assigned probabilities. There are, however, certain types of histories for which interference is negligible, and the candidate probabilities for histories do satisfy the sum rules. These histories may be found using the complex-valued *decoherence functional*:

$$D(\underline{\alpha}, \underline{\alpha}') = \mathrm{Tr}\left(P^n_{\alpha_n}(t_n) \cdots P^1_{\alpha_1}(t_1) \rho P^1_{\alpha_{1'}}(t_1) \cdots P^n_{\alpha_{n'}}(t_n)\right)$$

Here $\underline{\alpha}$ denotes the string $\alpha_1, \alpha_2, \cdots \alpha_n$. Intuitively, the decoherence functional measures the *amount of interference* between pairs of histories. It may be shown that the additivity requirement is satisfied for all coarse-grainings if and only if the *real part* of the complex-valued decoherence functional vanishes, that is $\mathrm{Re}D(\underline{\alpha}, \underline{\alpha}') = 0$, for all distinct pairs of histories $\underline{\alpha}, \underline{\alpha}'$. Such sets of histories are said to be consistent, or weakly decoherent. The consistency condition is typically satisfied only for coarse–grained histories, and this then often leads to satisfaction of the stronger condition of decoherence $D(\underline{\alpha},\underline{\alpha}')=0$ for $\underline{\alpha}\neq\underline{\alpha}'$. The condition of decoherence is associated with the existence of so-called *generalized records*, corresponding to the idea that information about the variables to follow is stored in the variables ignored in the coarse-graining procedure.

We may state concisely the key idea of the consistent histories approach to quantum mechanics by the realization that we can have additive probabilities only in case we *restrict* our resolution scale within particular sets of histories, called *consistent sets*, which are typically sets of appropriately coarse-grained histories. The main problems of the consistent-histories approach to coarse-graining induced decoherence may be summarized as follows:

[i] It uses an *external time* parameter to both formulate the notion of history, and express the process of coarse-graining by summation of projection operators at each moment of this external time;

[ii] It turns out that there exist *many consistent sets* of histories which are mutually *incompatible* such that they cannot be combined to give a maximal consistent set, and at the same time there is *no criterion* of selecting uniquely some consistent set among them; and

[iii] The criterion of consistency of an arbitrary set of histories is *not correlated* with the requirement of classical behavior of a large class of observables. This is due to the fact that the difference between the algebraic structure of quantum observables (non-Boolean and non-commutative) in comparison to the algebraic structure of macroscopic observables at suitable coarse-grained scales (Boolean and commutative) is essentially *ignored* in this formalism.

In order to cope with these problems and formulate a more satisfactory account of coarse-graining induced decoherence, we need to topologize quantum algebras of observables in the spirit of Stone and Gelfand topological representation theorems of Boolean and commutative algebras respectively.[7]

The link between the consistent histories approach and the topological approach is obtained in two steps:

[i] By considering explicitly the *difference in the algebraic structure* between an algebra of microscopic observables and various algebras of macroscopic observables at suitable coarse-grained measurement scales with respect to the property of *commutativity*, and

[ii] By *topologizing* this difference in terms of a local-to-global sheaf-theoretic construction, where the information of a global non-commutative algebra of quantum observables may be *localized* by *restriction* or *reduction* to *compatible families* of local commutative algebras of classical macroscopic observables topologically *covering* the former, and inversely, recovering its content by the topological method of *gluing*.

At the propositional level, that is at the level of projection operators, as we have already pointed out, it is a common realization that existing set-theoretic lattice axiomatizations of quantum observable structures hide the intrinsic fundamental significance of Boolean commutative localization systems in the formation of these structures. On the other side, topologically, we can explicitly associate them with the action of appropriate Boolean coverings of global quantum algebras. Consequently, the algebraic representation of a quantum observable structure in the relational topological terms of localization systems of commutative algebras of macroscopic Boolean observables, effectuates a *semantic transition* from the axiomatic set-theoretic context of orthocomplemented partially ordered sets, *à la* Birkhoff and Von Neumann, to the categorical sheaf-theoretic context of Boolean information sieves, *à la* Grothendieck.

In order to understand the semantic transition of quantum observable algebras from the set-theoretic level to the sheaf-theoretic level, as well as the significance of this transition in relation to the conceptual and technical modeling of decoherence, it is instructive to explain briefly the philosophy behind Stone and Gelfand topological representation theorems of Boolean and commutative algebras, respectively, being the precursors of Grothendieck-type sheaf-theoretic formulations expressed in terms of topological covering families (covering sieves) on categories. These theorems state that an arbitrary commutative algebra of observables $\mathcal{A}$ of a certain type (for instance a Boolean algebra of observables in Stone's case) may be represented as the *algebra of continuous* $\mathbb{F}$*-valued functions*, where $\mathbb{F}$ is a particular *standard algebra* of the given type, on a *topological space* associated with $\mathcal{A}$ and constituting in this way its $\mathbb{F}$ - *spectrum*. For example, in Stone's case, the standard algebra is the two-valued Boolean algebra $\{0,1\}$ and we make use of the discrete topology. The topological representation theorem is proved by the fact that a continuous function from the topological $\{0,1\}$-spectrum space X to $\{0,1\}$, that is $f : X \rightarrow \{0,1\}$ is completely determined by the clopen subset $f^{-1}(1)$ of X.

Now, the extension of this topological representation philosophy to larger classes of algebras of observables proves to be more complicated in the sense that a general algebra of observables $\mathcal{A}$ *cannot* be built up in this way from a single standard algebra like the two-valued Boolean algebra $\{0,1\}$ in Stone's case. This means that its spectrum is more complicated. An algebra of quantum observables falls into the latter category, and this is precisely the meaning of Kochen-Specker theorem, which essentially states that it is impossible to recover the global information content of a global quantum observable algebra using a single type of a Boolean measurement device globally. Of course, this can be done *locally at a certain coarse-grained resolution measurement scale*, but the global information content may be approximated only by employing simultaneously an appropriate *multitude of compatible Boolean devices* at different overlapping measurement scales *covering* its behavior completely. Mathematically, the inability to build up a global algebra of observables from a single standard algebra necessitates the introduction of *continuously variable* local standard algebras being capable of covering it completely. This pictorially means that we may identify the observables of $\mathcal{A}$ locally with *sections of a sheaf* defined on a spectrum X and taking values in one of a given class of standard local algebras, provided that the local algebra in which an observable takes values may vary continuously from point to point in the spectrum X.

The extrapolation of the sheaf-theoretic topological representation methodology, in the case of a quantum algebra of observables, sheds new light on the process of coarse-graining induced decoherence and the emergence of classical behavior. *In a nutshell it implies that quantum observables are not specified globally by their measured values, but they are specified by their Boolean observable germs*, viz., equivalence classes of locally compatible Boolean observables forming an inductive limit in a sheaf of macroscopic coarse-grained Boolean observable algebras covering the former at various resolution measurement scales. It is instructive to stress again that the concept of a categorical topology epitomizes the meaning of a *localization scheme* of a quantum algebra, whereas the covering sieves are interpreted as generalized measures of topological localization of quantum observables at various coarse-grained macroscopic (Boolean) levels. In the same vein of ideas the notion of a site generalizes the notion of a topological state space in categorical terms and constitutes the *relational variable topological background* for the effectuation of a localization scheme for coarse-graining the global information contained in a quantum algebra through localizing Boolean algebraic frames of macroscopic observables at various coarse-grained resolution scales. Thus, from the physical point of view, covering sieves of a global quantum algebra by local Boolean frames effectuate Boolean localization systems of the former. It is instructive to note that covering sieves of a quantum algebra L are actually subfunctors of the functor of Boolean frames of L, viz., *relational covering networks* of morphisms from local Boolean frames to L, expressing the generic morphology of the *internal mereotopological relations* constituting the process-theoretic scheme of extensive connection from the variable Boolean local to the quantum global level. These subfunctors are *partially ordered* by inclusion. The partial order relation among covering sieves of a quantum algebra L is interpreted as the *operation of coarse-graining* among Boolean localization systems. A Boolean localization system forces a transition in the semantics of quantum observables from the set-theoretic to the topological sheaf-theoretic level. The fact that the functor $\mathbf{R}_L$ of Boolean frames of L is a sheaf, means that there exists a gluing compatibility isomorphism of local Boolean frames on their overlap, meaning that these Boolean frames of L cover the same observable information content of a quantum algebra L in a *compatible* way.

The topological compatibility relation between local Boolean frames of a global quantum algebra sheds new light and forces a *strict criterion of consistency* in the process of coarse-graining as it is currently implemented in the context of the consistent histories approach. More concretely, as we have already mentioned, the major problem of this approach is that the criterion of consistency of an arbitrary set of histories is not correlated with the requirement of classical behavior of a large class of observables. This problem is due to the fact that the process of coarse-graining as it is currently implemented *does not have to satisfy any constraint*, it is just a summation over projection operators which describes the transition from a fine-grained description to some coarse-grained description. *We claim that an appropriate compatibility relation has to be imposed on the process of coarse-graining projection operators in order to obtain meaningful probabilities from the decoherence functional, and thus to explain the emergence of classical behavior at suitable coarse-grained macroscopic scales.* The Boolean localization scheme implies that the *coarse-graining process should respect the local Boolean structural rule as we pass from a local Boolean context to another*; that is, it should respect the morphology *of internal relation between local Boolean frames* covering a global quantum algebra.

In order to express this formally we need a compatibility condition which is given precisely by the *transition morphisms from a local Boolean frame to another local Boolean frame* as prescribed by the specification of the category of elements of the functor (sheaf) of Boolean frames of a quantum algebra. For example, referring to the same observable (for example, position) the proper coarse-graining procedure corresponds to a nesting of local Boolean frames, such that there exists compatibility on their overlaps preserving the local Boolean structural rule. In case we consider different observables and apply the coarse-graining process properly, we require partial compatibility on their overlapping coarse-grained Boolean frames (for instance, partial Boolean compatibility of coarse-grained position and momentum up to the limit of Heisenberg uncertainty relations) which is given by the topological gluing isomorphism. Then the formation of the inductive limit (which gives the transition from the local to the global) requires the formation of a set of equivalence classes (Boolean germs) of partially compatible Boolean observable information on all possible overlaps, glued topologically together in this way. This sheaf-theoretic methodology, based on the formation of Boolean localization systems of a quantum algebra, resolves completely the problem associated with coarse-graining in the consistent histories approach via the *idea of gluing on overlaps* such that the local Boolean structure rule is preserved via these transitions.

We may summarize the basic idea of the topological modeling approach to coarse-graining induced decoherence as follows: Although the global description of a quantum system should obey the laws of quantum logic and quantum probability, there exist coarse-grained descriptions at macroscopic levels, formulated for example in terms of *macroscopic hydrodynamic variables*, and expressed in terms of local Boolean subalgebras of a global non-Boolean quantum algebra, such that the attribution of properties by measurement valuations, effectuated by means of various distinct binary devices at these coarse-grained levels, takes place locally according to the laws of Boolean logic and classical probability theory. Essentially this means that *a local Boolean restriction or reduction of a quantum system with respect to a local Boolean context, or more generally with respect to a Boolean localization system, is conceived as a result of a decoherence process assuming validity with respect to an appropriate corresponding notion of coarse-graining subsumed by the function of covering sieves*. Only in this case, it is then meaningful to infer predictions about the behavior of the localized quantum system, via its local macroscopic reduction with respect to some local Boolean context in a Boolean localization system obtained by topological coarse-graining, using the methodology of classical probability theory.

The crucial point is that according to the general conceptual framework of topological localization, *decoherence makes sense only as an intrinsic, locally conceived, reductive process of the global-level description of a quantum system*. Now, recent experimental work seems to verify the above theoretical perspective.[8] More specifically, within the descriptive context of the environment-induced decoherence theories, it was assumed until now that the environment modeled by means of a bath of oscillators acts externally to a quantum system via an appropriate interaction coupling, and most importantly extends globally. These recent experiments show *on the contrary* that decoherence is caused by entities considered within a *localized environment*, and thus the global oscillator-bath picture is not appropriate to describe the phenomenon. More specifically, the role of such a locally conceived environment may be played by a nuclear spin bath, which gets internally entangled with a qubit (for example, a magnetically polarized molecule) and causing *intrinsic decoherence* without dissipation of energy.

10.6 Sheaf-Theoretic Quantum Relational Realism

The idea of a reality admitting *objective* existence is the starting point of our scientific investigations and the conceptual basis pertaining to the realistic interpretation of physics. We argue that our proposed sheaf-theoretic modeling scheme provides a *realistic* perspective on quantum structure, although *conceptually different* from the classical one. The classical view of reality presupposes that objects of our observation are simply individual entities in the world. Moreover an entity may be specified, both qualitatively and quantitatively, in a definite manner *independently* of procedures of observation, being capable of assuming individuality in isolation. According to the interpretative standards of classical realism, an objective description is one that determines the properties possessed by an independently real physical object, standing for an entity, by adopting a representation of that object as a physical system isolated from any observational interaction. Quantum theory does *not* conform to the descriptive ideals of classical realism. Thus, the proposed sheaf-theoretic localization scheme of quantum event algebras should *not* be judged by the descriptive ideals of the classical realism position, but instead should be considered as generating a generalization of the classical framework, mainly by initiating a *revision* of classical realism assumptions, concerning in particular the descriptive concepts involved in association with the objects they are used to describe. The basic claim is that the sheaf-theoretic descriptive framework makes good sense only on the basis of a *revised* realistic interpretation.

Our philosophical starting point is that a quantum event algebra reflects an *objective physical reality* at the quantum level having existence *independently* of some mind perceiving it. Evidently, such an interpretation viewpoint is compatible with the formalism of quantum theory if and only if a *non-Boolean* event structure is being manifested *globally*. Let us initially notice that we describe our *observations* using notions of validity adhering to *Boolean logic* as a consequence of the preparation of Boolean environments of measurement. This is due to the fact that only in such environments is it possible to *separate* elements and conceive of them as existing *in isolation* from the rest of the world. Indeed, the Boolean specification of environments of measurement incorporates the implicit assumption of a separative and compact underlying topology (Stone representation theorem).[9] Of course, such an assumption, in general, can be used as a methodological tool of enquiry, *only locally or partially*, and can never be claimed to assume global validity in the name of empirical findings, being themselves amenable to the specification of such global environments. The appropriate border in the domain of validity of such a Boolean description is decided by the relevant *abstractions* of a measurement situation conceived precisely as a *Boolean localizing context*.

It is instructive at this point to remind ourselves of Bohr's definition of the word *"phenomenon"* to refer exclusively to observations obtained under specific circumstances, including the account of the whole experiment. Thus, for methodological purposes, we may adopt Bohr's concept of phenomenon as a referent of the assignment of an observable quantity to a system, legitimately thought of as approximately isolated, *strictly* in the context of a local Boolean environment of measurement. Let us notice that this consideration is naturally required because of the globally non-Boolean event structure, necessitating a *multitude* of locally considered, isolated yet correlated manifestations with respect to distinct Boolean reference frames. In this sense, every Boolean filtering context corresponds to a particular *type* of observable phenomenon and every proposition of the universe of discourse *belongs to* at least one Boolean frame. Thus every proposition is in principle descriptive of a classically conceived observable, but *only under* the choice of a local physical context. It is also important to notice that the physical claim of partial descriptions embodies in an essential manner the claim that the description of a phenomenon, as above, necessitates a theoretical conception of the phenomenon as the referent of an interactive process structured in *two levels*:

The first level of this process is constitutive of the *generation* of a *localizing environment*, as a reality probing filter, endowed with an intentionally prepared mechanism of abstraction determined explicitly by the qualitative observable nature of the specific Boolean environment (Boolean reference frame).

The second level of this process is constitutive of the *actualization* of some observed event only after *interaction* with the relevant measuring apparatus attached as a binary code to the corresponding Boolean frame.

The conception of phenomena in this approach enforces reference to quantum systems as actually existing objects behind the phenomena, the latter being only their *local or partial* manifestations with respect to Boolean environments designed for that purpose. Evidently, each separate phenomenon *cannot* be regarded as a full representation of a quantum system. Only at the level of sheaves of Boolean frames, that is *Boolean localization systems*, constituting consistently *interconnected multiplicities* of phenomena in terms of Boolean observable germs (equivalence classes), can an *isomorphic representation* of the structure of events of a quantum system be reproduced by means of the *inductive limit* construction.

In the light of the above, it is legitimate to say that the sheaf-theoretic localization scheme replaces the classical static monolithic realist view with a form of *relational realism*, admitting *multiplicities in logical space*, but not some "actualized" space of existing objects, as an expression of qualitative structured *potential continuous variation* in the observable universe of discourse, which are simultaneously strictly conditioned to obey *collectively* the *global constraint* of global uniqueness of a sheaf. Hence, this explicitly *rules out* any interpretation in the classical realist sense of a one-to-one correspondence between concepts used to describe a phenomenal object and the presumed properties of an independent reality. Still, a revised realistic understanding of the description of quantum systems is retained.

Consequently, by virtue of the sheaf-theoretic scheme of interpretation, the *classical realist assumption* that knowledge of an object is achieved by forming a representation of that object as a substance possessing properties is *rejected* and subsequently *replaced* by the possibility of formulating local or partially compatible contextual Boolean descriptions with respect to Boolean localization systems. The systems are then *grounded* on the same actually existing quantum object, where the sameness is precisely determined by the possibility of *uniquely gluing* these descriptions globally via the synthetic ontological process of sheaf-theoretic local-to-global formation. In this way, the revised realist position justifies the usage of the adjective *complex* when we refer to quantum systems. Moreover, they are distinguished from simple systems by applying precisely *the norm* of this revised realist position. So, whereas classical systems are simple because they conform to the descriptive ideals of classical realism, quantum systems are *complex* because they conform to the more complex local-global requirements of revised relational realism.

In this way, we may argue that the quantum level of reality can be conceived only through a *relational sheaf-theoretic perspective*. It has now emerged that relations occupy substantial and remarkable territory in our description of nature, one which was once occupied exclusively by properties. An actually existing quantum system is *not* described through isolated properties, but only through its *relations* with localizing physical contexts interconnected together according to the *sheaf-theoretic* specifications of Boolean localization systems. More concretely, the topological covering process of a quantum event algebra associated with a Boolean localization system leads naturally to a contextual description of quantum events with respect to local Boolean reference frames of measurement. Stated more precisely, *quantum events are represented by evaluations of equivalence classes of Boolean coefficients (Boolean germs) with respect to local compatible Boolean contexts belonging to a Boolean localization system.* Thus, the global information content of a quantum event algebra remains *invariant* by the action of some topological covering system, if and only if that system forms a *Boolean localization system. Hence, the significance of a quantum event algebra is lifted from the orthoposet axiomatization at the level of events, to the sheaf-theoretic gluing conditions at the level of Boolean localization systems*. Eventually, the former axiomatization is fully and faithfully recaptured at the level of equivalence classes in these localization systems.

We argue that the relational sheaf-theoretic perspective on quantum reality provides a *viable mathematical model* to describe and understand Whitehead's conception of an *extensive event continuum.* Extension is fundamentally tied with the assumption of *potential divisibility* signifying a part-whole or local-global type of relation. In this way, the crucial role in the conceptualization of quantum event ontology is being played by the notion of a Boolean localization scheme, modeled in the *functorial/relational* terms of *sheaves.* The topological internal gluing relations characterizing sheaves force a consistency in the action of the inverse processes of extending from the local (Boolean) to the global (quantum) level and conversely reducing from the global to the local. Thus, *the real potentialities relative to all standpoints (local Boolean logical frames) are coordinated as diverse determinations of one extensive continuum, being a relational complex united by the sheaf-theoretic part-whole or local-global mereotopological relations*. Most importantly, the sheaf-theoretic extensive scheme provides the *generic morphology of the internal relations which bind the actual occasions into a nexus*. More concretely, the qualification of a uniform fibration expressing the potentialities relative to all Boolean frames, viz., of a presheaf of Boolean frames as a sheaf requires the explicit consideration of covering sieves. The covering sieves are functorial objects acting as the semantic carriers of networks of internal relations among Boolean frames with respect to which the *process-theoretic scheme of extensive connection* in the continuum is consistently realized. Notice that the sheaf-theoretic scheme of extensive connection is based on the *ontological primacy of events* or facts or actual occasions. This is the case because each actual occasion *self-determines its own subjective standpoint* (Boolean logical frame), and simultaneously instantiates a *bundle of potential relations* referring to this standpoint. It is crucial that all these potential relations, viz., all the relations among local observables, are *induced* by the internal relations among their underlying standpoints. Now, the idea of an actual occasion conceived in the sense of an actualized *concrescence*, via the evaluation of some Boolean germ expressing its internal constitution in this process, serves as a *datum* for subsequent becoming actual occasions in the extensive connection of the continuum and so on. Consequently, a global quantum event continuum is *not* conceptualized as a completed a priori existing set-theoretic structure but it is constituted in a *continuous* topological process of extension from the local to the global by *actualization* of new potential facts with respect to local Boolean frames.

The framework of category theory in its function to express the process of extensive connection in the quantum event continuum via the notion of Boolean localization systems, that is in the topological terms of sheaf theory, is indispensable in order to understand the revised relational realist scheme in connection to Whitehead's process theory. The most important aspect of its function in this context is the *conceptualization of the quantum event continuum via an inductive limit categorical construction in a sheaf of germs of local Boolean observable relations*. This is significant because the formulation of the mathematical procedure by which the *approximation to the global quantum level* is consistently achieved depends on the properties of the relation of *extension of Boolean germs* (equivalence classes), and most distinctively, this is implemented by *universal* means. This is the case because the inductive limit is described categorically via a *universal mapping property* and thus functions relationally as a *potentiality for actuality*.

Thus, from a generalized viewpoint, the congruence of the proposed sheaf-theoretic quantum relational realist position with Whitehead's process-theoretic scheme traces back to the semantic convergence between the category-theoretic notion of a *universal* (like the inductive limit) and Whitehead's notion of an *eternal object* or *pure potential for the specific determination of fact*. The theoretical basis of this semantic convergence stems from the fact that both of them act as *agents in objectification* under the operation of the *Category of Transmutation* which, in category-theoretic terms, is subsumed by the action of pairs of *adjoint functors*. In particular, keeping in focus quantum theory, the *mutually implicative dipolar functorial internal relation* between the multiplicity of local Boolean frames and the global quantum is expressed in terms of a pair of adjoint functors between the category of sheaves of Boolean event algebras and the category of quantum event algebras. As we have argued, this categorical adjunction formalizes the process-theoretic operation of the category of transmutation by relating internally and bidirectionally the variable and local Boolean with the quantum global level of structure.

A remarkable application of this scheme is the theoretical explanation of *coarse-graining induced decoherence* and the *emergence* of classical behavior. In this way, coarse-graining induced decoherence is understood and modeled as an *intrinsic* reductive global-to-local process, in the sense that classical behavior emerges intrinsically from the global quantum level via coarsening the scale of resolution of quantum observable behavior by means of macroscopic Boolean observables locally according to the sheaf-theoretic topological compatibility relation.

Notes

1. Elias Zafiris, "Boolean Information Sieves: A Local-to-Global Approach to Quantum Information," *International Journal of General Systems* 39, no. 8 (2010).

2. Alfred North Whitehead, *Process and Reality: An Essay in Cosmology*, Corrected Edition, ed. D. Griffin and D. Sherburne (New York: Free Press, 1978), 289.

3. Roland Omnès, *The Interpretation of Quantum Mechanics* (Princeton: Princeton University Press, 1994). See also: P. C. E. Stamp, "The Decoherence Puzzle," *Studies in History and Philosophy of Modern Physics* 37 (2006).

4. Elias Zafiris, "Quantum Observables Algebras and Abstract Differential Geometry: The Topos-Theoretic Dynamics of Diagrams of Commutative Algebraic Localizations," *International Journal of Theoretical Physics* 46, no. 2 (2007).

5. Omnès, *The Interpretation of Quantum Mechanics*. See also: J. J. Halliwell, "A review of the decoherent histories approach to quantum mechanics." In "Fundamental Problems in Quantum Theory - A Conference Held in Honor of Professor John A. Wheeler," *Annals of the New York Academy of Sciences*, Vol. 755 (1995): 726-40.

6. P. T. Johnstone, *Stone Spaces* (Cambridge: Cambridge University Press, 1986).

7. Johnstone, *Stone Spaces*.

8. S. Bertaina, S. Gamarelli, T. Mitra, B. Tsukerblat, A. Muller, and B. Barbara, "Quantum oscillations in a molecular magnet," *Nature* 453 (May 2008): 203-6.

9. Johnstone, *Stone Spaces*.

CHAPTER 11

Recapitulation: A Semantic Bridge Between Process Metaphysics and Quantum Theory via Sheaves

11.1 General Conceptual Dimension

A fundamental methodological concept of science is grounded on the idea that the advancement of our knowledge about the functioning of systemic wholes has to proceed by *reductive compartmentalization* or taxonomy. The classical mathematical implementation of this concept was based on the assumption that the understanding of a global system's behavior is achieved through the description of interactions of independently existing constituent objects, supplemented by the belief that this description is in accordance with the empirical facts. The extrapolation of the classical mathematical approach in the regime of complex and quantum systems was based historically on the further assumption that phenomena may be understood in terms of some individual elementary constituent objects localized in metrical space-time and their complex dynamical interactions. This classical mathematical approach was seriously challenged by the inadequacy to explain collective *emerging phenomena* within a single-layered metrical space-time background from the dynamical properties of the individual constituents. In this way, there arose the need for *multi-level* mathematical descriptions that would account for emergent phenomena. Mathematical tools employed for this purpose included the formalisms of quantum statistical physics and quantum field theory. For example, the methods of statistical physics account for the emergence of thermodynamic properties at the macroscopic scale from the non-trivial behavior of collections at the microscopic scale. On the other side, there is still missing a *unifying mathematical framework* that would provide a satisfactory logical explanation of multi-level emergent phenomena in the description of quantum systems going beyond or elaborating various assumptions, like the macroscopic scale statistical state formulations in terms of continuous probability distributions taken over the states of collections of individuals at some microscopic scale.

Closely related with the problem of modeling emergent phenomena in quantum systems' behavior are the problems of *structural organization* and *cohesion* of individuals into a variety of possible forms. These problems impose serious challenges for mathematical modeling that require revised conceptual, methodological and technical tools. On the conceptual and methodological state of affairs, the consideration of systemic wholes as aggregations of simple interacting individuals seems to be *simplistic* since it is based on a set-theoretic mode of thinking, which *ignores* the fundamental algebraic structural properties of collections. Put differently, the problem of decomposition of a quantum whole into subsystems should conform to *algebraic decomposition* methods. The relevance of decomposition methods referring to suitable algebraic structures of quantum observables may be made explicit if we think of them as *solutions* to various physical measurement problems. This is significant because it targets the possible *structural formation* of observable attributes of systems at *various scales* admitting a well-defined operational implementation in terms of corresponding measurement procedures. In this sense, the technical tools of decomposition of algebraic structures of observables into *prime* constituents or into *indecomposable* primary structures, as well as the techniques of *algebraic analysis*, should be employed for tackling the decomposition problem. Together with the relevance of algebraic decomposition methods comes the significance of *topological* modeling tools.[1] Topological ideas target the *localization properties of observables* of quantum systems at various measurement scales obtained, for instance, via *coarse-graining* observation processes from some microscopic level to various macroscopic levels. Moreover, topological ideas pertain to the problem of *organization* and *cohesion* of localized observable properties into global descriptions, to the problem of *form-invariance* under deformations, as well as to the problem of *structural order*. The topological modeling tools are locally, in a well-defined mathematical sense, *dual* to the corresponding algebraic ones. This is the main conclusion reached by the *deep topological representation theorems* of algebraic structures—for example, Stone's representation theorem for Boolean algebras and Gelfand's representation theorem for commutative algebras of observables.[2] The essence of these theorems is the *concrete local topological manifestation of some abstract algebraic structure of observables as an algebraic form-invariant structure of continuous functions defined over a topological space*. Most importantly, this topological space is neither postulated ad hoc, nor constructed by imposing metrical relations, but is defined by *prime or indecomposable constituents of the algebraic structure itself*, at least locally, forming in this sense its observable *spectrum*. Within this concrete topological framework of representation dynamical and invariant properties of quantum systems may be expressed by means of the methods of algebraic analysis and algebraic topology.

The extension of this topological representation philosophy to larger or more complicated classes of algebras of observables, implementing increased complexity levels, requires even more precise technical tools, like the ones provided by *sheaf theory*.[3] The concept of sheaf expresses essentially *gluing conditions*, namely, the way by which local data can be collated compatibly into global ones. Using the theory of sheaves we may *approximate* global observables of a quantum system by employing an appropriate multitude of locally compatible covers, referring to sufficiently understood local Boolean subsystems at overlapping measurement scales, reproducing its local-to-global topological behavior by consistent continuous local functional information amalgamation.

The unifying thread of these dual algebraic-topological modeling tools is provided by *category theory.*[4] The conceptual and technical cornerstone of category theory referring to systems is that the behavior of a system is solely determined by the *network of structural algebraic-topological relations* it bears with all other systems of the same categorical species. It is important to note that this does not exclude the inter-level relational determination (besides the intra-level one) by systems belonging to other categorical species of structure under the condition that there exists a *bidirectional functorial correlation* between them. This is the main methodological and semantical lesson provided by category theory, which is formulated in the syntactic language of *adjunctions*, formed by pairs of generalized *inverses*, called technically *adjoint functors.*[5] The crucial notion of adjunctions, which does not have a set-theoretic counterpart, provides the mathematical means of *relating relations*, and thus establishing partial or local structural *congruences* between categories of quantum information structures and categories of sufficiently understood classical Boolean information structures. Remarkably, the topological techniques of sheaf theory may be formulated in purely categorical relational terms providing a promising modeling approach for the formulation of a general theory of emergent behavior in quantum systems.

11.2 Subtleties of Quantum Theory

The idea of a reality admitting objective existence is the starting point of our scientific investigations and the conceptual basis pertaining to the realistic interpretation of our theories of physics. In this sense, the proposed framework of categorical sheaf theory in relation to the interpretation of quantum physics should be considered *not only* as a potent technical modeling tool of a purely mathematical nature but as a new conceptual development capable of initiating a *radical revision of the norms of classical realism.*

The classical view of reality presupposes that objects of our observation are simply individual entities in the world. Moreover an entity may be specified, both qualitatively and quantitatively in a definite manner, independently of procedures of observation, being capable of assuming individuality in isolation. According to the interpretative standards of classical realism, an objective description is one that determines the properties possessed by an independently real physical object, standing for an entity, by adopting a representation of that object as a physical system isolated from any observational interaction. Quantum theory *does not conform to the descriptive ideals of classical realism.* In sharp contrast, the establishment of quantum theory as the best fundamental theory of the material world has proved that the classical dogma is inadequate, since it predicts that individual objects cannot have existence in an absolute sense, or else do not exist in perfect isolation.

The non-separability of nature is described by certain correlations (Einstein-Podolsky-Rosen correlations), which exist even in absence of any interactions, making all subsystems of the universe inextricably entangled.[6] The phenomenon of *quantum entanglement*, which has been experimentally confirmed, precludes in a novel way the possibility of defining individual entities independently of the *local logical frames* (Boolean frames) under which their behavior is manifested. In this way, quantum theory incorporates as one of its basic features the idea that the *whole is non-trivially more than the sum of its parts.* In a well-defined sense, entanglement may be conceived as an *inseparable correlation of potentialities*, whose content is not exhausted by an aggregation of actual pre-existing values that may be assigned to the properties of parts considered disjointly and separately. *Thus, in considering any entangled compound quantum system the nature and properties of parts may only be determined from their specific localizing role within the whole, while the whole can only be inductively inferred via the interdependent amalgamated behavior of these relational parts.*

It is an astonishing realization that the essential aspect of quantum entanglement, abstracted in appropriate terms, lies at the heart of both *Whitehead's process theory*,[7] and *categorical sheaf theory*,[8] in relation to the conception and comprehension of the physical quantum event continuum. More concretely, this essential aspect of entanglement phenomena is besides the explication of a situation where the behavior of the whole is not reduced to the behavior of its parts, or else that the whole is more than the sum of its parts, that the parts do not assume an individuation or localization independently of the whole. Put differently, there exists a *mutually implicative bidirectional relation between the parts and the whole*, where the correlation of *potentialities* plays a crucial role. In this state of affairs, the *ontological primacy of objective events* or *facts* proves to be crucial not only to the relational realist interpretation of quantum theory proposed in this volume, but to *any* viable ontological interpretation.

In the framework of quantum theory *events* are identified as measurement outcomes referring to measurement procedures of corresponding *observables*.[9] The theory then provides the means of correlating these events. Observables are generally associated with physical quantities that, in principle, can be measured.[10] In this way, physical systems are completely described by the collection of all observed data determined by devices in corresponding measurement situations. The complexity of quantum theory in this respect stems from two factors:

First, the existence of an event referring to a quantum system can be inferred only *probabilistically* with respect to the context of measurement of a selected observable, and moreover, can be affirmed only after a measured result has been registered by a corresponding measuring device.

Second, if we agree that an event corresponds to an evaluation of an observable with respect to a measuring device as above, then all events related to the behavior of a quantum system *cannot be actualized within the same measurement context* due to the property of *non-commutativity* of quantum observables.

In more detail, the major foundational difference between classical and quantum physical systems from the perspective of the modeling scheme by observables is a consequence of a single principle, which may be called the principle of *potential simultaneous observability*. According to this, in the classical description of physical systems all their observables are theoretically compatible, or else they can potentially be simultaneously evaluated within a single measurement context. On the other side, the quantum description of physical systems is based on the assertion of incompatibility of all their theoretical observables within a single measurement context, and as a consequence quantum-theoretically the simultaneous specification of all observables is not possible. The conceptual roots of the violation of the principle of potential simultaneous observability in the quantum regime is associated with Heisenberg's uncertainty principle and Bohr's principle of complementarity of physical descriptions.

The first complexity factor bears on the prominent role that *potential relations* play in the process of determination of a quantum event, pertaining to both their qualification as probabilistic with respect to the context of measurement of a selected observable, and to their significance for correlating events, and thus forming histories. The second complexity factor bears on the *contextual* significance of preparation procedures for the evaluation of observables via measurement, and thus for the actualization of events. It is precisely these procedures which can be interpreted in a topological way as a means of localization of quantum observables with respect to *local Boolean* contexts.

These two factors taken together necessitate a thorough rethinking of our conceptual and mathematical representation of the notion of a *physical continuum* suited to the quantum domain of discourse. Most significantly, it turns out that these two factors are particularly important in relation to the foundations of quantum theory, both from the philosophical perspective of Whitehead's process theory, as it has been interpreted and implemented by Epperson, and from the mathematical perspective of sheaf theory, as it has been interpreted and implemented by Zafiris. The conceptual convergence of the underlying theoretical perspectives in relation to the interpretation of quantum theory has created the need for an *integrated framework* combining the relevant aspects of Whitehead's process theory, amenable to a *topological re-interpretation*, together with the relevant aspects of *sheaf-theoretic Boolean localization*, viewed from a process-theoretic standpoint, so as to provide the basis for development of a novel *relational realist ontological interpretation* of quantum theory synthesizing these aspects.

Epigrammatically, the major confluence of the advocated schemes—that is, the process-theoretic and the sheaf-theoretic—in relation to quantum theory, stems from the conception of a physical continuum of events being in a process of *continuous ontological formation via extensive connection*. This is made possible by considering explicitly the potential level, manifested in terms of predicative potential facts, or observables correspondingly, and making selections (for example, predication via measurement of an observable). The crucial requirement of sheaf theory in this conception of the quantum event continuum, being in agreement with Whiteheadian process theory, is that selection is local, conceived in a precise topological sense. With respect to quantum theory, this means that the selection of an observable to be measured (by the preparation procedure) locally fixes a context, which acts as a *Boolean logical frame* locally (that is, covering the continuum locally) for the individuation of events. In order to understand the significance of local Boolean logical frames for the individuation of events we have to explain in detail the notion of localization in the quantum domain.

11.3 Localization in the Quantum Domain

Procedures of physical measurement presuppose the existence of localization processes for extracting information related with the local behavior of physical systems. A localization process is implemented by the preparation of *local reference contexts* for measurement of observables. It is important to realize that the effectuation of a localization process is *not always* equivalent to conferring a numerical identity to observables, expressed in terms of some value corresponding to a physical attribute. On the contrary, the latter is only a *limited case* under the assumption that all local reference contexts can be *contracted* to points. In this limited case, points constitute *unique measures* of localization in the physical continuum. This is the crucial assumption underlying the employment of the set-theoretic structure of the real line as a model of the physical event continuum, understood as a set of independent points possessing the property of infinite distinguishability with absolute precision. The success of this localizing philosophy in classical physics is due to the association of the notion of physical continuum with the attribute of position and the fact that all classical observables can be potentially determined precisely and simultaneously at the unique measure of localization of this attribute—namely, at a spatial point parameterized by the field of real numbers. Nevertheless, the major foundational difference of quantum systems from classical ones is a consequence of the *incompatibility of precise determination* of all quantum observables within a *single* local measurement context, meaning that the simultaneous specification of all observables is impossible in a single measurement procedure. Thus, a foundational strategy implied by quantum theory should fulfill the following objectives:

First, it should disassociate the physical meaning of localization from its restricted spatial connotation context.

Second, it should enable the functional dependence of observables on generalized localization measures, not necessarily based on the existence of an underlying structure of points on the real line.

For implementing this strategy, it should be essential to interpret any local observable as a relational algebraic partial information carrier with respect to the corresponding local context of measurement. Next, it should be necessary to establish appropriate compatibility conditions for gluing the information content of local observables globally. Mathematically, the implementation of this localization strategy is precisely captured by the concept of a *categorical sheaf-theoretic localization structure referring to observables*.

The foundational nature of this research strategy lies on the following semantic elements:

[i] *It proposes a shift from point-set to topological localization models of quantum observable structures, which induce a semantic transition of events from a set-theoretic to a sheaf-theoretic level,* and

[ii] *As a consequence of (i), it proposes the replacement of the classical metrical gauge of point localization on a background manifold, with a sheaf gauge of Boolean topological localization on a categorical site capturing the relational information of observables in the quantum regime.*

In this way, the notion of local Boolean frames and their potential interrelations associated with local measurement contexts of observables quantum theoretically acquire a fundamental *topological significance* for the enunciation of the process of formation of the quantum event continuum. This is the case because the current set-theoretic axiomatizations of quantum event structures, or quantum logics,[11] underestimate or even suppress the functional role of *interconnected families of local Boolean frames* in the formation of these structures. This is entirely unsatisfactory because all operational procedures employed in quantum measurement are based on the preparation of appropriate Boolean environments. The implementation of these local physical contexts of measurement and their correspondence with local Boolean frames is tied with the particular formalization of the notion of an observable in quantum theory—namely, a self-adjoint operator.

More precisely, *each self-adjoint operator representing an observable has associated with it a Boolean frame which is identified with the Boolean algebra of projection operators belonging to its spectral decomposition.* In this way, Boolean algebras of projection operators can be metaphorically considered as local Boolean logical pattern recognition frames for the individuation of events. It is clear that there exists a variability of potential local Boolean frames employed to probe the quantum domain, which corresponds to a variability of the potential localization gauges of quantum events. Thus, the compatible partial amalgamation of observable information via an appropriate methodology of gluing local Boolean frames is of paramount importance for understanding the process of extensive connection in the quantum event continuum. This is precisely the functional role that categorical sheaf theory has to play for elucidating the completion of this process as we have already explained.

At this point, it is essential to clarify why a Boolean logical frame bears the topological semantics of a *local probe* or a *local cover* in a global non-Boolean quantum event structure. According to the Kochen-Specker theorem in quantum logic,[12] there does *not* exist a *global two-valued truth-functional assignment* pertaining to the global event structure of a quantum system. Notwithstanding this fact, there always exists a *local two-valued truth-functional assignment* with respect to a *complete Boolean algebra of projection operators* on the Hilbert space of a quantum system. This is identified with a complete Boolean subalgebra of the global non-Boolean event algebra of a quantum system. More precisely, each *self-adjoint operator* representing an observable has associated with it a *Boolean subalgebra* which is identified with the Boolean algebra of projection operators belonging to its *spectral decomposition*. Hence, given a set of observables of a quantum system, there always exists a complete Boolean algebra of projection operators, viz., a local Boolean subalgebra of the global non-Boolean event algebra of a quantum system with respect to which a local two-valued truth-functional assignment is meaningful. This is actually the case if and only if the given observables can be *simultaneously measurable*. Consequently, the possibility of local two-valued truth-functional assignments of the global non-Boolean event algebra of a quantum system points to the conclusion that *complete Boolean algebras of projections play the role of local Boolean logical frames for contextual true/false value assignments*.

At a logical level of interpretation, any meaningful proposition that an observer can actually state in the context of a local Boolean logical frame referring to the measurement of a quantum observable is being usually built, in the simplest case, in terms of some *binary schematism* objectified in terms of a corresponding *yes/no experiment.*[13] Through the functioning of such a binary schematism subsumed by a corresponding measuring device in the context of a local Boolean frame of measurement, it is left to the observed system the choice between two possibilities. In this simple case, the complexity of an observed quantum system is accepted insofar as the observer does not know which of the two possibilities the latter will choose, since before a particular measurement is performed the observer can infer only probabilities. In this sense, an observer can schematize the use of the indicated system's complexity in the corresponding Boolean logical frame as a *true/false proposition.* The schematism itself forces the observer to admit the contingency of the behavior, and thus the autonomy of the measured system in the context of the respective Boolean frame. Consequently, *all potential local Boolean logical frames play the role of selective local constraints on relational possibilities and constitute the precondition for the observation and description of a quantum system at the global level.* The locally Boolean logical structure associated to each local Boolean context of measurement may thus be effectively understood as a system of logical rules which *conditions* and *forbids* the constitution of contradictions within the boundaries of the corresponding frame covering locally the quantum domain. Nevertheless, from the global viewpoint of the totality of all different potential Boolean covering local frames, they function as a semantic tool for *combining features* under the condition of *partial compatibility* of observable information within the limits determined by the imposition of the uncertainty relations.

11.4 Relational Localization via Sheaves

Initially, it is important to indicate that concerning the theoresis of the physical continuum, both of our fundamental theories—namely, *general relativity* and *quantum mechanics*—can be characterized in general terms as special instances in *the process of replacement of the constant or rigid by the variable or elastic.*

More concretely, in the case of general relativity this process takes place through the *rejection of the fixed kinematical structure of spacetime*, by transforming the *metric* into a *dynamical* object determined solely by the solution of Einstein's field equations. This essentially means that the geometrical relations defined on a four-dimensional manifold, making it into a spacetime, become variable. The dynamically induced *infinitesimal variability* of the metric constitutes the means of making geometrical relations variable in the descriptive terms of general relativity. The intelligibility of this framework is enriched by the imposition of the principle of *general covariance* of the field equations under arbitrary smooth coordinate transformations of the points of the spacetime manifold. As an immediate consequence, the points of the manifold are *not* dynamically localizable entities in the theory. Thus, it is *illegitimate* to use the *set-theoretic point structure* of the manifold for individuating directly spacetime events, independently of the localization properties of the metric determined dynamically. In this way, it can be asserted that manifold points bear only an *indirect reference* as indicators of spacetime events, and most importantly, only after the dynamical specification of variable geometrical relations among them, obtained as particular *solutions* of the generally covariant field equations in terms of the metric. We conclude that *the conceptualization of the physical event continuum in general relativity requires explicit reference to the relational localization properties of the metric, which is determined dynamically in the context of this theory.*

In the case of quantum theory, the process of replacement of the constant by the variable is signified by the imposition of *Heisenberg's uncertainty relations*, which determine the limits for simultaneous measurements of certain pairs of *conjugate physical observables*, like position and momentum. In a well-defined sense, the uncertainty relations may be interpreted as measures of the *valuation vagueness* associated with the simultaneous determination of conjugate pairs of observables within the same experimental context. This is in *contradistinction* with the case of all classical theories, where the valuation algebra is *fixed* once and for all to be the *field of real numbers*, reflecting the fact that values admissible as measured results must be real numbers, *irrespectively* of the measurement context and simultaneously for all physical observables. The above problem of observable valuation vagueness in quantum theory can be *topologically resolved* via a relational process of topological localization of the global non-commutative algebra of quantum observables with respect to appropriately interlocking families of local commutative algebras of observables. Each *maximal local commutative algebra* contains all observables that can be *simultaneously* measured within the same experimental context. The set of *central idempotent elements* of such a local commutative algebra of observables, that is, the *projections*, form a *Boolean algebra*, which is locally identified as a *Boolean logical frame* of a global quantum event structure corresponding to the experimental context of measurement of these observables. Consequently, *any quantum event that happens is raised at the empirical level only in conjunction with the specification of an experimental context conforming to a Boolean domain of discourse, and corresponding to a local logical Boolean frame incorporating all observables which can be measured simultaneously in that context.* Thus, the quantum event continuum cannot be conceived independently of the localization properties of observables specified by the internal topological relations of variable local Boolean frames probing or covering the continuum.

It is instructive to indicate that the process of replacement of the constant or rigid by the variable or elastic is *deeper* and more *radical* in quantum theory in comparison to general relativity. This is the case because the relational variability induced in the process of individuation of events at the quantum domain refers to *topological localization* of observables via locally variable Boolean logical frames and *not* merely to *metrical localization* properties via infinitesimally variable geometric relations over the points of a spacetime manifold.

Put differently and more philosophically, concerning the *connectivity of the event continuum*, quantum theory touches upon the *logical order* of nature as this is induced by relational topological localization properties generated by interrelated variable Boolean frames, and thus *descends deeper* than the *causal order* of nature as this is induced by metrical localization properties generated by variable coordinate frames. The failure of *distinction* between the logical and causal order is a source of paradoxes and misconceptions like the ones associated with attempts to provide metrical explanations of entanglement correlations in the quantum domain. Whitehead's process theory incorporates an explicit distinction between the logical and causal orders of nature in relation to the elaboration of *extensive connection* in the conception of the continuum.

Parenthetically, it is illuminating to remark that the process of extensive connection should be of a *dynamical origin* and thus not dependent on the particular localization schemes used for probing the continuum. This is particularly relevant in the attempts to construct a background independent viable theory of *quantum gravity*. Essentially, the basic idea is that the usual analytic differential geometric framework of smooth manifolds employed for the formulation of general relativity provides a particular metric localization of the *universal algebraic homological process* of extensive connection, and thus it should be substituted appropriately in case a merging with quantum theory is being sought. This substitution is effectuated by *rejecting the fixed background manifold substratum* of the classical theory, while maintaining the *algebraic skeleton of extensive connection*, and subsequently instantiating it in the *topological site* reflecting the *relational localization* properties of observables in the quantum regime according to our previous analysis.[14]

From a mathematical point of view, the general process of semantic transition from the constant to the variable along the previous lines takes place by passing from *rigid set-theoretic structures* to *continuous sheaf-theoretic structures*. In general, a sheaf may be thought of as a continuously variable set, whose continuous variation is considered over specified local frames interlocking together non-trivially. The local frames are required to obey certain topological closure conditions, which generalize the definition of a topology formulated in terms of open regions in classical topological spaces.

We recall that a *topology* is defined by a collection of *open loci* (to be thought of as probes or covers of a space), which is closed by the formation of arbitrary unions and finite intersections of open loci. The definition of a topology on a space provides the means to talk about the notion of a *continuous function* between topological spaces. A function is said to be continuous if and only if the inverse image of every open locus of the range is an open locus of the domain's topological space. This is an attempt to capture the intuition that there are no breaks or separations in a continuous function. It is important to realize the following:

[i] *The definition of a topology on a space is solely used for the formalization of what a continuous function is on that space.*

[ii] *The continuity of a function is a property which is determined locally—that is, only by reference to open loci.*

This means that due to the variability or elasticity of open loci in the topology the property of continuity of a function should respect the inverse algebraic operations of *restriction* and *extension* (uniquely) with respect to open loci. Thus, a continuous function can be restricted consistently to open subloci of any open locus in the topology and inversely extended by gluing uniquely together all its local restrictions. This is the crucial conceptual insight which is incorporated in the technical definition of a sheaf.

This insight referring to the precise formulation of the property of continuity may be generalized in two directions:

First, instead of open loci or open regions covering a topological space we may consider generalized *local covering frames* (for example, local Boolean logical frames for probing or covering the quantum event continuum) under the constraint that they collectively obey topological *closure* conditions analogous to the ones used for open loci.

Second, instead of functions varying continuously over local frames we may consider *generalized functional relations*, called technically *sections* of a sheaf.

Conceptually speaking, local sections of a sheaf describe potential predicative facts relative to a local frame, which we call *local observables*. A sheaf is essentially the totality of its sections (both local and global if they exist) and their consistent interrelations with respect to the *local-global* distinctions subsumed by the interlocking properties of the underlying frames. In this sense, the operation of restriction of sections acquires a meaning with respect to *nesting* local frames, whereas the operation of extension or gluing of sections acquires a meaning with respect to compatibly *pairwise overlapping* local frames. We note that if only the operation of restriction for nested local frames is satisfied, then we obtain a structure called a *presheaf*. In case that the operation of extension is also satisfied, then we obtain a *separated presheaf*. Only in case that the operation of extension can be uniquely carried out—that is, local sections can be uniquely glued together—does a separated presheaf become a *sheaf*.

We stress that the notion of a sheaf depends on the fact that we require the *gluing condition* with respect to all local covering frames—that is, to all covers of any local frame. In principle, one could select some covers of a local frame and require the gluing condition only with respect to the *selected covers*. In this way, the notion of sheaf would be meant with respect to the selected family of covers. On the other hand, there is no restriction in considering only *hereditary covers*—that is, covers containing all their subcovers.

More precisely, any cover can be made hereditary (by adding to each cover all its considered subcovers) and compatible (uniquely glued) families of sections on the original cover are in bijective correspondence with compatible families of sections on the new one. This fact provides an important insight on the nature of the topological localization process of observables implicated by the concept of sheaf. Firstly, let us think of local frames as *partial information carriers* related to measurement of observables. For example, in the case of quantum theory, local frames are Boolean logical frames as we have explained previously. In their logical Boolean terms, potential observable quantum information is locally or partially reduced to probabilistic information referring to all the possible measurement outcome events.

Now, the idea of considering hereditary covers can be implemented physically via a procedure of *measuring by sieving*. More concretely, a *sieve* on a local frame can be thought of as consisting of *spectrum holes* distributed in different nested layers, such that every partial information carrier, which is in a sense "smaller" than one of its holes, passes through it but not otherwise. In this way, the notion of a sieve incorporates an *asymmetric logical order.* Moreover, since the partial information carriers should bear the semantics of local covers, certain conditions should hold distinguishing some sieves as *covering* ones within the class of all possible sieves. These conditions are the following:

[α] The covering sieves should be *stable under intersection*, meaning that the intersection of covering sieves should also be a covering sieve, for each local frame,

[β] The covering sieves should be *transitive*, such that covering sieves of subframes of a frame in covering sieves of this frame, should also be covering sieves of this frame themselves, and

[γ] The covering sieves should be stable under *change* of a local frame.

From a physical perspective, the consideration of covering sieves as hereditary topological covers used for localizing observable information with respect to local frames gives rise to *localization systems* of global observable structures which induce a semantic transition of events from a set-theoretic to a sheaf-theoretic level. The operation which assigns to each local frame a collection of covering sieves satisfying the stability and transitivity conditions described qualitatively above, defines a topology, which is technically called a *Grothendieck topology*. The notion of a Grothendieck topology formulated in terms of covering sieves is significant for the following reasons:

First, it elucidates precisely the topological concept of locality in relational logical order-theoretic terms, such that this concept becomes distinguished from its usual metrical spatiotemporal connotation.

Second, it permits the amalgamation of local observable information into global ones by utilization of the notion of sheaf with respect to a defined topology.

In more detail, the extension of observable information from the local to the global takes place via a *compatibly glued family of sections over a covering sieve* (*localization system*) constitutive of covering local frames (local covers). A sheaf assigns a set of sections to each local frame of a localization system, representing a set of observables which can be measured with respect to this local frame. A selection of sections from these sets, one for each local frame, forms a compatibly glued family with respect to a localization system, if the selection respects the operation of restriction, and additionally, if the sections selected agree whenever two local frames of the localization system overlap. If such a locally compatible selection of sections *extends uniquely* to a global one, then the sheaf conditions are satisfied. We note that in general, there will be more local sections than global ones (if they exist), since not all local observables need be extendible to global ones, but a compatible family of local observables uniquely extends to a global one with respect to a localization system.

We conclude from the previous analysis that *sheaf-theoretic gluing provides the means of extensive connection in the physical continuum*, that is, extending a local section over a local frame to a higher level local frame and so on towards approximating the global with respect to some covering sieve. In the case of quantum theory, a covering sieve stands for a *Boolean localization system* composed of covering local Boolean logical frames (Boolean covers of the quantum continuum). We think of a *becoming actual occasion* as the process of evaluating a section with respect to a local Boolean frame, where the result of the evaluation is given by the measurement outcome or actualized event. Notice that in quantum theory due to the Kochen-Specker prohibition, sections cannot be evaluated globally, but only locally with respect to a Boolean frame. So how is the process of extensive connection taking place in the quantum continuum from the local to the global incorporating the *contextual information* of actualized events with respect to Boolean frames? The answer provided by the sheaf-theoretic scheme is by the formation of *Boolean germs*. Intuitively, a Boolean germ should be thought of as an appropriate *equivalence class of compatible local sections* contextualizing the information of an actualized event. In order to understand the role of germs (*seeds of contextualized potentiality*) sheaf-theoretically, it is easier to explain their functional role in a spatial topological environment first, where local frames stand for open covers of a topological space; then, building on this intuition, one can analogously explain the functional role of Boolean germs in the quantum domain.

11.5 Germs and Extensive Connection

Let us consider a point of a topological measurement space. Moreover, let us consider a basis of open covers of this topological space containing the point, such that the following condition holds: For any two open covers containing the point, there exists another open cover in this basis contained in their intersection. Now, we proceed as follows:

First, we form the *disjoint union* of all variable sets of sections defined over all these open covers in the basis containing the considered point.

Second, we define an *equivalence relation* in the above disjoint union of sets of local sections, by requiring that two sections are equivalent at the considered point if they have the *same restriction to a smaller open cover* contained in this basis.

Third, we define the *quotient* of the disjoint union of sets of local sections modulo the above defined equivalence relation of local sections at the considered point.

The quotient set obtained is *independent* of the choice of basis of open covers containing the point and is called the *stalk of the sheaf* at the considered point. The elements of the stalk, that is, the equivalence classes of local sections containing the point, are called the *germs of the stalk* at this point.

Notice that there is no definite domain of definition for a germ at a point, since its domain corresponds to an open cover (containing the point) specified by the requirement of identical restriction of local sections, and thus it is *intrinsically variable*. Hence, a germ at a point is only accessed through its *representatives*, that is, by selecting some local section from the corresponding equivalence class of local sections. *The ultimate significance of germs is that they allow us to consider local sections as genuine functions*. More precisely, each local section corresponds to the function which assigns to each point of its domain of definition (open cover) its germ (equivalence class) over that point. Hence, all local sections admit a *functional representation* via their germs. Then, the space of all germs at all points may be endowed with the finest topology (the discrete topology) such that this functional representation of sections is continuous.

The construction of the set of germs at a point, that is, the stalk at a point, is an instance of an *inductive logical* construction. This is because the set of germs at a point constituting the stalk is formed by induction on equivalent local sections at the *limit* of diminishing extent. For this reason, the process of forming the stalk of a sheaf at a point is actually an inductive limit procedure on sets of local sections over a basis of open covers containing that point, which form a *filter* of open covers around the considered point. We may think of this filter of open covers as a *partial order of refinement* among open covers containing the point which is the inverse of the inclusion partial order of these open covers; that is, it approximates closer and closer the point by *converging* to it as we move to higher and higher and thus more refined levels in this order. This constitutes a topological formulation of the process-theoretic notion of *extensive abstraction*.

In the physical state of affairs, we remind that we have identified a local section with a local observable defined over an open cover. Then, the equivalence relation used in the definition of an observable germ at a considered point is interpreted as follows:

Two local observables induce the same *contextualized potential information* at a point if they have the same restriction to a smaller (more refined) open cover contained in a filter of open covers around this point.

Then, the stalk is the set incorporating all contextual information at this point—that is, the set of all germs at this point. Moreover, the image of a local observable at the stalk over this point is the germ of the local observable at this point.

Every local observable admits a functional representation; namely, it corresponds to the continuous function from the open cover over which it is defined to the space of all observable germs equipped with the discrete topology. Thus, the functional representation of a local observable assigns continuously to each point its germ at this point.

It is clear that the notion of a germ at a point is *not* defined *independently* from its topological domain (open cover), since its very definition makes sense only with respect to such a local frame over which local observables have identical restrictions. Notice also that in the conception of a germ, a local frame is never considered apart from the local information content which is contextualized with respect to it.

We interpret the evaluation of a germ of the stalk at a point as an event (becoming actual occasion) corresponding to a measurement outcome in the context of measurement depicted by all equivalent local observables forming this germ. In the proposed scheme events have ontological primacy and the actualization of any potential fact is an objective augmentation of the already formed global event structure. From the conceptual perspective of *event-based process theory*, each actual occasion self-determines its own *subjective standpoint*—that is, the pertinent local frame together with the local information content contextualized with respect to it—and simultaneously instantiates a bundle of potential contextualized relations referring to this standpoint.

Thus, it is appropriate to think of a germ as representing the *internal constitution* of an actual occasion in its process of *concrescence*. *A germ integrates compatible contextualized information at higher and higher (more refined) local frames in the form of an equivalence class*. The formation of this equivalence class constituting a germ takes place *relative* to the concrete standpoint of an actual occasion. Equivalently put, the standpoint of an actual occasion instantiates the local frame, which is correlated bijectively with the local frame of a germ tracing in this way the *genetic process* of actualization of some potential contextualized fact. Thus a germ provides the internal constitution of a becoming actual occasion, whereas an actual occasion is specified by the evaluation of this germ in its concrescence.

Concomitantly, a sheaf-theoretic explication of Whitehead's dictum that the "many become one and are increased by one" is formulated as follows: The many become one by compatible interconnection or gluing of their contextualized information content relative to the local frame of an actual occasion, and thus by forming a germ, but concurrently they are increased by one because the already integrated totality serves as a new context for the integration of compatible observable information related with a subsequent actual occasion and so on. Thus, an actualized event being contextualized by some germ serves as a datum for subsequent becoming actual occasions, and thus besides its *"subject"* aspect, it incorporates a *"superject"* aspect as well. Now, we understand the continuous extension of a germ from event to event as a *historical process* which is mathematically depicted by the continuous functional representation of a section in sheaf-theoretic terms.

The previous analysis targeting the explanation of the functional role of germs in a topological spatial environment can be analogically transferred to the *quantum domain*. In order to perform this analogy we consider instead of open covers of a topological space playing the role of local frames for contextualizing observable information referring to events, *logical Boolean covers*—that is, local Boolean logical frames covering the quantum event continuum. In the spatial topological case, we observe that the role of points can be completely substituted by filters of open covers around them. In the quantum logical case the role of filters of open covers is subsumed by *covering sieves*—that is, by Boolean localization systems.

Employing a Boolean localization system we can form *Boolean logical germs* which contextualize the information of quantum events and provide their internal logical constitution. In this way, the sheaf-theoretic conceptualization of the quantum event continuum interprets it not as a completed already existing set-theoretic structure but in terms of a *continuous topological process of extensive connection* conducted by actualizations of contextualized potential facts with respect to local Boolean frames in covering sieves, and inversely by contextualization and integration of actualized event information in terms of Boolean germs.

Most significantly, this process of ontological formation is coherent in the sense of respecting the rules of transition from the local to the global and conversely. We stress that the actualization of any potential fact is an *objective augmentation* of an already formed global event structure. The functional dependence implicated by sheaf-theoretic localization *relativizes* the presupposed rigid relations between quantum events with respect to variable local Boolean frames in covering sieves conditioning and contextualizing the actualization of events. In this way, each new actualized event creates a *novelly integrated whole*, and this is due to the fact that its relational integration to the previously established totality is possible only via the gluing and compatibility properties of Boolean germs.

11.6 The Road Ahead: Differential Extensive Connection and Dynamics with Applications to Quantum Gravity

We have emphasized in our previous discussion in section 11.4 that the continuous sheaf-theoretic process of extensive connection should be of a dynamical origin and thus exhibit a functorial behavior with respect to all particular localization schemes used for probing the physical event continuum. More precisely, although a localization scheme is indispensable for a proper formalization of the mechanism of observability suited to the operational means (reference frames) of a physical theory (for instance, quantum observability), the process of extensive connection initiated dynamically by the actualization of events should be universal, and hence covariant with respect to these localization schemes, under the proviso that it is enacted naturally in terms of germs—that is, sheaf-theoretically according to our previous analysis. This is of particular significance in the context of the current research programs targeting the construction of a tenable dynamical theory of quantum gravity, where the basic conceptual and technical issue to be faced refers to the problem of independence from a fixed metrical space-time manifold background.[15] From our perspective, the smooth metrical manifold substratum represents a particular type of a metrical localization scheme, which is determined dynamically by solving Einstein's field equations, and is employed for expressing the connectivity of gravitational interactions by means of the differential geometric framework of smooth four-dimensional semi-Riemannian point manifolds.

Differential geometry, which provides the means of formulating the process of extensive connection in the context of general relativity, is traditionally associated with the notion of smooth manifolds. Smooth manifolds are geometric spaces which are locally homeomorphic to Euclidean spaces of an appropriate dimension. The topological (actually metrical) linear vector space structure of a Euclidean space in the neighborhood of a manifold point is used for transferring to smooth manifolds the methods of calculus, thus generating locally the whole differential geometric framework. Conclusively, this local linear structure provides the seed of generation of spatially determined differential analysis on a smooth manifold, which is interpreted physically as a process of dynamical extensive connection caused by a field.

The localization scheme provided by the notion of a smooth manifold is very restrictive not only for the incorporation of quantum phenomena (where the local Euclidean linear space structure is not tenable) but even for an efficient treatment of spacetime singularities obstructing the validity of the field equations formulated in terms of the corresponding algebra of smooth functions. So the pertinent question is the following: Is the generation of local differential analysis, and thus differential geometry, solely dependent, and thus spatially determined, by the local linear space structure of Euclidean spaces? Or is there a universal algebraic mechanism of generation of differentiability not dependent on this particular metrical Euclidean localization scheme? If such a functorial algebraic mechanism of differential analysis exists, then the whole differential geometric framework of smooth manifolds would constitute a particular manifestation shaped by the modeling Euclidean linear space localization scheme. Again from our perspective, the prospect of dependence of the process of continuous extensive connection on some local linear Euclidean spatial structure seems entirely unnatural, since such a structure has not been used at all, either for the development of the notion of a localization scheme, or for the explication of the connective process in terms of germs.

A major step towards providing a satisfactory answer to the pertinent question posed above comes from commutative algebra and algebraic topology.[16] Kaehler's purely algebraic theory of generation of differential forms does not involve any assumptions involving Euclidean spaces and is solely based on the properties of the inverse processes of extension and restriction of commutative algebras. De Rham's algebraic-topological theory of cohomology is based on the existence of algebraic modules of exterior differential n-forms (admitting an appropriate characterization as closed or exact by intrinsic means), and the formation of complexes by means of differential operators acting on these modules. The Kaehler-de Rham homological mechanism of algebraic differential analysis is not dependent on any local Euclidean vector space structure for its function, and thus provides an excellent candidate for the generalization of the differential geometric framework of smooth manifolds to more general localization schemes specified by dynamical means (in analogy to the case of general relativity).[17]

In order to be applicable for modeling the process of extensive connection dynamically (that is via the action of a physical field) in the observability scale of a physical theory (defined by the corresponding reference frames) it should be susceptible to sheaf-theoretic localization. It is instructive to think of the method of sheaf-theoretic localization of the homological Kaehler-de Rham algebraic differential mechanism as the dynamically induced germification of the process of extensive connection initiated in the physical continuum after the actualization of every event. The fact that the localized (or sheafified) Kaehler-de Rham universal differential mechanism is functorial with respect to any localization scheme (modulo the satisfaction of the appropriate cohomological conditions) is a strong indication that it should be carefully considered for instantiating the differential geometric dynamics of a tenable theory of quantum gravity, independently of any smoothness assumption—that is, independently of any locally Euclidean smooth manifold background. Clearly, in this case the Kaehler-de Rham differential mechanism should be functional on the Grothendieck site reflecting the relational localization properties of observable events in the quantum regime under the action of gravitational interactions. It is only under this dynamically induced localization scheme that the Kaehler-de Rham differential mechanism becomes covariant with the fact that quantum events are raised at the empirical level only in conjunction with the specification of a local logical frame conforming to a Boolean domain of discourse (Boolean information sieves[18]) corresponding to a local commutative algebra of all observables which can be measured simultaneously with respect to this logical frame.

The fact that the homological Kaehler-de Rham algebraic differential mechanism actually admits a sheaf-theoretic localization with respect to appropriate localization schemes is not a trivial issue. What it shows is that the concepts of differential geometry which remain logically invariant under abstraction from the smooth manifold scaffolding are the ones which can be localized by sheaf-theoretic means. Accordingly, the sheaf gauge of localization of observables shows that the intrinsic mechanism of differential observability induced by the action of a physical field is of a sheaf cohomological nature. Cohomology can be conceived as an effective method of assigning global invariants to a localization scheme (for example to a topological space, or more generally to a categorical site) in a homotopy-invariant way. The cohomology groups measure the global obstructions for extending sections from the local to the global level. From a physical perspective, these obstructions have global observable effects and may be associated with field sources preventing the uniform and maximally undisturbed dynamical process of extensive connection. It is precisely these obstructions detected by cohomology groups, which make sheaf cohomology the proper measurement tool of global observability.

Regarding the non-triviality of sheaf-theoretic localization of the Kaehler-de Rham algebraic differential mechanism, it is instructive to clarify the following: Since it is possible to interpret standard algebraic notions such as homomorphism, kernel, image, subobject and quotient object for sheaves, in such a way that these concepts have essentially the same meaning as in algebra, one can consider them from a categorical standpoint and apply all the constructions of homological algebra to sheaf theory. The resulting category of sheaves thus has the same classical properties as the category of Abelian groups or the category of modules; in particular, one can define for sheaves direct sums, direct products, tensor products, inductive limits, and other concepts.

In this way, the apparatus of sheaf theory has penetrated into various fields of mathematics providing an effective algebraic tool, especially in those areas which ask for global solutions to problems whose hypotheses are local. This is due to the fact that there is a natural definition of the cohomology of a site (for example, a topological space) with coefficients in a sheaf. In particular, the development of algebraic geometry by Grothendieck led to the crystallization of the idea that the natural argument of a cohomology theory is a pair consisting of a topological space (or more generally a site) together with a sheaf of commutative algebras defined over it, rather than just a space. This realization has been non-trivially and meticulously transferred to the mathematical field of complex analysis, and more recently, to the field of differential geometry by the development of the geometric theory of vector sheaves equipped with a connection. In this way, the differential-analytic tools of

Euclidean spaces and smooth manifolds have been substituted, abstracted, and generalized by sheaf cohomology.

The interpretation of the Kaehler-de Rham differential mechanism on sheaf-theoretic localization schemes defined over an abstract topological space has been accomplished by Anastasios Mallios, who developed the theory of Abstract (Modern) Differential Geometry (ADG)[19] and reformulated it in terms of the Gauge Theories (Maxwell and Yang-Mills) of physics.[20] This theory is conceptualized as a framework of relational differential calculus[21] which refers directly to the geometric objects (sheaves) without the intervention of any locally Euclidean space. ADG generalizes the differential geometric mechanism of smooth manifolds by proving explicitly that most of the usual differential geometric constructions on manifolds can be carried out by purely sheaf-theoretic means without any use of any sort of smoothness or any of the conventional analytic methods of spatial differential calculus associated with it.[22]

The technique used for the generalization of the analytic methods of Classical Differential Geometry (CDG) consists of the following: Initially, a concept of CDG is suitable for extension to a broader differential context (beyond the smooth context of manifolds) if it is liable to a process of localization. Then, this concept is expressed by means of a natural transformation of sheaves. The prominent role is played by the sheaf of observable algebras, which is required to be commutative. This sheaf plays the role of a functional arithmetic, meaning that all geometric objects involved in the formalism are locally expressed in terms of its sections. Thus, all vector sheaves (of states), group sheaves of automorphisms (symmetries) and sheaves of differential forms partaking in ADG are typically required to be locally free modules and of finite rank over the sheaf of observable algebras.

The concept of infinitesimal differentiability of the sections of vector sheaves (covariant derivative of sheaf sections) is formulated in terms of connections. A connection on a vector sheaf bears the dynamical semantics of a potential causing the differentiability of its sections (states). In this way, a vector sheaf equipped with a connection may be thought of as a generalization of the notion of a differential equation. The integrability properties of a connection under extension of the differential mechanism from the local to the global level are considered from the prism of sheaf cohomology and not by the typical analytic methods. The curvature of a connection is measured by the strength of the associated potential covariantly with respect to the corresponding sheaf of observable algebras.

One of the most important didactics crystallized by ADG is a new semantic content attributed to the notion of a geometric space.[23] A geometric space is merely a localization basis for the formulation of the differential mechanism, whereas the mechanism itself is not subordinate to this basis, meaning that the process of differential extensive connection is not organically dependent on any particular localization basis chosen experimentally, although it should respect it. Thus, the object of primary significance in ADG is not the space itself but the sheaf of observable algebras localized over it. Put differently, the process of differential extensive connection is enacted and completely expressed locally in terms of the sections of the sheaf of observables, whereas the base space itself is only the carrier of the sections defined over it, or else it provides only the means of localization of these sections.

Thus, it is precisely the covariance of the differential extensive process with respect to the employed sheaf of observables which permits the detection of dynamical interactions caused by some physical field.[24] This is a crucial insight and provides a justification of the gauge principle used successfully in theoretical physics. Moreover, it suggests an efficient method for overcoming the problem of singularities which might be associated with a geometric space relative to an insufficient choice of some sheaf of algebras (for example smooth functions) without breaking down the validity of the dynamical law. This method is based on the substitution of the pathological sheaf by a new one being able to engulf algebraically (in the form of ideals) the former singularities while leaving the localization basis unreduced. The fact that the differential mechanism of ADG remains invariant and effective under the substitution of the sheaf of smooth functions (which is used to model a smooth manifold) by singular sheaves of generalized functions,[25] including distributions (defined by Rosinger[26]), proves the validity and power of this method.

Finally, contemplating further the nature of the process of differential extensive connection, and using the conceptual paradigm of ADG, stressing its universality with respect to any sheaf-theoretic localization scheme, we reach the following conclusion: The dynamical underpinning of this connection process is always due to the existence of an adjoint pair of functors, expressing the inverse algebraic procedures of infinitesimally extending and restricting the observables compatibly, induced by the actualization of every event due to the presence of a field, so that all the events form a connected nexus. In particular, the notion of a gravitational connection is the semantic information carrier of the connectivity caused by gravitational interactions. Moreover, the development of the gravitational dynamical connection process involves the satisfaction of appropriate global requirements referring to the transition from the infinitesimal to the global level. These requirements are of a cohomological nature and characterize the integrability property of the infinitesimal dynamical variation process induced by a gravitational connection. More concretely, they are properly addressed by the construction of the De Rham complex associated to an integrable connection. Now, the non-integrability of a connection is characterized by the notion of curvature bearing the semantics of observable disturbances to the process of cohomologically unobstructed variation induced by the corresponding connection.[27]

The appealing aspect of modeling the process of differential extensive connection along these lines is that the induced differential geometric mechanism is universal and thus functorially expressed with respect to both [a] sheaves of observable algebras used as functional arithmetics (subject to the appropriate global cohomological requirements for the formation of generalized de Rham complexes); and [b] the localization schemes used as a means of empirical observability and logical classification (subject to the requirement that they are compatible with the formation of germs). Therefore, referring in particular to the connection induced by gravitational interactions, it can be transferred functorially to the quantum regime by utilization of the notion of a sheaf of commutative observable algebras with respect to the Grothendieck topology of epimorphic families[28] (based on logical classification with respect to Boolean germs according to the central thesis of our interpretation scheme).

In more intuitive terms, a global non-commutative algebra of quantum observables can be localized sheaf-theoretically with respect to its commutative subalgebras. Thus, the derived localization scheme, defined by the category of sheaves of commutative subalgebras of a quantum observable algebra, constitutes a legitimate universe of discourse for the formulation of a theory of quantum gravity, obtained by transferring functorially the classical differential geometric framework of modeling gravitational interactions in the quantum regime. In this way, given that the proposed quantum localization schemes are illuminating for a proper formalization and understanding of quantum measurement and observability, the dynamical process of gravitational extensive connection is functorial with respect to these localization schemes, since it is enacted naturally in terms of germs—i.e., sheaf-theoretically according to the conceptual framework of quantum relational realism.

Notes

1. H. Weyl, "Part I. Topology and Abstract Algebra as Two Roads of Mathematical Comprehension," *The American Mathematical Monthly* 102, no. 5 (1995).

2. P. T. Johnstone, *Stone Spaces* (Cambridge: Cambridge University Press, 1986).

3. A. Mallios, *Geometry of Vector Sheaves: An Axiomatic Approach to Differential Geometry,* Vol. 1 (Dordrecht: Kluwer Academic Publishers, 1998). See also: A. Mallios, *Geometry of Vector Sheaves: An Axiomatic Approach to Differential Geometry*, Vol. 2 (Dordrecht: Kluwer Academic Publishers, 1998) and S. Mac Lane and I. Moerdijk, *Sheaves in Geometry and Logic* (New York: Springer-Verlag, 1992).

4. S. Awodey, *Category Theory*, Oxford Logic Guides (Oxford: Oxford University Press, 2006).

5. D. Ellerman, "A Theory of Adjoint Functors with Some Thoughts on Their Philosophical Significance," in *What Is Category Theory?* ed. G. Sica (Milan: Polimetrica, 2006). See also: Elias Zafiris, "Rosen's Modeling Relations via Categorical Adjunctions," *International Journal of General Systems* 41, no. 5 (2012).

6. Roland Omnès, *The Interpretation of Quantum Mechanics* (Princeton: Princeton University Press, 1994).

7. Alfred North Whitehead, *Process and Reality: An Essay in Cosmology*, Corrected Edition, ed. D. Griffin and D. Sherburne (New York: Free Press, 1978).

8. S. Mac Lane and I. Moerdijk, *Sheaves in Geometry and Logic* (New York: Springer-Verlag, 1992).

9. V. S. Varadarajan, *Geometry of Quantum Theory* (New York: Springer, 1985).

10. Elias Zafiris, "Interpreting Observables in a Quantum World from the Categorial Standpoint," *International Journal of Theoretical Physics* 43, no. 1 (2004).

11. G. Birkhoff and J. von Neumann, "The logic of quantum mechanics," *Annals of Mathematics* 37 (1936).

12. S. Kochen and E. Specker, "The problem of hidden variables in quantum mechanics," *Journal of Applied Mathematics and Mech*anics 17 (1967).

13. V. S. Varadarajan, *Geometry of Quantum Theory* (New York: Springer, 1985). See also: Omnès, *The Interpretation of Quantum Mechanics.*

14. A. Mallios and E. Zafiris, "The Homological Kaehler-De Rham Differential Mechanism part I: Application in General Theory of Relativity," *Advances in Mathematical Physics*, doi:10.1155/2011/191083 (2011); and A. Mallios and E. Zafiris, "The Homological Kaehler-De Rham Differential Mechanism part II: Sheaf-Theoretic Localization of Quantum Dynamics," *Advances in Mathematical Physics*, doi:10.1155/2011/189801 (2011). See also: E. Zafiris, "Quantum Observables Algebras and Abstract Differential Geometry: The Topos-Theoretic Dynamics of Diagrams of Commutative Algebraic Localizations," *International Journal of Theoretical Physics* 46, no. 2 (2007); A. Mallios, "Geometry and Physics Today," *International Journal of Theoretical Physics* 45, no. 8 (2006): 1557; A. Mallios, "A-Invariance: An Axiomatic Approach to

Quantum Relativity," *International Journal of Theoretical Physics* 47, no. 7 (2008); and S. A. Selesnick, *Quanta, Logic and Spacetime*, 2nd ed. (New York: World Scientific, 2004).

15. L. Smolin, "The Case for Background Independence," pre-print (2005) arXiv:gr-qc/0507235; C. J. Isham, "Some Reflections on the Status of Conventional Quantum Theory when Applied to Quantum Gravity," in *The Future of Theoretical Physics and Cosmology: Celebrating Stephen Hawking's 60th Birthday*, ed. G. W. Gibbons, E. P. S. Shellard and S. J. Rankin (Cambridge: Cambridge University Press, 2003) and arXiv:quant-ph/0206090; D. R. Finkelstein, *Quantum Relativity: A Synthesis of the Ideas of Einstein and Heisenberg*, (New York: Springer-Verlag, 1996); S. A. Selesnick, *Quanta, Logic and Spacetime*, 2nd ed. (New York: World Scientific, 2003).

16. D. Eisenbud, *Commutative Algebra: With a View Toward Algebraic Geometry*, Graduate Texts in Mathematics, Vol. 150 (New York: Springer, 1995).

17. A. Mallios and E. Zafiris, "The Homological Kähler-De Rham Differential Mechanism—Part I: Application in the General Theory of Relativity," *Advances in Mathematical Physics* (2011), doi:10.1155/2011/191083; A. Mallios and E. Zafiris "The Homological Kähler-de Rham Differential Mechanism—Part II: Sheaf-Theoretic Localization of Quantum Dynamics," *Advances in Mathematical Physics* (2011). doi:10.1155/2011/189801.

18. E. Zafiris, "Boolean Information Sieves: A Local-to-Global Approach to Quantum Information," *International Journal of General Systems* 39, no. 8 (2010).

19. A. Mallios, *Geometry of Vector Sheaves: An Axiomatic Approach to Differential Geometry*, Vols. 1-2 (Dordrecht: Kluwer Academic Publishers, 1998); E. Vassiliou, *Geometry of Principal Sheaves* (Dordrecht: Kluwer Academic Publishers, 2004).

20. A. Mallios, *Modern Differential Geometry in Gauge Theories, Vol. 1: Maxwell Fields* (Boston: Birkhäuser, 2006); A. Mallios, *Modern Differential Geometry in Gauge Theories: Vol. 2: Yang-Mills Fields* (Boston: Birkhäuser, 2009).

21. A. Mallios, "Geometry and Physics Today," *International Journal of Theoretical Physics* 8, no. 45 (2006): 1557; A. Mallios, *Relational Mathematics: A Response to Quantum Relativity* (Paris: Publ. Ecole Normale Superieure, 2010).

22. A. Mallios, "Quantum gravity and 'Singularities,'" *Note di Matematica* 25, no. 57 (2005-2006) and arXiv:physics/0405111; A. Mallios, "On Algebra Spaces," *Contemporary Mathematics* 263 (2007): 427.

23. I. Raptis, "A Dodecalogue of Basic Didactics from Applications of Abstract Differential Geometry to Quantum Gravity," *International Journal of Theoretical Physics* 12, no. 46 (2007); A. Mallios, "On Algebra Spaces," *Contemporary Mathematics* 263 (2007): 427; A. Mallios and E. Zafiris, "The Homological Kähler-De Rham Differential Mechanism—Part I: Application in the General Theory of Relativity," *Advances in Mathematical Physics* (2011), doi:10.1155/2011/191083

24. A. Mallios, "A-invariance: An Axiomatic Approach to Quantum Relativity," *International Journal of Theoretical Physics* 7, no.47 (2008).

25. A. Mallios and E. E. Rosinger, "Abstract Differential Geometry, Differential Algebras of Generalized Functions and de Rham Cohomology," *Acta Applicandae Mathematicae* 55, no. 231 (1999); A. Mallios and E. E. Rosinger, "Space-Time Foam Dense Singularities and de Rham Cohomology," *Acta Applicandae Mathematicae* 67, no. 59 (2001).

26. E. E. Rosinger, *Non-Linear Partial Differential Equations: An Algebraic View of Generalized Solutions* (Amsterdam: North-Holland, 1990).

27. E. Zafiris, "A Sheaf-Theoretic Topos Model of the Physical Continuum and its Cohomological Observable Dynamics," *International Journal of General Systems* 38, no. 1 (2009); E. Zafiris, "A Categorical Perspective on Connections with Application in the Formulation of Functorial Physical Dynamics," in *Emerging Topics on Differential Geometry and Graph Theory*, ed. Lucas Bernard and Francois Roux (New York: Nova Science 2010); A. Mallios and E. Zafiris, "The Homological Kähler-De Rham Differential Mechanism—Part I: Application in the General Theory of Relativity," *Advances in Mathematical Physics* (2011), doi:10.1155/2011/191083.

28. E. Zafiris, "Quantum Observables Algebras and Abstract Differential Geometry: The Topos-Theoretic Dynamics of Diagrams of Commutative Algebraic Localizations," *International Journal of Theoretical Physics* 46, no. 2 (2007); E. Zafiris, "A Categorical Perspective on Connections"; A. Mallios and E. Zafiris, "The Homological Kähler-De Rham Differential Mechanism—Part I."

BIBLIOGRAPHY

Aspect, A., J. Dalibard, and G. Roger. "Experimental Test of Bell's Inequalities Using Time-Varying Analyzers." *Physical Review Letters* 49, no. 25 (1982): 1804-7.

Aspect, A., et al. "Experimental Tests of Realistic Local Theories via Bell's Theorem." *Physical Review Letters*. 47, no. 460 (1981).

Awodey, S. *Category Theory*, Oxford Logic Guides. London: Oxford University Press, 2006.

Bain, Jonathan. "Category-theoretic Structure and Radical Ontic Structural Realism." *Synthese* 0039-7857 (2011): 1-15.

Barry, Patrick. "What's done is done. . ." *New Scientist* 191, Iss. 2571 (2006): 36.

Bell, J. L. "From Absolute to Local Mathematics." *Synthese* 69 (1986).

Bell, J. L. "Categories, Toposes and Sets." *Synthese* 51, no. 3 (1982).

Bell, J. S. "On the Einstein Podolsky Rosen Paradox." *Physics* 1, no. 3 (1964): 195-200.

Bertaina, S., S. Gamarelli, T. Mitra, B. Tsukerblat, A. Muller, and B. Barbara. "Quantum oscillations in a molecular magnet." *Nature*, 453 (May 2008): 203-6.

Birkhoff, G., and J. Von Neumann. "The logic of quantum mechanics." *Annals of Mathematics* 37 (1936): 823-43.

Bohm, David. "A Suggested Interpretation of the Quantum Theory in Terms of 'Hidden' Variables." *Physical Review* 85 (1952): 166-93.

Bohm, David. "Time, the Implicate Order, and Pre-space." In *Physics and the Ultimate Significance of Time*, edited by David R. Griffin, 177-208. New York: State University of New York Press, 1986.

Bohm, David, and B. J. Hiley. *The Undivided Universe: An Ontological Interpretation of Quantum Theory*. London: Routledge, 1993.

Bohm, David, and Y. Aharonov. "Discussion of Experimental Proof for the Paradox of Einstein, Rosen and Podolski." *Physical Review* 108 (1957): 1070-76.

Bohr, Niels. "Can quantum-mechanical description of physical reality be considered complete?" *Physical Review* 48 (1935): 696-702.

Bohr, Niels. *Atomic Physics and Human Knowledge.* New York: Wiley, 1958.

Bohr, Niels. "Discussion with Einstein." In *Albert Einstein: Philosopher-Scientist,* edited by Paul Arthur Schilpp, 210. New York: Harper, 1959.

Brumfiel, Geoff. "Scientists Supersize Quantum Mechanics." *Nature* (March 17, 2010), doi:10.1038/news.2010.130.

Bub, Jeffrey. "Quantum Logic, Conditional Probability, and Interference." *Philosophy of Science* 49 (1982): 402-21.

Bub, Jeffrey. "The problem of properties in quantum mechanics." *Topoi* 10, no. 1 (1991): 27-34.

Caldeira, A., and A. Leggett. "Quantum Tunneling in a Dissipative System." *Annals of Physics* 149 (1983): 374-456.

Cox, R. T. "Probability, Frequency and Reasonable Expectation." *American Journal of Physics* 14 (1946): 1-13.

Eastman, T. E. "Duality without Dualism." In *Physics and Whitehead*, edited by T. Eastman and H. Keeton. Albany: State University of New York Press, 2004.

Eddington, A. S. "The Theory of Groups." In *The World of Mathematics*, Volume 3, edited by J. R. Newman. London: Dover, 2003.

Einstein, A., B. Podolsky, and N. Rosen. "Can Quantum-Mechanical Description of Physical Reality Be Considered Complete?" *Physical Review* 47, no. 10 (1935): 777-80.

Eisenbud, David. *Commutative Algebra: With a View Toward Algebraic Geometry*. Graduate Texts in Mathematics, Vol. 150. New York: Springer, 1995.

Ellerman, D. "A Theory of Adjoint Functors with Some Thoughts on Their Philosophical Significance." In *What Is Category Theory?* edited by G. Sica. Milan: Polimetrica, 2006.

Epperson, Michael. *Quantum Mechanics and the Philosophy of Alfred North Whitehead*. New York: Fordham University Press, 2004.

Epperson, Michael. "Whitehead and Modern Physics." In *Handbook of Whiteheadian Process Thought*, edited by Nicholas Rescher, Johanna Seibt, and Michel Weber. Berlin: Ontos-Verlag, 2008.

Epperson, Michael. "Quantum Mechanics and Relational Realism: Logical Causality and Wavefunction Collapse." *Process Studies* 38, no. 2 (2009): 340-67.

Epperson, Michael. "Relational Realism: The Evolution of Ontology to Praxiology in the Philosophy of Nature." *World Futures* 65, no. 1 (2009): 19-41.

Epperson, Michael. "The Mechanics of Concrescence: Quantum Theory and Process Metaphysics." *Studia Whiteheadiana* 4 (2010): 159-90.

Everett, Hugh. "'Relative State' Formulation of Quantum Mechanics." *Reviews of Modern Physics* 29 (1957): 454-62.

Finkelstein, D. R. *Quantum Relativity: A Synthesis of the Ideas of Einstein and Heisenberg*. New York: Springer-Verlag, 1996.

French, S. "On the Withering Away of Physical Objects." In *Interpreting Bodies: Classical and Quantum Objects in Modern Physics*, edited by E. Castellani. Princeton: Princeton University Press, 1998.

French, S., and J. Ladyman. "Remodeling structural realism: Quantum physics and the metaphysics of structure." *Synthese* 136 (2003): 31-56.

Gelfand, Sergei, and Yuri Manin. *Methods of Homological Algebra*, 2nd ed. New York: Springer, 2003.

Gell-Mann, M., and J. Hartle. "Quantum Mechanics in the Light of Quantum Cosmology." Proceedings of the 3rd International Symposium on the Foundations of Quantum Mechanics in the Light of New Technology, Tokyo, Japan, August 1989, 321-43.

Ghirardi, G. C. *Sneaking a Look at God's Cards: Unraveling the Mysteries of Quantum Mechanics*. Princeton: Princeton University Press, 2004.

Ghirardi, G. C., A. Rimini, and T. Weber. "Unified dynamics for microscopic and macroscopic systems." *Physical Review D* 34 (1986): 470-91.

Griffiths, Robert. "Consistent Histories and the Interpretation of Quantum Mechanics." *Journal of Statistical Physics* 36, no. 1 (1984): 219-72.

Halliwell, J. J. "A review of the decoherent histories approach to quantum mechanics." In "Fundamental Problems in Quantum Theory—A Conference Held in Honor of Professor John A. Wheeler." *Annals of the New York Academy of Sciences*, Vol. 755 (1995): 726-40.

Halliwell, J. J., and E. Zafiris. "Decoherent Histories Approach to the Arrival Time Problem." *Physical Review D* 57 (1998): 3351

Halliwell, J. J., J. Pérez-Mercader, and W. H. Zurek. *Physical Origins of Time Asymmetry*. Cambridge: Cambridge University Press, 1996.

Hartle, James, and Stephen Hawking. "The Wave Function of the Universe." *Physical Review D* 28 (1983): 2960-75.

Heisenberg, Werner. *Physics and Philosophy*. New York: Harper Torchbooks, 1958.

Heisenberg, Werner. "The Development of the Interpretation of Quantum Theory." In *Niels Bohr and the Development of Physics*, edited by Wolfgang Pauli. New York: McGraw-Hill, 1955.

Henry, Granville, and Robert Valenza. "Idempotency in Whitehead's Universal Algebra." *Philosophia Mathematica* 1, no. 2 (1993): 171.

Isham, C. J. "Some Reflections on the Status of Conventional Quantum Theory when Applied to Quantum Gravity." In *The Future of Theoretical Physics and Cosmology: Celebrating Stephen Hawking's 60th Birthday*, edited by G. W. Gibbons, E. P. S. Shellard, and S. J. Rankin. Cambridge: Cambridge University Press, 2003 and arXiv:quant-ph/0206090.

Johnstone, P. T., *Stone Spaces*. Cambridge: Cambridge University Press, 1986.

Karakostas, V. "Nonseparability, Potentiality and the Context-Dependence of Quantum Objects." *Journal for General Philosophy of Science* 38 (2007).

Kochen, Simon, and Ernst Specker. "The Problem of Hidden Variables in Quantum Mechanics." *Journal of Mathematics and Mechanics* 17 (1967): 59-87.

Ladyman, J., and D. Ross. *Every Thing Must Go: Metaphysics Naturalized*. Oxford: Oxford University Press, 2009.

Lewis, David. "How Many Lives Has Schrödinger's Cat?" In *Lewisian Themes: The Philosophy of David K. Lewis*, edited by Frank Jackson and Graham Priest, 4-23. Oxford: Oxford University Press, 2004.

Lewis, David. *On the Plurality of Worlds*. Oxford: Basil Blackwell, 1986.

Lombardi, Olimpia, S. Fortin, and M. Castagnino. "The Problem of Identifying the System and the Environment in the Phenomenon of Decoherence." In *The European Philosophy of Science Association Proceedings: Amsterdam 2009*, edited by H.W. de Regt, S. Hartmann, and S. Okasha. Berlin: Springer, 2012.

Mac Lane, Saunders. *Categories for the Working Mathematician*. Graduate Texts in Mathematics, 2nd ed. New York: Springer-Verlag, 1998.

Mac Lane, S., and I. Moerdijk. *Sheaves in Geometry and Logic*. New York: Springer-Verlag, 1992.

Mallios, A., and E. Zafiris. "The Homological Kähler-De Rham Differential Mechanism—Part I: Application in the General Theory of Relativity." *Advances in Mathematical Physics* (2011), doi:10.1155/2011/191083.

Mallios, A., and E. Zafiris. "The Homological Kähler-de Rham Differential Mechanism—Part II: Sheaf-Theoretic Localization of Quantum Dynamics." *Advances in Mathematical Physics* (2011), doi:10.1155/2011/189801.

Mallios, A. *Relational Mathematics: A Response to Quantum Relativity*. Paris: Publ. Ecole Normale Superieure, 2010.

Mallios, A. *Modern Differential Geometry in Gauge Theories: Vol. 2: Yang-Mills Fields*. Boston: Birkhäuser, 2009.

Mallios, A. "A-Invariance: An Axiomatic Approach to Quantum Relativity." *International Journal of Theoretical Physics* 47, no. 7 (2008).

Mallios, A. "On Algebra Spaces." *Contemporary Mathematics* 263 (2007): 427.

Mallios, A. *Modern Differential Geometry in Gauge Theories, Vol. 1: Maxwell Fields*. Boston: Birkhäuser, 2006.

Mallios, A. "Geometry and Physics Today." *International Journal of Theoretical Physics* 45, no. 8 (2006).

Mallios, A. "Quantum gravity and 'Singularities.'" *Note di Matematica*, 25 no. 57 (2005-2006) and arXiv:physics/0405111.

Mallios, A. "On Localizing Topological Algebras." *Contemporary Mathematics* 341, no. 79 (2004).

Mallios, A., and E. E. Rosinger. "Space-Time Foam Dense Singularities and de Rham Cohomology." *Acta Applicandae Mathematicae* 67 (2001).

Mallios, A., and E. E. Rosinger. "Abstract Differential Geometry, Differential Algebras of Generalized Functions and de Rham Cohomology." *Acta Applicandae Mathematicae* 55 (1999).

Mallios, A. *Geometry of Vector Sheaves: An Axiomatic Approach to Differential Geometry*, Vol. 1. Dordrecht: Kluwer Academic Publishers, 1998.

Mallios, A. *Geometry of Vector Sheaves: An Axiomatic Approach to Differential Geometry*, Vol. 2. Dordrecht: Kluwer Academic Publishers, 1998.

Mazzola, G. "Towards Big Science: Geometric Logic of Music and Its Technology." Paper presented at KlangArt-Kongress 1995, Universität Osnabrück, 1995.

Mohanty, Pritiraj, E. M. Q. Jariwala, and R. A. Webb. "Intrinsic Decoherence in Mesoscopic Systems." *Physical Review Letters* 78, no. 17 (1997): 3366.

Moore, G. E. "External and Internal Relations," *Proceedings of the Aristotelian Society* 20 (1919-1920): 40-62.

O'Connell, A. D., et al. "Quantum ground state and single-phonon control of a mechanical resonator" *Nature* 464, no. 7289 (2010): 697-703.

Omnès, Roland. "Logical Reformulation of Quantum Mechanics I.-IV." *Journal of Statistical Physics* 53, nos. 3-4 (1988): 893-957.

Omnès, Roland. *Philosophie de la Science Contemporaine*. Paris: Editions Gallimard, 1994. English translation published as *Quantum Philosophy: Understanding and Interpreting Contemporary Science*. Princeton: Princeton University Press, 2002.

Omnès, Roland. *The Interpretation of Quantum Mechanics*. Princeton: Princeton University Press, 1994.

Poincare, H. *The Foundations of Science: Science and Hypothesis, The Value of Science and Methods*. New York: The Science Press, 1921.

Popper, Karl. *Quantum Theory and the Schism in Physics: From the Postscript to the Logic of Scientific Discovery*. London: Routledge, 1992.

Raptis, I. "A Dodecalogue of Basic Didactics from Applications of Abstract Differential Geometry to Quantum Gravity." *International Journal of Theoretical Physics* 12, no. 46 (2007).

Rescher, N. *Studies in Leibniz's Cosmology*. Berlin: Ontos-Verlag, 2006.

Rosinger, E. E. *Non-Linear Partial Differential Equations: An Algebraic View of Generalized Solutions*. Amsterdam: North-Holland, 1990.

Rovelli, C. "Relational quantum mechanics." *International Journal of Theoretical Physics* 35 (1996): 1637-78.

Russell, Bertrand. "Mathematical Logic as Based on the Theory of Types." *American Journal of Mathematics* 30 (1908): 224.

Schlosshauer, Maximilian. *Decoherence and the Quantum-to-Classical Transition*. New York: Springer, 2007.

Schlosshauer, Maximilian. "Decoherence, the Measurement Problem, and Interpretations of Quantum Mechanics." *Reviews of Modern Physics* 76, no. 4 (2005): 1267

Schrödinger, Erwin. "The Present Situation in Quantum Mechanics" (translated by J. Trimmer). *Proceedings of the American Philosophical Society* 124 (1980): 323-38.

Selesnick, S. A. *Quanta, Logic and Spacetime*, 2nd ed. New York: World Scientific, 2003.

Sheehan, Daniel., ed. *Frontiers of Time: Retrocausation—Experiment and Theory*. Presented at the Pacific Division of the American Association for the Advancement of Science, American Institute of Physics, San Diego, California, June 20-22, 2006.

Smolin, L. "The Case for Background Independence." pre-print (2005) arXiv:gr-qc/0507235.

Stamp, P. C. E. "The Decoherence Puzzle." *Studies in History and Philosophy of Modern Physics* 37 (2006): 467-97.

Stapp, Henry. *Mindful Universe: Quantum Mechanics and the Participating Observer*. Berlin: Springer, 2007.

Takahashi, S. et al., "Decoherence in Crystals of Quantum Molecular Magnets." *Nature* 476 (July, 2011): 76-79.

Varadarajan, V. S. *Geometry of Quantum Theory*. New York: Springer, 1985.

Von Neumann, John. *Mathematical Foundations of Quantum Mechanics*. Princeton: Princeton University Press, 1955.

Walborn, S., et al. "Double-slit Quantum Eraser." *Physical Review A* 65, no. 3 (2002): 033818.

Weber, R. *Dialogues with Scientists and Sages: The Search for Unity*. London: Routledge, 1986.

Weyl, H. "Part I. Topology and Abstract Algebra as Two Roads of Mathematical Comprehension." *The American Mathematical Monthly* 102, no. 5 (1995).

Whitehead, Alfred North. *Process and Reality: An Essay in Cosmology*. Corrected Edition, edited by D. Griffin and D. Sherburne. New York: Free Press, 1978.

Whitehead, Alfred North. *The Principle of Relativity with Application to Physical Science*. New York: Dover, 2004.

Whitehead, Alfred North. *A Treatise on Universal Algebra, With Applications*. Cambridge: Cambridge University Press, 1898.

Whitehead, Alfred North. "The Philosophy of Mathematics." Review of *Mysticism in Modern Mathematics* by Hastings Berkeley. *Science: Progress in the Twentieth Century* 5 (October 1910): 234-39.

Whitehead, Alfred North. "Mathematics and the Good." In *The Philosophy of Alfred North Whitehead,* edited by Paul Arthur Schilpp, 671-72. New York: Tudor, 1951.

Whitehead, Alfred North. "On Mathematical Concepts of the Material World." In *Alfred North Whitehead: An Anthology*, edited by F. S. C. Northrup and M. W. Gross. New York: MacMillan, 1953. Originally published in *Philosophical Transactions of the Royal Society of London, Series A* 205 (1906): 465-525.

Whitehead, Alfred North. *Personal Letter to Henry Leonard* (January 10, 1936). United States Library of Congress, Manuscript Division (http://www.loc.gov/today/pr/2011/11-020.html).

Whitehead, Alfred North, and Bertrand Russell. *Principia Mathematica to *56.* London: Cambridge at the University Press, 1913.

Zafiris, Elias. "Rosen's Modeling Relations via Categorical Adjunctions." *International Journal of General Systems* 41, no. 5 (2012).

Zafiris, Elias. "A Categorical Perspective on Connections with Application in the Formulation of Functorial Physical Dynamics." In *Emerging Topics on Differential Geometry and Graph Theory*, edited by Lucas Bernard and Francois Roux. New York: Nova Science 2010.

Zafiris, Elias. "Boolean Information Sieves: A Local-to-Global Approach to Quantum Information." *International Journal of General Systems* 39, no. 8 (2010).

Zafiris, Elias. "A Sheaf-theoretic Topos Model of the Physical 'Continuum' and its Cohomological Observable Dynamics." *International Journal of General Systems* 38, no. 1 (2009): 1-27.

Zafiris, Elias. "Quantum Observables Algebras and Abstract Differential Geometry: The Topos-Theoretic Dynamics of Diagrams of Commutative Algebraic Localizations." *International Journal of Theoretical Physics* 46, no. 2 (2007).

Zafiris, Elias. "Sheaf-Theoretic Representation of Quantum Measure Algebras." *Journal of Mathematical Physics* 47 (2006).

Zafiris, Elias. "Topos-Theoretic Classification of Quantum Events Structures in Terms of Boolean Reference Frames." *International Journal of Geometric Methods in Modern Physics* 3, no. 8 (2006).

Zafiris, Elias. "Generalized Topological Covering Systems on Quantum Events' Structures." *Journal of Physics A: Mathematical and General* 39 (2006).

Zafiris, Elias. "Interpreting Observables in a Quantum World from the Categorial Standpoint." *International Journal of Theoretical Physics* 43, no. 1 (2004).

Zafiris, Elias. "Quantum Event Structures from the Perspective of Grothendieck Topoi." *Foundations of Physics* 34, no. 7 (2004).

Zafiris, Elias. "Boolean Coverings of Quantum Observable Structure: A Setting for an Abstract Differential Geometric Mechanism." *Journal of Geometry and Physics* 50, no. 99 (2004).

Zafiris, Elias. "Probing Quantum Structure Through Boolean Localization Systems." *International Journal of Theoretical Physics* 39, no. 12 (2000).

Zurek, Wojciech. "Decoherence and the Transition from the Quantum to the Classical." *Physics Today* 44, no. 10 (1991): 36-44.

Zurek, Wojciech. "Letters." *Physics Today* 46, no. 4 (1993): 84.

Zurek, Wojciech. "Decoherence, Einselection and the Existential Interpretation (the Rough Guide)." *Philosophical Transactions of the Royal Society A: Mathematical, Physical and Engineering Sciences* 356, no. 1743 (1998): 1793-1821.

Zurek, Wojciech. "Decoherence, Einselection, and the Quantum Origins of the Classical." *Reviews of Modern Physics* 75, no. 3 (2003): 715-75.

Index

Abstract (Modern) Differential Geometry (ADG), 381–84
actualities/actualization: Boolean logic and quantum mechanics and, 43–51; category theory and, 119–32; compatibility condition and, 58–63, 148–56; EPR and decoherence, 106–10; implicate order of, 9; internal local-to-global ontological formation, 269–70; internal relation and, 52–58, 140–47; local to global transformation, 63–68; logically related actual occasions, 168–69; mutual implicative relata and, 7–8, 24–26, 163–68; of physical state, 15–16, 33–37; of potentia, 34–35; potential-to-probable evolution, 157–63; quantum events interpretation, 305–7; quantum measurements, xvi–xviii, 37–38; quantum nonlocality and, 71–80; relational realism and, 86–95, 133, 202–9, 347–52; self reference problem, 113–15; set theory/sheaf theory semantics and, 255–57; topological *vs.* metrical relations, 80–85; of universe, 11
actual occasions: quantum mechanics and, 163–68, 202–9; topological localization and genetic analysis, 333–34
adjoint factors: Boolean topological significance, 266–67; category theory and, 229–32; functorial Boolean-quantum internal relation, 274–77; localization theories, 362–64
adjunction: category theory, 223; partial congruence and, 328–30
Aharonov-Bohm effect, 215–16
algebra: decoherence analysis, 340–45; differential extensive connection, 378–84; functorial entanglement, 295–303; Whitehead's discussion of, 120–32
angular momentum, EPR and quantum nonlocality, 71–80
aphairein (Aristotle), 228–32
Aristotle, 5, 86–87, 180, 333–34
arithmetization, localization theories, 213–16
arrow-theoretic formulation: arrow reversal, 228; category theory, 222

assimilation: fallacy of misplaced concreteness and, 13–21; mutually exclusive relata and, 12
asymmetrical relationships: compatibility condition for logical causality, 58–63; internal relations in quantum mechanics and, 140–47; philosophy of mathematics and, 119–21, 130–32; quantum mechanics and Boolean logic and, 44–51
atomic topology, category theory and, 221–22

Bacon, Francis, 30
base categories, localization topology and, 235–40
Berry phase, 215–16
bipolar dualism: mutually exclusive relata and, 5–10; relational realism and, 3–27; scientific method and, 10–12
Bohm, David, 8–10
Bohm-Hiley non-local hidden variables, quantum mechanics and, 9–10
Bohm's correlations, entanglement, 307–13
Bohr, Niels, 111–15, 346–47; complementarity principle of, 5–7
Boolean algebra, 261–65, 327–28
Boolean cover, physical contexts, quantum localization, 327–28
Boolean frames-quantum adjunction: functorial entanglement, 300–303; quantum events interpretation, 303–7
Boolean germs, 194; decoherence and, 337–45; entanglement, EPR correlations, 311–13; extensive connection theory and, 372–76; localization systems, 279–81; local quantum events and, 206–9; measurement events classification, 323–26; mereotopology and, 187–90; quantum events interpretation, 304–7; quantum logical event algebras and observables, 261–65; relational realism and, 349–52; topological relations, 261–84; truth valuation, Boolean localization system, 321–23
Boolean localization system: classical to quantum transition, 330–32; decoherence and, 334–45; epimorphic family formation, 279–81; truth valuation, 318–23
Boolean logic: category-sheaf theory and, 200–209; compatibility condition for logical causality and, 59–63, 149–56; composite system entanglement, 288–94; dipolar duality and, 23; dipolar quantum units, 163–68; functorial entanglement, 295–303; functor of Boolean reference frames, 270–73; global induction from local conditions, 63–68; internal relations and, 52–

58, 140–47; local Boolean reference frames, gluing isomorphism, 281–84; localization, 214–16, 279–81; objective local contextuality and, 39–43; potentiality-probability evolution and, 158–63; quantum event classification, 313–17; quantum localization, sheaf of Boolean reference frames, 277–79; quantum mechanics and, 43–51; reference frame, sheaf localization, 277–79; relational realism and, 92–95, 346–52; self reference in quantum mechanics and, 111–15; separation of science from, 32–39; topological localization, 236–40; truth valuation, localization systems, 318–23; Whitehead's philosophy of mathematics and, 126–32

Boolean measurement context: EPR and decoherence and, 107–10; fallacy of misplaced concreteness and, 15–21; predication in quantum mechanics and, 103–4; quantum theory and, xvi–xx

Boolean realization functor, 274–77

Borel algebra, 263–65

Born's rule: compatibility condition for logical causality and, 60–63; quantum mechanics and Boolean logic and, 43–51

Boyle, Robert, 30

Bub, Jeffrey, 106

Burali-Forti paradox, 117–19

candidate probability, decoherence, 339–45

canonical injections, category theory, 226

Cartesian coordinates, quantum mechanics and Boolean logic and, 47–51

Cartesian dualism, relational realism and, 5

categorical topology: localization and, 235–40; quantum observables, 263–65

categories, properties of, 223

category-sheaf theory, 64–68; EPR experiments and, 99n46; extensive relations and, 192–94; mereotopology and, 185–90; philosophy of mathematics and, 126–32; quantum mechanics and spacetime and, 83–85; relational realism and, 199–209

category theory: conceptual framework for, 220–32; extensive connection theory and, 185–90; mutual relations and transformations, 220–22; partial congruence and adjunction, 328–30; principles of, 127–32, 222–27; process metaphysics, 356; terminology, 222–27; universality and equivalence in, 227–32

Cauchy sequence, 180–84

causal-physical pole, 119–20, 131–32
causal relations: decoherence and, 68–70; EPR and quantum nonlocality, 75–80; one-way efficiency in, 198n13; predication in quantum mechanics and, 104–5; relational localization, 366–72
Classical Differential Geometry (CDG), 381–84
classical physics: decoherence and, 335–45; localization theories and, 213–16; quantum mechanics and, 356–60; quantum theory and, xiv–xx; reference frames, 330–32
classification: measurement events, Boolean germ classification, 323–26; quantum events, 313–17; truth valuation, Boolean localization system, 318–23
coarse-graining: decoherence, 335–45; relational realism, 351–52
cocones, category theory, 225
coequalizer condition, functorial entanglement, 297–303
coherence, dipolar duality and, 23–27
coincidence measurement, entanglement, EPR correlations, 310–13
colimits: category theory, 226; functorial entanglement, 298–303
commutativity: decoherence, 340–45; quantum mechanics and Boolean logic and, 45–51
compatibility condition for logical causality: entanglement, EPR correlations, 309–13; functorial entanglement, 296–303; internal local-to-global ontological formation, 267–70; quantum mechanics and, 58–63, 148–56
complementarity: fallacy of misplaced concreteness and, 13–21; mutually exclusive relata and, 6–8
composite systems, entanglement and, 287–94
conceptual inverses, category theory and, 229–32
concrescence: philosophy of mathematics and, 133–34; potentiality-probability evolution and, 157–63; relational realism and, 89–95, 350–52; set theory/sheaf theory semantics and, 255–57
conditional probability, composite system entanglement, 289–94
congruence: internal local-to-global ontological formation, 267–70; partial congruence, 328–30; topological localization and genetic analysis, 333–34
consistent sets, 339–45; decoherence and, 340–45
contextual dependency: internal relations in quantum mechanics, 143–47; measurement events,

Boolean germ classification, 324–26; sheaf germs and stalks, 248–51
contextualized measurement: internal relations and quantum mechanics and, 54–58; quantum mechanics and Boolean logic and, 47–51; relational realism and, 92–95
continuity, in quantum theory, xiv–xx
coordinate division, 196–97
coordination, localization theories, 213–16
Copenhagen Interpretation, 112–16
correlation events: entanglement, EPR correlations, 308–13; topological localization and genetic analysis, 333–34
cosmology: bipolar dualism and, 10–12; dipolar duality and, 24–27; fallacy of misplaced concreteness and, 14–21; relational realism and, 86–95; Whitehead's essay on, 23
covariant representational factor, category theory, 230–32
covering sieve: Boolean germs and extensive connection, 375–76; localization topology, 235–40; mereotopology and, 187–89, 193–94; relational localization, 369–72; representable functors as, 257–59
covering sites, localization topology, 233–40
decoherence: Boolean localization and, 334–45; causal relation and logical integration, 68–70; compatibility condition for logical causality and, 151–56; histories, 19; localization theory and, 215–16; potentiality-probability evolution and, 159–63; predication in quantum mechanics and, 105–10; relational realism, 351–52
Dedekind real numbers, set theory/sheaf theory semantics, 254–57
density matrix: decoherence and, 160–62, 334–45; off-diagonal terms in, 98n34, 110, 152, 159, 171–72; PNC violations and, 98n34, 101n.65, 107
density operator, composite system entanglement, 287–94
De Rham's algebraic-topological theory, differential extensive connection, 378–84
Descartes, Rene, 5
detectors: compatibility conditions and, 148–56; Copenhagen formalism in, 114; dipolar unit quantum events, 163–68; double-slit experiments, 33–34, 95n4, 144–45, 145–47; environmental relations, 69–70; indexical eventuality in, 158–63; local relations, 60–61; logically related actual occasions, 168–73; nonlocal relations, 62, 71–

80; objective local contextuality, 39–40; orthonormal basis for, 192–93; predication and, 109, 133, 142; quantum measurements, 33–35; relational realism and, 88–95; self reference in quantum systems and, 111; topological *vs.* metrical relations, 82–85
diagrams, category theory, 225
differential extensive connection, 376–84
dipolar duality: functorial Boolean-quantum internal relation, 273–77; mutually implicative relata and, 22–27; predication and, 119–32; relational realism and, 351–52
discreteness, in quantum theory, xiv–xx
disjoint relations: Boolean germs and extensive connection, 372–76; EPR and quantum nonlocality, 77–80; quantum mechanics, 52–58; sheaf germs and stalks, 249–51; uniform localization of observables, 245–46
dissipative energy transfer, compatibility condition for logical causality and, 151–56
double-slit experiments, 33–34, 95n4, 145–47
duality, quantum events interpretation, 305–7

Eddington, Arthur, 237–40
eigenvalues/eigenstates, quantum measurement and, 134n1
Eilenberg, Samuel, 127, 182–84
Einstein, Albert: field equations, 365; theories of, xiv, 197n4
Einstein, Podolsky and Rosen (EPR) correlations: compatibility condition for logical causality and, 62–63; experimental arrangements for, 99n44; functorial entanglement, 307–13; internal relations in quantum mechanics and, 55, 144–47; predication in quantum mechanics and, 105–10; quantum nonlocality and, 71–80, 98n40, 99n44; quantum theory and, 357–60; relational realism and, 99n46; spin systems, 204–9
Eleatic School, 8–10
elementary objects: quantum mechanics and spacetime and, 81–85; uniform localization of observables, 244–46; Whitehead's philosophy of mathematics and, 125
empirical adequacy, dipolar duality and, 23–27
empirical applicability, dipolar duality and, 23–27
empiricism, mutually exclusive relata and, 12
entanglement: composite systems and, 287–94; EPR correlations, 307–13; functorial entanglement, 295–303; quantum events

interpretation, 303–7; quantum theory and, 357–60
environmental decoherence. *See also* system-environment relations: Boolean localization and, 335–45; compatibility condition for logical causality and, 151–56
Epimenides paradox, 21, 31
epimorphic family formation: Boolean localization systems and germs, 279–81; gluing isomorphism, local Boolean reference frames, 281–84; reference frame, sheaf localization, 277–79
equalizers, category theory, 226–27
equivalence: category theory, 227–32; functorial entanglement, 300–303; quantum events interpretation, 305–7; uniform equivalence, fibration methodology, 232–33
erasure procedure, double-slit experiments, 145–47
Euclidean space, 377–84
Everett, Hugh, 8, 42
Excluded Middle, Principle of. *See* Principle of the Excluded Middle (PEM)
exclusive disjunction: Boolean logic and, 48, 135n11; outcome states, 41, 51
extensive abstraction theory, 179–80; categories and, 228–32, 235–40
extensive connection theory, 185–90; Boolean germs and, 372–76; classical to quantum transition, 332; localization topology and, 234–40; mereotopology and, 190–94; quantum gravity applications, 376–84; relational localization, 367–72; relational realism and, 194–96, 349–52
external objects, category theory and, 228–32

fibration: functor of Boolean reference frames, 272–73; split discrete fibration, 245–46; uniform methodology, 232–33
fine-graining, decoherence, 338–45
formalism, limitations of, 120–32
functions: category theory, 227–32; uniform fibration methodology, 232–33
functorial tensor product composition, 299–303, 310–13
functors: Boolean-quantum internal relation, 273–77; Boolean reference frames, 270–73, 277–79; composite system entanglement, 291–94; covering sieves as, 257–59; entanglement in, 295–303; quantum events interpretation, 303–7; subobject functor, 314–17

Gelfand topological representation theorem, 337–45
generalized records, 339–45

genetic analysis: coordinate division and, 196–97, 208; internal relations and, 52–58, 173–74, 179–84; mereotopological extensive relations and, 192–95; philosophy of mathematics and, 120–21, 131, 139; topological localization and, 333–34
global relations: compatibility condition for logical causality and, 60–63, 148–56; dipolar quantum units and, 163–68; global-to-local compatibility, gluing operations, 246–48; internal relations and quantum mechanics and, 54–58; local contextuality and, 63–68, 143–47; mereotopology and, 187–90; potentiality-probability evolution and, 158–63; predication in quantum mechanics and, 104–5; self reference in quantum mechanics and, 112–15; sheaf theory and, 348–52; uniform localization of observables, 244–46
global-to-local transformation: actualities, 63–68; genetic analysis and, 333–34; gluing operations, 246–48; internal relations, 269–70; quantum nonlocality, 71–78; relational realism, 351–52
gluing axiom, 67–68; category-sheaf theory, 194; classical to quantum transition, 332; global-to-local compatibility, 246–48; internal local-to-global ontological formation, 269–70; isomorphism, local Boolean reference frames, 281–84; local Boolean reference frames, 281–84; localization theories and, 215–16; localization topology and, 234–40; mereotopology and, 203–9; process metaphysics, 355–56; relational localization, 369–72
Gödel, Kurt, incompleteness theorems of, 18–19, 31, 124
Griffiths, Robert, 153–56
Grothendieck topology, 64–68; Boolean localization systems and germs, 279–81; covered sieves, 258–59; decoherence, 337–45; differential extensive connection, 380–84; functorial Boolean-quantum internal relation, 276–77; gluing isomorphism, local Boolean reference frames, 281–84; localization and, 235–40; local topological region and, 201–9; quantum localization, sheaf of Boolean reference frames, 277–79; quantum mechanics and spacetime and, 81–85; reference frame, sheaf localization, 277–79; relational realism and, 90–95

Hartle, James, 15
Hausdorff topological spaces, 305–7
Hawking, Stephen, 15

Heisenberg, Werner, 133, 137n52, 160
Heisenberg uncertainty relations, 365–72
Henry, Granville, 126–27, 129
Heraclitus, 5–6, 22
hidden variables theory, quantum mechanics and, 9, 104
Hilbert, David, 47
Hilbert space formalism: compatibility condition for logical causality and, 149–56; composite system entanglement, 287–94; objective local contextuality and, 39–43; quantum mechanics and Boolean logic and, 46–51; quantum observables, 264–65; relational realism and, 194–96
histories: of actual occasions, 25–26, 35, 164–68, 187–94; decoherence, 337–45; of decoherence, 19, 152–56; of facts, 74–75, 133, 159–62; predication and, 115; of quantum events, 206–8
homomorphisms: category theory, 227–32; functorial entanglement, 301–3; quantum event classification, 314–17; quantum logical event algebras and observables, 261–65; sheaf germs and stalks, 250–51

idealization, functorial entanglement, 295–303
idempotency: mathematics and, 129–32; relational localization, 366–72
identity, mathematics and, 129–32
implicative relations: EPR and quantum nonlocality, 77–80; quantum mechanics, 52–58
incompleteness, Gödel's theorems of, 18–19, 31, 124
inductive limit: Boolean localization systems and germs, 279–81; functorial entanglement, 296–303; global-to-local compatibility, 64–68; partial congruence and adjunction, 330; sheaf germs and stalks, 248–51; truth valuation, Boolean localization system, 318–23
infinitesimal differentiability, 381–84
infinitum actu non datur (Aristotle), 180
information carriers, relational localization, 369–72
information classification, partial congruence and adjunction, 328–30
ingression, category theory and, 228–32
interference fringe/effect: decoherence, 335–45; quantum superposition and, 33–39
internal relations: category theory and, 127–32; covering sieves as representable functors, 257–59; elemental units of, 119–20,

132–34; entanglement, EPR correlations, 310–13; functorial Boolean-quantum internal relation, 273–77; global induction from local and, 63–68; localization schemes and, 219; local-to-global ontological formation, 267–70; logical asymmetry in, 198n13; mereotopology and, 185–90; potentiality-probability evolution and, 157–63; quantum mechanics and, 52–58, 140–47; relational realism and, 92–95
intrinsic relativity: Boolean germs and extensive connection, 373–76; functorial entanglement, 296–303
isomorphisms: canonical isomorphisms, 228–29, 249; functorial entanglement, 296–303; gluing, local Boolean reference frames, 281–84; natural isomorphisms, 224, 230–31; partial congruence and adjunction, 329–30; quantum topology, 275–77; relational realism and, 348–52; set theory/sheaf theory semantics, 253–57; sheafification, 253

joint probability distribution, composite system entanglement, 287–94

Kaehler-de Rharm homological mechanism, 378–84
Kaehler's theory of generation, differential extensive connection, 378–84
Kant, Immanuel, 12
Kauffman, Stuart, 93
Kochen-Specker theorem, 56–58, 64–68, 134n3; localization topology and, 236–40; predication in quantum mechanics and, 103–4; quantum localization, 363–64; quantum observables, 264–65; relational realism and, 202–9

Leibniz, Karl, 59
Leonard, Henry, 91–93, 120
Lewis, David, 42–43
local Boolean frames, 281–84; quantum theory and, 357–64; relational realism and, 347–52
localization: Boolean covers and physical contexts, 327–28; Boolean systems and germs, 279–81; Boolean topological significance, 266–67; category theory and, 220–32; classical to quantum transition, 330–32; decoherence, Boolean localization and, 334–45; differential extensive connection and, 377–84; functorial entanglement, 295–303; genetic analysis, topological localization, 333–34; measurement events, Boolean germ classification, 324–26; partial congruence and adjunc-

tion, 328–30; in physical theories, 213–16; quantum events interpretation, 303–7, 306–7; quantum mechanics and, 360–64; relational sheaf-based theories, 364–72; schemes for, 216–19; sheaf of Boolean reference frames, 277–79; sheaf-theoretic quantum relational realism, 345–52; topological space, 243–44; topology and, 233–40; truth valuations in, 318–23; uniform fibration methodology, 232–33; uniform observables localization, 244–46
local relations: compatibility condition for logical causality and, 60–63, 148–56; contextuality of, 39–43; global conditions and, 63–68; global-to-local compatibility, gluing operations, 246–48; mereotopology and, 187–90; uniform localization of observables, 244–46
local-to-global transformation: actualities, 63–68; genetic analysis and, 333–34; gluing operations, 246–48; internal relations, 269–70; quantum nonlocality, 71–78; relational realism, 351–52
local topological region, defined, 200–209
Locke, John, 5, 12, 59
logical causality: compatibility condition for, 58–63, 148–56; EPR and quantum nonlocality, 71–80; logically related actual occasions, 168–69; mereotopology and, 185–90; potentiality to probability evolution, 157–63; predication and, 105; quantum logical event algebras and observables, 261–65; in quantum mechanics, 58–63, 139–72; relational localization, 366–72
logical-conceptual pole, 119–20, 131–32
logical integration, decoherence and, 68–70
logical structure, dipolar duality and, 23–27
logical types, theory of: philosophy of mathematics and, 119–32; quantum mechanics and, 116–19; set-theoretic framework for, 182–84

Mac Lane, Saunders, 127, 182–84
macroscopic superposition, 32–39; decoherence, 335–45
Mallios, Anastasios, 381–84
Many Worlds Interpretation (MWI), objective superposition of states and, 8–9, 39–43
Mathematical Foundations of Quantum Mechanics (von Neumann), 112–15
mathematics: mereotopology and, 179–84; philosophy of, 18–21, 119–32; process metaphysics and, 354–56

"Mathematics and the Good" (Whitehead), 120, 122
maximal sieves, 258–59
Mazzola, Guerino, 231–32
measurement: Boolean germ event classification, 323–26; localization topology and, 236–40; quantum actual occasions, 170–73; quantum mechanics and Boolean logic and, 45–51; in quantum theory, xiii–xx, 134n1
mereotopology: compatibility condition for logical causality and, 148–56; extensive relations in, 190–94; global induction from local and, 65–68; internal relation and logical implication in, 185–90; logical-extensive relation integration in, 179–96; quantum mechanics and spacetime and, 82–85
metrical relations: classical to quantum transition, 330–32; EPR and quantum nonlocality, 75–80; quantum mechanics and spacetime and, 80–85; relational realism and, 89–95
microscopic observables, decoherence, 336–45
misplaced concreteness, fallacy of, 13–21
mixed state, composite system entanglement, 288–94
Möbius strip, 67–68
momentum, bipolar dualism and, 6–7
mutually exclusive relata: bipolar dualism and, 5–10; cosmology and, 10–12; fallacy of misplaced concreteness and, 13–21
mutually implicative relata: composite system entanglement, 291–94; dipolar duality and, 22–27; dipolar quantum units, 163–68; entanglement, EPR correlations, 308–13; functorial Boolean-quantum internal relation, 273–77; internal local-to-global ontological formation, 268–70; relational realism and, 209

natural isomorphisms, category theory, 223–24, 230–32
natural philosophy, EPR and quantum nonlocality, 76–80
natural transformation, category theory, 222–27
Nature (journal), 32–34
negative selection, 88–89, 107–9; integration of relations and, 128–29; internal relations, 171–73; logical causality and, 151–53; potential relations and, 161–62; transmutation mechanism and, 162–63
nested composite inclusions, covering sieves as representable functors, 257–59
New Organon (Bacon), 30
non-dissipative decoherence, 152–56

nonlocality. *See* quantum nonlocality: quantum events interpretation, 304–7
non-separability, EPR correlations, 307–13

objective indeterminacy, 39–43
objective local contextuality, 39–43
objective superposition of states, 40–43
objects and relations, quantum theory and, xiv–xx
observables: decoherence, 335–45; entanglement, EPR correlations, 308–13; functorial entanglement, 295–303; localization theories, 213–16; quantum logical event algebras and observables, 261–65; quantum mechanics and, 358–60; relational localization, 365–72; set theory/sheaf theory semantics and, 254–57; uniform localization, 244–46
"Of the Excellency and Grounds of the Corpuscular or Mechanical Philosophy" (Boyle), 30
Omnès, Roland, 38–43, 90–95
one-dimensional projectors, composite system entanglement, 288–94
"On Mathematical Concepts of the Material World" (Whitehead), 125
On the Plurality of Worlds (Lewis), 42–43
open neighborhoods, localization topology and, 237–40
operational procedures, category theory, 221–22
orthogonal projections, quantum observables, 264–65
overlapping relations: EPR and quantum nonlocality, 77–80; quantum mechanics, 52–58; relational localization, 369–72

Parmenides, 5–6, 8–9, 22
partial Boolean algebra, 261–65
partial states: composite system entanglement, 292–94; functorial entanglement, 295–303
partitioning: functor of Boolean reference frames, 272–73; uniform fibration methodology, 232–33
phenomenon, relational realism and, 346–52
physical context, Boolean covers, quantum localization, 327–28
physical continuum: category theory, 220–22; localization schemes and, 216–19
physics, localization theories, 213–16
Plato, 5, 17–18; Protagoran doctrine and, 30–31
Platonic dualism, relational realism and, 5
Podolsky, B. *See* Einstein, Podolsky and Rosen (EPR) correlations

Poincare, Henry, 216
pointer basis, 335
positivism, relational realism and, 91–95
potentia: category theory and, 228–32; causal relation and logical integration, 68–70; EPR and decoherence and, 107–10; evolution to probability, 157–63; global induction from local conditions, 63–68; Heisenberg's concept of, 137n52; logically related actual occasions, 168–69; quantum mechanics and Boolean logic and, 47–51, 359–60; quantum theory and, xvi–xx; relational realism and quantum mechanics and, 86–95
potential outcome states, quantum superposition and, 33–39
potential simultaneous observability, 359–60
predication: EPR and decoherence and, 105–10; logical types theory and, 116–19; in quantum mechanics, 103–34; self reference in quantum mechanics and, 110–15; Theory of Logical Types and, 119–32
presheaf sets: display bundles, completion and functionalization, 252–53; functor of Boolean reference frames, 270–73; global-to-local compatibility, gluing operations, 246–48; localization topology and, 234–40; mereotopology and, 189–94; relational localization, 369–72; uniform fibrations methodology, 232–33; uniform localization of observables, 244–46
pre-Socratic philosophy, mutually exclusive relata and, 8–10
Principia Mathematica (Russell and Whitehead), 90–95, 117–21, 130
Principle of Identity (PI), 30–31; category theory and, 127; quantum actual occasions, 170–73; relational realism and, 92–95
Principle of Non-Contradiction (PNC), 25–27, 31; Boolean logic and, 32–39; causal relation and logical integration, 68–70; compatibility condition for logical causality and, 60–63, 151–56; EPR and decoherence, 105–10; EPR and quantum nonlocality, 73–80; exclusive disjunction and, 41, 51, 135n11; internal relations and quantum mechanics and, 57–58; objective superposition of states and, 40–43; potentiality-probability evolution and, 159–63; quantum actual occasions, 170–73; quantum mechanics and Boolean logic and, 43–51, 139; relational realism and, 88–95, 101n65
Principle of the Excluded Middle (PEM), 31; EPR and decoherence, 105–10; EPR and quan-

tum nonlocality, 73–80; exclusive disjunction and, 41, 51, 135n11; internal relations and quantum mechanics and, 57–58; internal relations in quantum mechanics and, 142–47; objective superposition of states and, 40–43; potentiality-probability evolution and, 159–63; quantum actual occasions, 170–73; quantum mechanics and Boolean logic and, 43–51; quantum superposition and, 38–39; relational realism and, 88–95, 101n65
probability: compatibility condition for logical causality and, 60–63; potentiality evolution to, 157–63; quantum mechanics and Boolean logic and, 45–51; quantum theory and, xvi–xx, 37–39; quantum topology, 264–65
Process and Reality (Whitehead), 23–27, 87–95; extensive connection theory in, 187–92; localization schemes and, 217–19; mereotopology in, 179–84; philosophy of mathematics and, 119–21, 123, 125, 130
process metaphysics: basic concepts, 353–56; differential extensive connection, quantum gravity applications, 376–84; germs and extensive connection, 372–76; localization in quantum domain, 360–64; quantum theory and, 356–60; relational localization, 364–72
product states, composite system entanglement, 290–94
projection operator: entanglement, EPR correlations, 310–13; localization theories, 362–64
property-indicating facts, quantum event classification, 313–17
Protagoras, 22, 30–31
pullback: category theory, 227; covering sieves as representable functors, 258–59; localization topology and, 238–40; quantum event classification, 317; quantum events interpretation, 305–7; truth valuation, Boolean localization system, 318–23
pure state: composite system entanglement, 288–94; entanglement, EPR correlations, 308–13
Pythagorean theorem, 21, 48

quantum gravity, differential extensive connection and, 376–84
quantum logical event algebra: Boolean germs and, 261–65; classification, 313–17; internal local-to-global ontological formation, 268–70; quantum events interpretation, 303–7; truth valuation, Boolean localization system, 318–23
quantum mechanics: Boolean logic, 43–51; category theory and,

127–32; compatibility condition for logical causality and, 148–56; cosmology and, 10–12, 15–20; decoherence and, 68–70; events interpretation, 303–7; historic interpretation, xv–xviii; internal relations in, 52–58, 140–47; interpretive problems in, xiii–xx; localization theories and, 213–16; local topological region and events in, 201–9; logical causality in, 58–63, 139–72; logical-extensive relation integration in, 179–96; logically related actual occasions, 168–69; logical types theory and, 116–19; objective indeterminacy and, 39–43; objective local contextuality and, 39–43; objective superposition of states, 40–43; potentiality-probability evolution and, 158–63; potentiality to probability evolution in, 157–63; predication in, 103–34; process metaphysics and, 353–56; reference frames, 330–32; relational localization, 364–72; relational realism and, 86–95, 163–68, 345–52; self reference problem and, 110–15; separation of Boolean logic from, 32–39; sheaf theory and, 345–52, 356–60; substance and logic in, 29–95; topological *vs.* metrical relations in, 80–85

quantum nonlocality, xiii–xx; decoherence and, 109–10; EPR correlations and, 71–80, 109–10

quantum superposition, 32–39; composite system entanglement, 288–94; internal relations and, 145–47; objective superposition of states and, 40–43; potentiality-probability evolution and, 160

rationalist philosophy, mutually exclusive relata and, 12

real numbers: localization theories, 213–16; relational localization, 366–72

recoherence, Boolean localization, 336–45

reduction: fallacy of misplaced concreteness and, 13–21; mutually exclusive relata and, 12

reductive-deductive epistemology, fallacy of misplaced concreteness and, 16–21

reference contexts, localization topology and, 239–40

reference frames: Boolean functors, 270–73; classical to quantum transition, 330–32; gluing isomorphism, 281–84; sheaf localization, 277–79

relata, relations as, 119–32; logically related actual occasions, 168–69; quantum mechanics and, 179–84

relational decomposition, quantum event classification, 313–17
relational localization, 364–72
relational realism: category-sheaf theory and, 199–209; category theory and, 228–32; dipolar duality and, 26–27; EPR and quantum nonlocality, 71–80, 99n46; logical causality in quantum mechanics and, 139–72; mereotopology and, 194–96; partial congruence and adjunction, 328–30; potentiality-probability evolution and, 157–63; quantum dipolar events, 163–68; quantum mechanics and spacetime and, 84–85, 359–60; quantum theory and, xvi–xx, xxn3; sheaf theory and, 345–52; speculative philosophy and, 3–27; Whitehead's cosmology and, 86–95
relativistically restricted data, 119–20, 131–32
relativity theory, 197n4; quantum theory and, xiv–xx
Rescher, Nicholas, 233
restriction operations, global-to-local compatibility, gluing operations, 247–48
rotational invariance, entanglement, EPR correlations, 309–13
Russell, Bertrand, 18, 91, 116, 182–84; Theory of Logical Types, 119–32
Russell's paradox, 18, 116–19; philosophy of mathematics and, 130–32

Schrödinger's Cat: EPR and decoherence, 106–10; objective superposition of states and, 41–43; quantum mechanics and Boolean logic and, 43–51; quantum superposition and, 33–39; quantum theory and, xvi–xx
scientific method: bipolar dualism and, 10–12; Boolean logic separation from, 32–39, 146–47; fallacy of misplaced concreteness and, 13–21
seed of extension, set theory/sheaf theory semantics, 256–57
self-determination, of actual occasions, 163–68
self reference: quantum mechanics and, 110–15; relational realism and, 208–9
semantic transition, decoherence, 340–45
set theory, 136n31; category theory and, 131–32; localization theories and, 213–16; relational localization, 365–72; topological relations and, 253–57
sheaf theory. *See also* category-sheaf theory: Boolean germs and extensive connection, 372–76; classical to quantum transition, 332; decoherence, Boolean localization, 337–45; differen-

tial extensive connection, 379–84; display bundles, completion and functionalization, 252–53; entanglement, EPR correlations, 307–13; functorial entanglement, 296–303; global-to-local compatibility, gluing operations, 246–48; gluing axiom in, 67–68; internal relations in, 191–92; localization topology and, 234–40, 360–64; overlapping structures, 150–51; partial congruence and adjunction, 328–30; process metaphysics, 355–56; quantum events and, 199–208; quantum localization, sheaf of Boolean reference frames, 277–79; quantum mechanics and, 356–60; quantum relational realism and, 345–52; relational localization, 364–72; set theory semantics and, 253–57; sheaf of germs, contextuality, 248–51; sheaf of germs, uniform localization of observables, 244–46; topology and, 182, 186

singlet pure state, entanglement, EPR correlations, 308–13

spacetime continuum (spatiotemporal extension): classical to quantum transition, 330–32; differential extensive connection and, 377–84; dipolar predication and, 119–20, 131–32; localization topology and, 234–40; relational realism and, 202–9; in thermodynamics, 198n13; topological *vs.* metrical relations in, 80–85

spectral decomposition: Boolean topological significance, 266–67; composite system entanglement, 288–94

spectrum holes, relational localization, 370–72

speculative philosophy: mathematics and, 120–32; relational realism and, 3–27

Spinoza, Baruch, 12, 59

spin particles: entanglement, EPR correlations, 307–13; EPR and quantum nonlocality, 71–80, 98n40, 99n41; internal relations and, 145–47; local quantum events and, 204–9; properties of, 99n41

split discrete fibration, 245–46; functor of Boolean reference frames, 270–73

state, composite system entanglement, 287–94

Stone representation theorem, 305–7, 337–45

Stone space, 305–7

structure-preserving mappings: category theory, 222; process metaphysics, 354–56

subject-superject concept, Whitehead's actual occasion and, 209

subobject functor: measurement events, Boolean germ classifica-

tion, 323–26; partial congruence and adjunction, 330; quantum event classification, 314–17
substance, relational realism and quantum mechanics and, 86–95
subsystems: EPR correlations, 307–13; functorial entanglement, 295–303
superluminal energy transfer, 99n48
system-environment relations: compatibility condition for logical causality and, 151–56; potentiality-probability evolution and, 159–63

temporal antecedence, EPR and quantum nonlocality, 73–80
tensor product, 287–94
terminal object, category theory, 226
Theaetetus (Plato), 17–18
Theory of Extension (Whitehead), 89–95
Theory of Logical Types. *See* logical types, theory of
Theory of Prehension (Whitehead), 89–95
topological relations: asymmetrical internal structure, 137n49, 150–56; basic principles, 100n50; Boolean contexts, 266–67; Boolean germs, 261–84; category-sheaf theory and, 127–32; contextuality, sheaf germs and stalks, 248–51; covering sieves as representable functors, 257–59; decoherence analysis, 340–45; differential extensive connection, 377–84; display bundles, completion and functionalization, 252–53; genetic analysis, 333–34; gluing sections and global-to-local compatibility, 246–48; internal local-to-global ontological formation, 267–70; localization theories and, 215–16, 233–40, 360–64; local topological region, 200–209; process metaphysics, 354–56; quantum events interpretation, 305–7; quantum mechanics and, 182–84; space localization, 243–44; spacetime in quantum mechanics and, 80–85
topos structures, set theory/sheaf theory semantics and, 254–57
total spin conservation, entanglement, EPR correlations, 309–13
trace function, 98n34
transcendental philosophy, complementarity and, 12
transformations, classical to quantum transition, 330–32
transitivity, localization topology and, 238–40
translational coding, functorial Boolean-quanter internal relation, 275–77
transmutation mechanism: concrescence and, 162–63; functorial

Boolean-quantum internal relation, 273–77; relational realism and, 351–52
Treatise on Universal Algebra, A (Whitehead), 120–23, 126, 129
truth valuation: Boolean localization system, 318–23; localization theories, 363–64; measurement events, Boolean germ classification, 324–26; partial congruence and adjunction, 330

uniform fibration methodology: functorial entanglement, 296–303; localization of observables, 245–46; variable set presheaves, 232–33
unit of relation, dipolar duality and, 25–27
universality, category theory, 227–32

Valenza, Robert, 126–27, 129
variable references, category theory, 220–22
vector sheaves, 381–84
Von Neumann, John: bipolar dualism and, 5; internal relations and quantum mechanics and, 54–58; projection postulate of, xviii–xx; quantum superposition and, 37–39; self reference in quantum mechanics and, 111–15
Von Neumann chain, 164–68
Von Neumann universe (set theory), 136n31
Von Neuman's spectral theorem, quantum observables, 264–65

wavefunction mechanics, fallacy of misplaced concreteness and, 15–21
wave-particle complementarity, 6–7
"The Wave Function of the Universe" (Hawking and Hartle), 15
Whitehead, Alfred North: category theory and, 227–32; extensive connection theory of, 185–90, 234–40, 332, 350–52, 372–76; fallacy of misplaced concreteness and, 13–21; genetic analysis concept of, 333–34; internal local-to-global ontological formation, 269–70; localization schemes and, 216–19; mereotopology theory of, 179–84; philosophy of mathematics and, 119–34; process theory of, 199–209, 358–60; quantum actual occasions of, 163–68, 202–9; quantum theory and work of, xvi–xx; relational realism and philosophy of, 86–95; on speculative philosophy, 3; Theory of Logical Types, 119–20; topological localization and, 333–34

Yoneda Lemma, 231–32
Young, Thomas, 95n4

Zafris, Elias, 64, 67–68, 81–85; quantum theory and, xviii–xx
Zermelo-Fraenkel (ZFC) set theory, 136n31
Zurek, Wojciech, 160–63

About the Authors

Michael Epperson did his doctoral work in philosophy of science and philosophy of religion at The University of Chicago, and earned his Ph.D. there in 2003. His dissertation, *Quantum Mechanics and the Philosophy of Alfred North Whitehead*, was written under the direction of philosopher David Tracy and physicist Peter Hodgson, Head of the Nuclear Physics Theoretical Group at the University of Oxford. It was published the following year by Fordham University Press, and re-released in paperback by Oxford University Press in 2012.

He is currently the founding director of the Center for Philosophy and the Natural Sciences in the College of Natural Sciences and Mathematics at California State University, Sacramento, where he is a Research Professor and Principal Investigator.

Elias Zafiris holds an M.Sc. (Distinction) in Quantum Fields and Fundamental Forces from Imperial College at the University of London, and a Ph.D. in theoretical physics from Imperial College. He has published numerous papers on category-theoretic methods in quantum physics and complex systems theories, topological localization and modern differential geometry in quantum field theory and quantum gravity, generalized spacetime quantum theory, decoherence, and many other topics in the foundations of physics.

He is a research professor in theoretical and mathematical physics at the Institute of Mathematics at the University of Athens, and is currently a visiting professor in the Department of Logic, Institute of Philosophy, Eötvös Loránd University in Budapest. He is also a senior research fellow and Principal Investigator at the Center for Philosophy and the Natural Sciences in the College of Natural Sciences and Mathematics at California State University, Sacramento.